HANDBUCH DER UROLOGIE

ENCYCLOPEDIA OF UROLOGY

ENCYCLOPÉDIE D'UROLOGIE

HERAUSGEGEBEN VON · EDITED BY
PUBLIÉE SOUS LA DIRECTION DE

C. E. ALKEN V. W. DIX H. M. WEYRAUCH
HOMBURG (SAAR) LONDON SAN FRANCISCO

E. WILDBOLZ
BERN

VI

Springer-Verlag Berlin Heidelberg GmbH 1959

ENDOSCOPY

BY

ROGER W. BARNES

R. THEODORE BERGMAN · HENRY L. HADLEY

LOS ANGELES

WITH 184 FIGURES

Springer-Verlag Berlin Heidelberg GmbH 1959

ISBN 978-3-642-86573-2   ISBN 978-3-642-86571-8 (eBook)
DOI 10.1007/978-3-642-86571-8

© by Springer-Verlag Berlin Heidelberg 1959
Ursprünglich erschienen bei Springer-Verlag OHG. Berlin · Göttingen · Heidelberg 1959
Softcover reprint of the hardcover 1st edition 1959

This volume is dedicated to the Urologists of India
who are struggling for recognition of the speciality of Urology

Contents

Chapter III — Postendoscopic care, reactions and complications

Chapter IV — The normal bladder and prostatic urethra

Chapter V — Abnormal ureteral orifices

Chapter X — Abnormal bladder contents

Chapter XI — Abnormalities of the bladder neck and posterior urethra in the male

Part II — Endoscopic Surgery

Chapter XIV — Miscellaneous endoscopic surgical procedures and treatments

Contents XVII

Chapter XVI — Electrosurgical units

Chapter XVII — Indications for endoscopic surgery

Chapter XX — Variations in technique of endoscopic prostatic resection

Contributors to volume VI

ROGER W. BARNES, M. S., M. D., D. Sc. (Med.), F. A. C. S., F. I. C. S. Professor of Surgery (Urology) and Chairman of the Division of Surgery, School of Medicine, College of Medical Evangelists; Chief of Urology Service, Los Angeles County Hospital; Senior Attending Surgeon, White Memorial Hospital; Attending Staff, Glendale Sanitarium and Hospital and Good Samaritan Hospital. Consultant in Urology to Christian Medical College, Vellore, South India, 1956 to 1957 and

R. THEODORE BERGMAN, B. Sc., M. D., D. N. B., F. A. C. S., F. I. C. S. Clinical Professor of Surgery (Urology), School of Medicine, College of Medical Evangelists; Chief of Urology Service, White Memorial Hospital; Senior Attending Surgeon, Los Angeles County Hospital; Attending Staff, Glendale Sanitarium and Hospital and Good Samaritan Hospital and

HENRY L. HADLEY, B. A., M. D., D. N. B., F. A. C. S. Assistant Professor of Surgery (Urology), School of Medicine, College of Medical Evangelists; Attending Staff, White Memorial Hospital, Los Angeles County Hospital, Glendale Sanitarium and Hospital, Good Samaritan Hospital. Consultant, Southern Pacific Company

Part I. Diagnostic Endoscopy

Chapter I

Endoscopic armamentarium

Endoscopy has elevated the modern urologist from the status of a venereal disease doctor to that of a highly trained and skillful specialist (McCARTHY 1951). It has put the urinary bladder and urethra on the outside of the body, as it were, for those who are experienced in the use of the endoscopic armamentarium. Intricate diagnostic and surgical procedures which would otherwise be impossible, are made available by means of endoscopy.

A. Endoscopes

Endoscopes, the instruments with which visual examination of the urinary bladder and urethra is accomplished, are many and varied. There are two general types — direct vision, and indirect vision utilizing a lens system.

The indicated size of endoscopes used in this text is the French Charrière scale which is the diameter in one-third millimeters; for example, a No. 15 Fr. is 5 mm in diameter. In Europe the same scale is sometimes indicated by the abbreviation Charr.

I. Direct vision endoscopes

1. Advantages

There are some advantages of direct vision endoscopes for cystoscopic and urethroscopic examination (RIDLEY). The view obtained through them is the actual picture of the area under observation; there is no distortion by magnification nor by diminution. A lesion can sometimes be more accurately identified than through a lens. Rigid instruments for manipulation within the bladder and urethra can more easily be used through the direct vision endoscope, although rigid catheters are not necessary as the catheter guides are small tubes which may be placed directly in front of the ureteral orifice. Catheters passed through them cannot loop or buckle. There is no friction to their passing and an extremely delicate sense of touch and feel is attained. It is therefore sometimes possible to catheterize ureters which are difficult to catheterize through a lens instrument. Bleeding points in the bladder are more easily identified because there is less clouding of the distention medium when air is used to distend the bladder; when water is used, the direct view penetrates the cloudiness better than does the view through a lens.

2. Cystoscopes

The Braasch is the most commonly used direct vision instrument. Modifications by BUMPUS, THOMPSON, and others have improved the original. The model (Fig. 1) which is most widely used at the present time is size 24 Fr.; size 28 Fr. is sometimes used. The beak is convex and the fenestra circular and located at the distal end of the cystoscope sheath. The light bulb is in the beak and the

Table 1. *Summary of available endoscopes*
(Some available endoscopes are not included in this list)

Name	Use and special features	Optical system and telescope		Sheath size, catheter capacity, type, attachments	Manufacturer[1] and/or distributor (see p. 10 for list of manufacturers)
		standard equipment	others available		
Alcock Lithotrite	Visual lithotripsy in adults	Right angle carries light	None	Motive power; rack and pinion on a wheel. Large stopcocks. Jaws open parallel to shaft	A.C.M.I.
Ballenger Urethroscope	Posterior urethroscopy. Direct view light on carrier	Single magnifying lens mounted on light carrier	None	24 Fr. open tube. Convex beak	A.C.M.I.
Beer Infant Cystoscope	Cystoscopy and catheterization of one ureter in infants	Rt. angle size 10 Fr. Used for examination only without a sheath. Detachable from catheterizing sheath	None	15 Fr. with catheterizing sheath which is detachable from telescope. 1—4 Fr. Concave beak	A.C.M.I.
Braasch-Bumpus Cystoscope	Routine cystoscopy and urethroscopy. Direct vision and water distention	Magnifying lens in viewing window	Right angle	24 Fr.: 3 — No. 6 Fr. 1 — No. 10 Fr. 28 Fr.: 2 — No. 6 Fr. 1 — No. 14 Fr. Convex beak. Light in beak.	Electro-Surg.
Braasch-Bumpus-Thompson Resectoscope	Endoscopic prostatic surgery. Direct vision water distention	Magnifying lens in viewing window	Right angle	28 Fr. Cold punch. Fulgurating electrode. Large stopcock	Br. cysto. Electro-Surg. End. Inst. G.U. Mfg.
Brown-Buerger Cystoscopes (Any combination of telescopes and sheaths is available)	Complete for adult routine cystoscopy	Right angle examining, operating, double catheterizing	Obliquely forward for convex sheath. Retrograde convertible (both operating & catheterizing)	16 Fr.: 2 — No. 5 Fr. 21 Fr.: 2 — No. 5 Fr. 1 — No. 7 Fr. 24 Fr.: 2 — No. 7 Fr. 1 — No. 9 Fr. Convex and concave beak	A.C.M.I. Br. Cysto. Inst. End. Inst. G.U. Mfg.
Brown-Buerger Female Convertible	Shorter and larger for universal use in women	Rt. angle. Examining. Convertible (both operating and catheterizing)	None	28 Fr. Short 2 — No. 8 Fr. 1 — No. 10 Fr. Concave beak. Light in beak	A.C.M.I. Br. Cysto
Buerger Cysto-Urethroscope	Routine cystoscopy and posterior urethroscopy in adults	Rt. angle. Examining. Convertible (both operating and catheterizing)	None	21 Fr.: 2 — No 6 Fr. 1 — No. 7 Fr. 24 Fr.: 2 — No. 7 Fr. 1 — No. 9 Fr.	A.C.M.I.

[1] Some endoscopes might be available from other manufacturers or distributors than those listed.

Table 1. (Continued)

Name	Use and special features	Optical system and telescope		Sheath size, catheter capacity, type, attachments	Manufacturer and/or distributor (see p. 10 for list of manufacturers)
		standard equipment	others avaiable		
Buerger Universal Urethroscope	Examining bladder and posterior and anterior urethra	Direct vision. Catheter channel. Light carrier separate. May also be used without. Water or air distension	None	25 Fr. Window for water distention without telescope. Large stopcock. Also used as open tube. Obturator has flexible beak	A.C.M.I.
Butterfield Double Catheterizing Children's Cysto-urethroscope	Examining bladder and urethra and double catheterizing in children	Rt. angle. Double catheterizing	None	15 Fr. 2 — No. 4 Fr. Convex beak. Small stopcock. Light in beak	A.C.M.I.
Butterfield Female Urethroscope and Infant Vaginoscope	Examining, operating and catheterizing in women. Urethroscopy in women. Vaginoscopy in infants	Foroblique[1]	None	18 Fr. Oval. Short. 2 — No. 4 Fr. (through one channel) 1 — No. 6 Fr. Catheter channel on sheath	A.C.M.I.
Campbell Dilating Cysto-urethroscope	Examining, operating, catheterizing and urethroscopy in adults	Foroblique interchangeable between 17 Fr. and 21 Fr. sheaths	None	17 Fr. 21 Fr. 1 — No. 10 Fr. Catheter channel on sheath. Open end. Concave beak	A.C.M.I.
Campbell Miniature Operating and Double Catheterizing	Operating and catheterizing in infants	Rt. angle examining	None	13½ Fr. Operating. 1 — No. 4 Fr. 15 Fr. Double catheterizing. 2 — No. 3½ Fr. Concave beak. Light in beak. Operating and catheter channel on separate sheaths. Fixed inclined plane catheter deflector	A.C.M.I.
Campbell Universal Infant Cystoscope	Examining and double catheterizing in infants	Rt. angle; light carried separately. Light and telescope project into bladder	Foroblique. Direct vision	9½ Fr. Examining only 15 Fr. Double catheterizing. 2 — No. 4 Fr. Catheter channel on sheath. Straight. Open end	A.C.M.I.
Deming Bladder Tumor Forceps	Removing biopsy specimens from bladder in adults	Rt. angle Carries light	None	21 Fr. Jaws activated by forceps handle and motion parallel to shaft.	A.C.M.I.
Gentile endoscopes	Complete assortment of endoscopes for diagnosis and surgery, including bright light source through quartz tube which can be used for photography				Gentile

[1] Foroblique is the American Cystoscope Makers Inc. registered trade mark to designate their obliquely forward vision telescope.

1*

Table 1. (Continued)

Name	Use and special features	Optical system and telescope		Sheath Size, catheter capacity, type, attachments	Manufacturer and/or distributor (see p. 10 for list of manufacturers)
		standard equipment	others available		
Hendrickson Lithotrite	Visual lithotripsy in adult	Foroblique	None	Forceps grip motivation of jaws which open parallel to shaft. Large lumen stopcock	A.C.M.I.
J. E. Semple Rotatable Resectoscope	Endoscopic surgery one hand operation. Scissor type grip for index and second fingers	Foroblique		26 Fr. Bakelite. Sheath telescope and cutting loop rotate	End. Inst.
Kelly Cystoscope	Air distention open tube for women	None		Various sizes. External light reflected by head mirror	
Kirwin Cystoscope	Routine cystoscopy. Double fenestra and flexible beak	Rt. angle. Examining. Convertible. (Both operating and catheterizing)	None	15, 20 & 24 Fr. Carries light. Fenestra on each side of sheath. Flexible jointed beak. Large stopcock	A.C.M.I.
Kirwin Lithotrite	Visual lithotripsy in adults	Rt. angle (Young) Carries light	A.C.M.I.	Forceps grip motivation of jaws which open at right angle to shaft	A.C.M.I.
Kirwin Rongeur	Visual removal of biopsy specimens	Same as above except jaws are smooth and keen edged			A.C.M.I.
Kirwin Rotary Resector	Endoscopic surgery. Rotating loop electrode makes transverse cut. Motivated by gears and wheel	Foroblique	Right angle Retrograde Direct	28 Fr. Metal covered. Short straight beak obliquely open end. Hinged obturator	A.C.M.I.
Laidley 16 Fr. Double Catheterizing Children's Cystoscope	Examining and double catheterizing in children	Right angle. Examining. Double catheterizing	None	16 Fr.: 2—No. 5 Fr. Concave and convex beaks. Light in beaks	A.C.M.I.
Lowsley Grasping Forceps	Visual removal of biopsy specimens and foreign bodies	Foroblique	None	22 Fr.: Forceps grip motivates jaws. Lower jaw hinged and serrated. Open parallel to shaft. Stopcock	A.C.M.I. Br. Cysto. End. Instr.
Lowsley-Peterson Universal Endoscope	Anterior and posterior urethroscopy. Ejaculatory duct and ureteral catheterization. Fulguration in bladder	Direct vision. Separate light. Carrier and two catheter channels	None	24 Fr.: 2—No. 4 Fr. Partial oblique fenestra. Straight end. Hinged obturator	A.C.M.I.

Table 1. (Continued)

Name	Use and special features	Optical system and telescope		Sheath size, catheter capacity, type, attachments	Manufacturer and/or distributor (see p. 10 for list of manufacturers)
		standard equipment	others available		
Lowsley Urethroscope	Anterior and posterior urethroscopy and operating in adult	Foroblique. Separate light carrier and catheter guide	None	24, 26, 28 Fr. Partial oblique fenestra. Straight end. Hinged obturator	A.C.M.I.
McCarthy Convertible Close Vision Cysto-urethroscope	Routine cystoscopic and posterior urethroscopic operating and catheterizing	Obliquely forward special close vision objective lens. Catheter guides and deflector on telescope convertible (both operating and catheterizing)	None	21 Fr. 2—No. 5 Fr. 24 Fr. 2—No. 7 Fr. 1—No. 9 Fr. 27 Fr. 2 — No. 8 Fr. 1 — No. 10 Fr. Carries light. Fenestra on side. Slight convex beak	A.C.M.I. Br. Cysto. End. Instr.
McCarthy Deflecting Foroblique Panendoscope	Routine cystoscopic and posterior urethroscopic operating and catheterizing	Special foroblique Carries light. Catheter guides and deflector on telescope. Convertible	None	24 Fr. 2 — No. 7 Fr. 1 — No. 10 Fr. Obliquely open straight end	A.C.M.I.
McCarthy Ejaculatory Duct Catheterizing Instrument	Catheterizing ejaculatory ducts	Foroblique. Catheter tunnels and roller deflecting mechanism	None	24 Fr. 2 —No 4 Fr. Ejaculatory duct catheters. Obliquely open straight end	A.C.M.I.
McCarthy Foroblique Panendoscope	Routine cystoscopy and posterior urethroscopy. Catheterizing and operating. First instrument to use McCarthy foroblique optical system	Foroblique. Detachable catheter guides. Interchangeable catheter washers and rigid deflectors	Foroblique examining (larger field of vision) Rt. angle Retrograde	Even numbers 16 Fr. to 30 Fr. All use same telescope and bridge assembly 16 Fr. 2—No. 4 Fr. 1—No. 6 Fr. 30 Fr.: 2 — No. 9 Fr. 1 — No. 14 Fr. Straight obliquely open end. Also available: Extra long sheath. Convex open end beaked sheath. Light carrier to use sheath as open tube urethroscope. Multiple catheter adapters	A.C.M.I. Br. Cysto. G.U. Mfg.
McCarthy Infant Cystoscope	Examination, single catheterizing and posterior urethroscopy in infants	Foroblique. Single catheter guide or deflector	None	10 Fr. Oval. 1 — No. 3½ Fr. Convex beak. Obliquely open end	A.C.M.I. G.U. Mfg.

Table 1. (Continued)

Name	Use and special features	Optical system and telescope		Sheath size, catheter capacity, type, attachments	Manufacturer and/or distributor (see p. 10 for list of manufacturers)
		standard equipment	others available		
McCarthy Miniature Cystoscope	Examining, catheterizing and posterior urethroscopy in infants	Foroblique	None	11 Fr. Examining only. 12 Fr. 1—No. 4 Fr. 14 Fr. Slightly beaked 2 — No. 4 Fr. Catheter channel in sheath. Inclined plane deflector	A.C.M.I.
McCarthy Periscopic Cystoscope	Examining and single catheterizing or operating	"Periscopic" foroblique with pivoted movable mirror giving 170° field	None	24 Fr. 1—No. 8 Fr. Straight obliquely open end	A.C.M.I.
McCarthy Routine Cystoscope	Routine examining, catheterizing, operating and posterior urethroscopy in adults	Foroblique. Three catheter guides. Convertible (operating and catheterizing). Rigid deflectors	Rt. angle Retrograde	Even numbers 18 to 28 Fr. 18 Fr. 2—No. 5 Fr. 1—No. 6 Fr. 28 Fr. 2 — No. 8 Fr. 1 — No. 13 Fr. Obliquely open end convex beak	A.C.M.I.
McCarthy Electrotome (Resectoscope) (Stern-McCarthy visual Prostatic Electrotome)	Endoscopy surgery. Most widely used resectoscope. Loop controlled by rack and pinion attached to handle. Cutting loop activated by damped current	Foroblique removable from working element	Rt. angle Retrograde Direct	24, 26, 28 Fr. Bakelite, metal covered or plastic. Straight obliquely open end. Long or short straight beak. Straight or hinged (Timberlake) obturator	A.C.M.I. Endo. Inst. G.U. Mfg.

Modifications of McCarthy Electrotome (A.C.M.I.)

BAUMRUCHER: One hand operation. Forefinger pulls loop against spring.

CREEVY: One hand operation. Two finger grips push loop against spring.

Foley Rotatable: One hand operation. Thumb pushes loop against spring, with a disc surrounding telescope. Supply lines remain stationary while remainder of resectoscope rotates.

GIBSON: One hand operation. Forefinger pulls loop against spring. Pistol grip.

IGLESIAS: One hand operation. Thumb pushes loop against leaf type spring. Two finger rests on sheath.

McCarthy Convertible: Adjustment to make telescope move with loop.

McCarthy Infant: No. 12 Fr. and short. Rotating water inlet. Rack and pinion attached to knob.

McCarthy Remote Control: Switch for current to loop is on a knob handle which moves loop.

McCarthy of Smaller Calibre: No. 16 and 20 Fr. Standard length. Rotating water inlet. Rack and pinion attached to knob.

MILLER: One hand operation. Thumb on a ring surrounding telescope pushes loop against spring.

NESBIT: Original one hand operation modification. Thumb in thumb rest pushes loop against spring which returns loop to original position. All sizes corresponding to McCarthy except the 12 Fr. infant.

Nesbit Perineal: One hand operation 33 Fr. and short (6½").

SCOTT: One hand operation. Finger pulls loop against spring. Pistol grip. Sheath telescope and loop rotate within housing for supply lines and handle.

Table 1. (Continued)

Name	Use and special features	Optical system and telescope		Sheath size, catheter capacity, type, attachments	Manufacturer and/or distributor (see p. 10 for list of manufacturers)
		standard equipment	others available		
McCarthy Urethroscope	Anterior and posterior urethroscopy in adults	Magnifying lens mounted on light carrier	None	Even numbers 22 to 28 Fr. Straight and beaked open tube. Separate light carrier to distalend	A.C.M.I. G.U. Mfg.
MILLIN's Cystoscopes and Resectoscope	For routine diagnostic and surgical endoscopy				Br. Cysto.
McCrea Infant Cystoscope	Examining, single catheterizing and operating in infants	Right angle examining	None	Oval 13 Fr.: 1 — No. 4 Fr. Catheter guides on sheath. Inclined plane deflector. Slightly convex beak. Light in beak	A.C.M.I.
Otis-Brown Cystoscope	Examination only in adults and children. First electrically lighted cystoscope made in U.S.A.	Right angle examining	None	13, 15, 18, 20 Fr. Convex and concave beaks. Light in beak	A.C.M.I.
Ravich Convertible Cystoscope	Routine examining, catheterizing and operating in adults	Right angle examining. Convertible	None	21 Fr.: 2—No. 6 Fr. 1—No. 8 Fr. Slight convex beak. Light in beak	A.C.M.I.
Ravich Lithotriptoscope	Visual lithotripsy in adults	Foroblique	None	27 Fr. Forceps grip motivation of jaws open parallel to shaft stopcock	A.C.M.I.
Ravich Urethroscope	Posterior urethroscopy in adults	Magnifying lens mounted on sheath	None	22 Fr. Open tube Fenestra in end Convex beak carries light	A.C.M.I.
Squire Urethroscope	Anterior and posterior urethroscopy in adults	Magnifying lens mounted on sheath	None	Anterior tube 22, 24, 26 and 28 Fr. Straight open end. Posterior tube 24, 26, 28 Fr. Convex beak oblique fenestra in end. Light mounted externally	A.C.M.I.
Stern-McCarthy visual prostatic electrome — same as McCarthy electrotome					
Swift Joly Aero-urethroscope (Harkness)	Anterior urethroscopy in adults	Magnifying lens		Closed tube for air distension. Light source incorporated in ocular end	Br. Cysto. G.U. Mfg.

Table 1. (Continued)

Name	Use and special features	Optical system and telescope		Sheath size, catheter capacity, type, attachments	Manufacturer and/or distributor (see p. 10 for list of manufacturers)
		standard equipment	others available		
Swift Joly's Cystoscope	Routine cystoscopy and posterior urethroscopy in adults and older children. Diathermy attachment	Right angle		15 Fr. to 24 Fr. Double Catheterizing	Br. Cysto. End. Inst. G.U. Mfg.
Vest Dilating Cystoscope	For large size operating and dilating instruments	Foroblique. One catheter guide	None	28 Fr.: 1 — No. 17 Fr. Straight, obliquely open end	A.C.M.I.
Wilhelm Insulated Urethroscope	Anterior and posterior urethroscopy in men and women. Especially for fulguration	Magnifying lens mounted on sheath		Anterior 21, 24, 28 Fr. Posterior 24 Fr. Female 24, 28 Fr. Bakelite open tube. Light mounted externally	A.C.M.I.
Wolf adult cystoscopes (any combination of telescopes and sheaths available)	Complete for adult cystoscopy and urethroscopy	Rt. angle 90°. Obliquely forward 135°. Obliquely backward 60°. 2 slightly obliquely forward 100°, 110°. Telescope removable from catheter guides single and double		12 Fr. (Charr.) examining 15 Fr.: 1 — No. 5 Fr. 17 Fr.: 2 — No. 5 Fr. Various sizes up to No. 24 Fr. which takes 1 — No. 10 Fr. or 2 — No. 7 Fr. Concave ⎫ Fenestra Convex ⎬ on Straight ⎭ side Light on sheath. Operating sheath contains guide	Richard Wolf G.m.b.H. (Germany)
Wolf Infant and Children's Cystoscope	Examination, catheterizing and operating	Rt. angle. 7 Fr. telescope curved beak without sheath	Obliquely forward	10 Fr. — examining 12 Fr.: 1 — No. 5 Fr. 13 Fr.: 2 — No. 4 Fr. Concave beak. Light in beak	Richard Wolf G.m.b.H. (Germany)
Wolf Resectoscope	Endoscopic prostatic surgery. Rack and pinion motivation of loop	Rt. angle. Direct (almost)	Obliquely retrograde	27 Fr. Slightly convex beak. Fenestra on side of sheath	Richard Wolf G.m.b.H. (Germany)
Wolf Resectoscope Small	Endoscopic surgery. Trigger manipulation of loop	Rt. angle. Obliquely forward		16, 20 Fr. Straight beak. Fenestra in side	Richard Wolf G.m.b.H. (Germany)
Wolf-Hosel Resectoscope	Endoscopic prostatic surgery. Pistol grip and trigger manipulation of loop.	Obliquely forward		24, 27 Fr. Obliquely open end. Handle and connections rotate around sheath, loop and telescope	Richard Wolf G.m.b.H. (Germany)
Wolf Lithotriptor	Visual lithotripsy in adults	Rt. angle. Moves with jaws. Carries light	None	Jaws move parallel to shaft. Rock and pinion on a wheel	Richard Wolf G.m.b.H. (Germany)

Table 1. (Continued)

| Name | Use and special features | Optical system and telescope | | Sheath size, catheter capacity, type, attachments | Manufacturer and/or distributor (see p. 10 for list of manufacturers) |
		standard equipment	others available		
Wolf Stone and Foreign Body Forceps	Visual lithotripsy, foreign body and biopsy in adults	Rt. angle. Carries light	None	Interchangeable serrated and sharp jaws	Richard Wolf G.m.b.H. (Germany)
Young Adult Cystoscope	Routine examining, operating, catheterizing and posterior urethroscopy in adults	Rt. angle. Examining. Convertible. Protecting disc	Retrograde Obliquely Forward	24 Fr.: 2 — No. 7 Fr. 1 — No. 9 Fr. Convex and concave. Light in beak. Rotating large stopcocks	A.C.M.I.
Young Cystoscopic Ronguer	Foreign bodies and biopsy specimens in adults	Rt. angle carries light	None	Transverse forceps handle motivation of sharp jaws which open at right angles to shaft	A.C.M.I. G.U. Mfg.
Young Cystoscopic Ronguer Improved	Foreign bodies and biopsy specimens in adults	Rt. angle carries light	None	Forceps handles in plane of shaft	A.C.M.I.
Young Infant Cystoscope	Examining and catheterizing in infants	Rt. angle carries light	None	9½ Fr. Examining. 12 Fr. Double catheterizing. 2 — No. 3 Fr. Inclined plane deflector. Convex beak	A.C.M.I.
Young Urethroscope	Anterior and posterior urethroscopy in adults	Magnifying lens mounted on sheath	None	Anterior 22, 24, 26, 28 Fr. Straight open tube. Posterior 24, 26, 28 Fr. Convex open tube. Light mounted externally	A.C.M.I.
May (Ferd.) Strahl-Cystoscope (no beak)	Routine cystoscopy and posterior urethroscopy, while washing the field of vision	System Zeiss-Kollmorgen-Heynemann boilable, 135⁰ forward	Panor (panorama, prograde and retrograde)	No. 21 or 16 Fr. boilable	Heynemann
May (Ferd.) Strahl-Cystoscope (no beak)	Routine cystoscopy and posterior urethroscopy, while washing the field of vision, fixed catheter deflector, examining and single catheterizing	System Zeiss-Kollmorgen-Heynemann boilable, 135⁰ forward	Panor (panorama, prograde and retrograde)	No. 22 or 17 Fr. boilable 1 — No. 5 Fr.	Heynemann
May (Ferd.) Strahl-Cystoscope (no beak)	Routine cystoscopy and posterior urethroscopy, while washing the field of vision, fixed catheter deflector, examining and double catheterizing	System Zeiss-Kollmorgen-Heynemann boilable, 135⁰ forward	Panor (panorama, prograde and retrograde)	No. 23,5 or 20 Fr. boilable 2 — No. 5 Fr.	Heynemann

Table 1. (Continued)

Name	Use and special features	Optical system and telescope		Sheath size, catheter capacity, type, attachments	Manufacturer and/or distributor *
		standard equipment	others available		
May (Ferd.) Strahl-Cystoscope (no beak)	Routine cystoscopy and posterior urethroscopy, while washing the field of vision, fixed catheter deflector, examining, single catheterizing and operating	System Zeiss-Kollmorgen-Heynemann boilable, 135° forward	Panor (panorama, prograde and retrograde)	No. 24 or 20 Fr. boilable 1 — No. 8 Fr.	Heynemann
May (Ferd.) Strahl-Cystoscope (no beak)	Childrens' cystoscopy and posterior urethroscopy, while washing the field of vision, fixed catheter deflector, examining, single catheterizing and operating	System Zeiss-Kollmorgen-Heynemann boilable, 135° forward	None	No. 15 Fr. boilable 1 — No. 5 Fr.	Heynemann
May (Ferd.) Strahl-Cystoscope	Infants' cystoscopy while washing the field of vision, fixed catheter deflector, examining, single catheterizing and operating	System Zeiss-Kollmorgen-Heynemann 135° forward	None	No. 11 Fr. 1 — No. 4 Fr.	Heynemann
Fischer (Karl S.) Urethroscope with May Haywalt dilators	Anterior and posterior urethroscopy, cystoscopy of small bladders, single catheterizing and operating in urethra and bladder, especially narrow strictures, combined with Heywalt-May dilators No. 9, 28.5 Fr.	System Zeiss-Kollmorgen-Heynemann boilable, 172° forward	None	No. 19,5 or 22 Fr. 1 — No. 5 Fr. 2 — No. 5 Fr.	Heynemann
Mauermayer Resector	Endoscopic surgery of all tumors (prostate and bladder), double illumination system: 1 lamp fixed on sheath, 1 lamp movable with loop, one hand operation, the cutting force is executed by the spring; thumb controls irrigation even while forefinger assists in rectum; one two-ways-stopcock	System Zeiss-Kollmorgen-Heynemann boilable, 172° forward	135° forward Panor (panorama, prograde and retrograde)	No. 27 Fr. (metal sheath)	Heynemann

* See below for list of manufacturers.

Availability of endoscopes
The endoscopes listed in this table are available from the following companies

Abbreviation	Manufacturing or Distributing Company
A.C.M.I.	American Cystoscope Makers, Inc., 1241 Lafayette Ave., New York, 59, New York
Br. Cysto.	British Cystoscope Company, 44 Clerkenwell Road, E.C. 1, London
Electro-surg. . . .	Electro Surgical Instrument Company, Rochester, New York, U.S.A.
End. Inst.	Endoscopic Instrument Company, Ltd., 52 Shirland Road, London W. 9
G.U. Mfg.	Genito-urinary Manufacturing Company, Ltd., 28 a, 33 and 32 Devenshire Street, London W. 1
Gentile	P. Gentile and Cie, Société à Responsabilité Limitée au Capital de 18,000,000 de Fr., 49, rue Saint-André des Arts, Paris, VI
Greenwald	Greenwald Company, 2688 Dekalb Street, Gary, Indiana, U.S.A.
Heynemann . . .	C. G. Heynemann, München 8, Germany
Takei	Takei Company, Tokyo, Japan
Nat. Elect.	National Electric Instrument Company, Elmhurst, New Hampshire, U.S.A.
Wolf	Richard Wolf Instrument Company, Germany

ocular end is covered by a window which may be either plain glass or a magnifying lens. Catheter guides are inserted through the sheath and direct the catheters

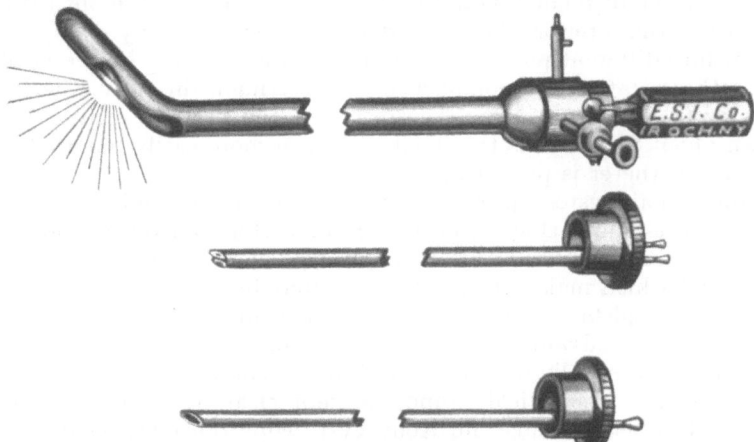

Fig. 1. Braasch direct vision cystoscope

straight forward within the bladder. A single, larger catheter guide is used when larger instruments are passed. A right angle lens system is available to pass through the sheath for examination purposes.

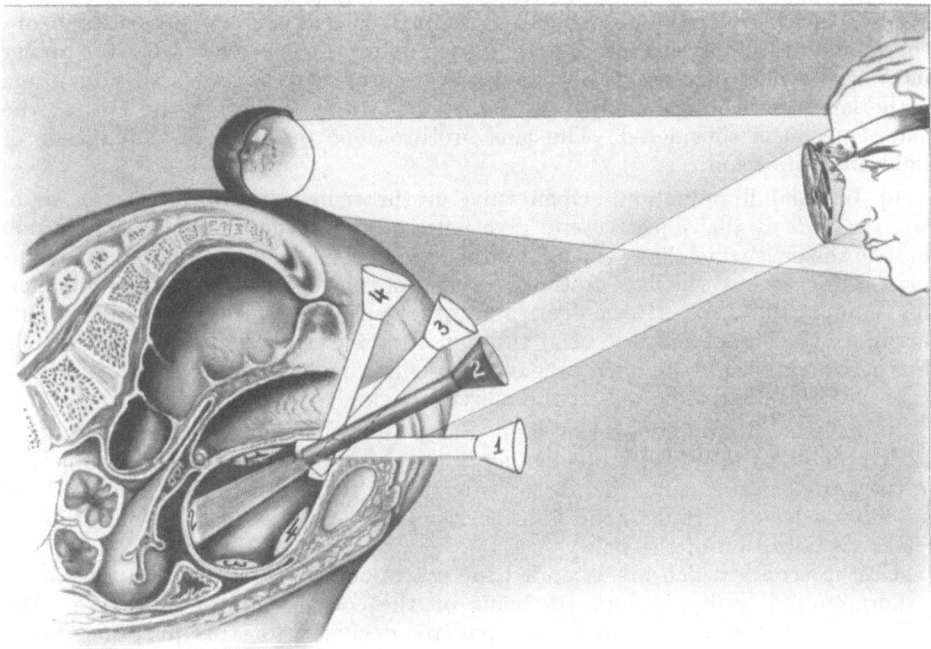

Fig. 2. Cystoscopy with Kelly cystoscope showing reflected light from head mirror and various positions of instrument which permit visualization of most of interior of bladder (adapted from KELLY and BURNHAM)

One of the disadvantages of the direct vision cystoscope is that a smaller than 24 Fr. size does not provide adequate vision.

The Kelly cystoscope (Fig. 2) is used in women. The patient is placed in the knee chest position. Negative intraabdominal pressure in this position allows the bladder to become distended with air which enters through the open tube cystoscope. Most of the interior of the bladder can be surveyed by manipulating the instrument into different positions. Illumination is reflected from a head mirror. Ureteral catheterization is accomplished by inserting a fine wire stilette through the catheter to stiffen it. After the tip enters the ureteral orifice, the wire is withdrawn about 2 cm.; thus the flexible tip can more easily follow the ureteral curves as the catheter is passed upward.

An improved air cystoscope has recently been developed in France (GODDARD). The ocular end is funnel shaped, providing for better controlled vision and facilitating the passing of instruments through it. There is a handle attached near the ocular end which makes it easier to manipulate.

Some gynecologists who also practice female urology find the Kelly cystoscope to have some advantages over others. This instrument, however, permits of only a small and usually dark field of vision. Debris and clots cannot be easily evacuated and the visual field cannot be kept clear by irrigating fluid during cystoscopy. Physicians using the Kelly cystoscope routinely claim good visualization.

3. Urethroscopes

Open tube urethroscopes are useful because they give direct access to the urethral mucosa. Silver nitrate stick or solutions on an applicator can be applied directly to lesions in the anterior or posterior urethra. The field of vision through an open tube urethroscope is nearly as large as that obtained through a lens urethroscope because the area being examined is always very close to the objective end of the instrument; it is only by moving the lens away from the object that a larger field of vision is obtained. Therefore, there is very little advantage to the lens urethroscope compared to the open tube one when the size of the field of vision is considered. The lens urethroscope provides the advantage of slight magnification.

a) **Internal illumination.** Open tube urethroscopes in which the *source of light is inside* at the objective end give better illumination of the field of vision than do the ones which have the light outside the tube. A lamp mounted on a light carrier which brings the light to the distal end of the tube is used in the Ballenger, the Young, the Wolf (Germany), and other urethroscopes. A similar light-carrying attachment may be used through the McCarthy panendoscope sheath, thus converting it into an open tube urethroscope. The Ravich open tube urethroscope has the light mounted in the convex beak.

b) **External illumination.** The light *source is outside* the tube in the Squier, the Young and the Wilhelm open tube urethroscopes. A strong light is mounted on the flange at the ocular end of the tube and is focused into the tube. A small magnifying lens is fitted to the light carrier and can be swung into position for better visualization of the field.

Urethroscopes which are intended for use in the prostatic urethra only have a short curved beak, the fenestra being on the convex side of the curve. The Ballenger and the Ravich urethroscopes are designed for this purpose. Some instruments such as the Squier and Young have both the straight and the curved tubes. The straight tube of the Wilhelm instrument is made of nonconducting bakelite which facilitates the use of electrodes through it. Some urethroscope tubes are supplied in several sizes, usually from 22 Fr. to 28 Fr. A short tube for use in the female urethra is supplied with the Wilhelm instrument.

II. Lens endoscopes

1. Advantages

Cystoscopes with lens systems are much more widely used and have numerous advantages over the open tube instrument. A larger and brighter field of vision is obtained. The bladder can be more thoroughly examined; all of its interior can be surveyed clearly when the different optical systems — forward vision, obliquely forward, right angle, and retrograde — are used. When the inner lens is close to the tissue being viewed, there is magnification. The field of vision can be kept clear by allowing fluid to flow in through the instrument during the examination. Blood clots and debris can be evacuated through the sheath.

Lens urethroscopes provide a slightly larger field of vision than open tube instruments, but have the disadvantage of not permitting the use of medication through them because water is used as a distending medium.

2. Optical systems used in endoscopes

The lenses in modern endoscopes are as accurately ground, as complicated and as numerous as those in a microscope (Fig. 3). The image enters through the objective lenses, is reflected into the tube of the telescope where it passes

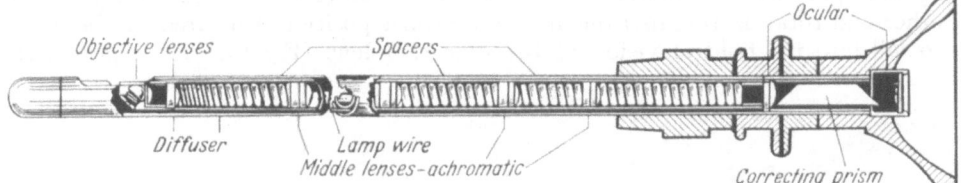

Fig. 3. Optical system in right angle lens telescope for Brown-Buerger instrument (adapted from A.C.M.I. catalogue)

through several lenses accurately spaced; it is corrected by a prism and emerges from the ocular end in parallel rays which need no focusing by the eye. Optical systems are constructed in endoscopic telescopes to give different views such as directly forward, obliquely forward, right angle and retrograde. Most endoscopes utilize one of these systems as standard equipment, but telescopes with the other views are also available for most of them. For example, the Brown-Buerger cystoscope utilizes the right angle optical system as standard equipment, but any of the other telescopes are available for use in this instrument. Similarly the obliquely forward optical system is standard for the McCarthy cystoscope, but the other telescopes are also made to use in it.

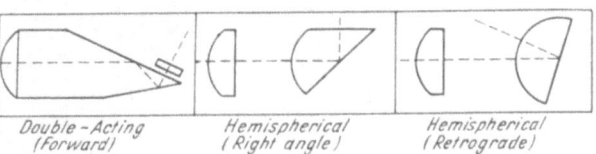

Fig. 4. Schematic view of construction and position of objective lenses to give different directional views (adapted from A.C.M.I. catalogue)

The position and construction of the objective lenses determines the direction of the view through the telescope (Fig. 4). The hemispherical prism developed by REINHOLD H. WAPPLER (American Cystoscope Makers) for Dr. WILLIAM K. OTIS in 1900 (OTIS), has largely taken the place of the right angle prism which was found unsatisfactory for clear wide vision and durability. The hemispherical prism is now used in most right angle and retrograde objectives. The double acting prism was also developed by WAPPLER and is used for vision slightly forward of the right angle.

a) Right angle. The *right angle* view through the telescope is the most widely used for routine cystoscopies. Most of the telescopes containing this optical system have a 60 degree field of vision which is projected at right angles (90⁰) to the shaft of the telescope (Fig. 5). It is the type used as standard equipment for the Brown-Buerger, the Young, the Gentile, the Wolf, and is available for

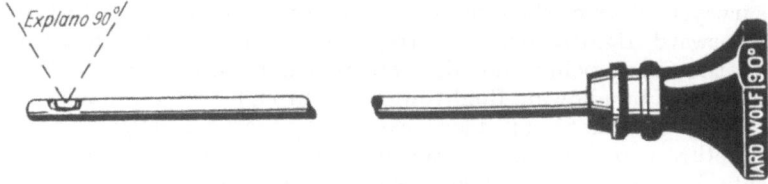

Fig. 5. Right angle view telescope Wolf

most other lens cystoscopes. With the right angle view telescope, the entire interior of the normal bladder can be visualized by rotating and moving the instrument into different positions. The most difficult area to see is the portion of the fundus opposite the vesical orifice, the so-called "blind spot". However, by manipulating the ocular end extremely to one side, then extremely to the other and rotating the instrument to the proper position, this area can be made to fall into the field of vision of the objective lens (Fig. 6). The right angle

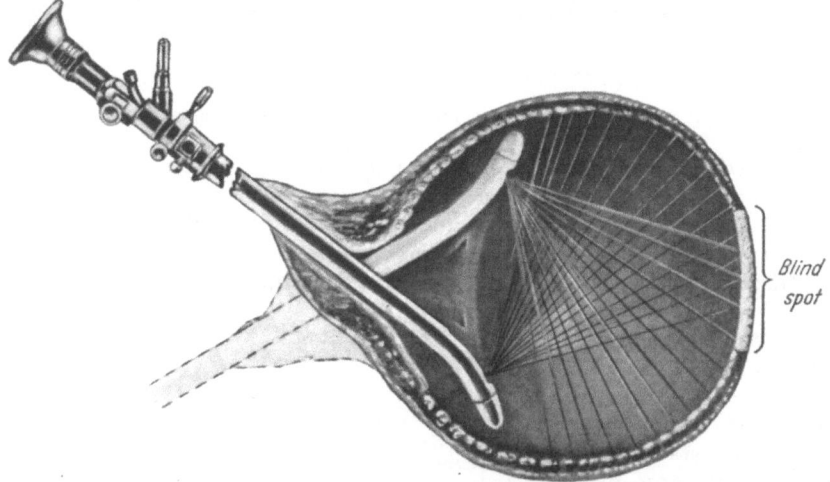

Fig. 6. Manipulation of right angle telescope to visualize area of bladder opposite vesical orifice-"blind spot" of bladder

lens provides a broad "straight down" vision of the floor of the bladder. This area, where the ureteral orifices are located and where most of the pathology occurs in the bladder, is the most important to visualize clearly. The right angle view does this better than the views in other directions, and is therefore the most widely used for routine cystoscopy (Fig. 7A).

b) Obliquely forward. The *obliquely forward* view through the telescope is used by some urologists for routine diagnostic endoscopy. It is admirably suited for visualization during endoscopic surgery and for viewing the prostatic urethra.

The McCarthy foroblique[1] optical system is used in the McCarthy panendoscope and all of its modifications including the Stern-McCarthy electrotome, and is also available for other instruments. The center of the visual field extends forward to a 27.5 degree angle. The upper edge of the field is straight forward, and the entire field encompasses 56 degrees (Fig. 7 B). A similar optical system is made by RICHARD WOLF. Besides being advantageous for use in the prostatic urethra, this optical system more clearly brings into view the area of the bladder wall opposite the vesical neck which is the "blind spot" through the right angle telescope. The distal tip of rigid instruments used through the cystoscope can be seen with the foroblique lens but would be out of the right angle field of vision. It is, however, difficult to visualize the bladder wall near the vesical orifice through the obliquely forward optical system, and the view of the base of the bladder is from an oblique angle rather than straight down as through the right angle telescope. For some urologists it is easier to catheterize the ureters while using the

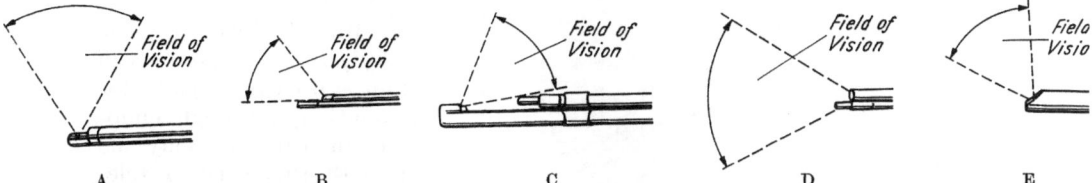

Fig. 7 A—E. Fields of vision of different optical systems. A Right angle view examining telescope. B McCarthy foroblique[1] Obliquely forward view telescope. C Retrograde view examining telescope. D Direct vision examining telescope. E Obliquely forward vision telescope

foroblique lens because there is a more direct approach to the orifice — the catheter can be inserted from a more nearly parallel approach.

A field of vision which is obliquely forward at its upper margin and right angle at its lower margin (Fig. 7 E) is useful for examination purposes. A good view of the prostatic urethra is obtained with this optical system and the "blind spot" of the right angle lens is brought into view. Telescopes with this angle vision are available for use in the Brown-Buerger convex sheath, the Young, the Ravich, the Wolf and the Gentile cystoscopes.

c) Retrograde. The *retrograde* view is used chiefly for examination of the bladder neck from within the bladder and is especially useful in estimating the amount of thickening or intravesical protrusion of the edges of the vesical orifice or prostate. The field of vision is obliquely backward and encompasses about 70 degrees (Fig. 7 C). The trigone may not be visible through any other telescope when there is an overhanging intravesical middle lobe of the prostate. An instrument such as a fulgurating tip can be used through some retrograde telescopes for treatment of the area immediately inside the vesical orifice. A retrospective optical system is made to fit most of the lens cystoscopes, including the Brown-Buerger, the Young, the McCarthy, the Wolf, the Gentile, and their modifications. It is valuable for examination of the prostatic border just before transurethral prostatectomy and for survey of the bladder neck at the conclusion of the operation. When thus used through the resectoscope sheath it gives the surgeon knowledge of the relation of the ureteral orifices to the prostatic border, in addition to the amount of intravesical protrusion of the gland. The cystoscopist who is not accustomed to using the retrograde telescope has difficulty in orientation when employing this visual system. One edge of the field of vision is backward along the shaft of the telescope. The portion of the bladder neck closest to the

---

[1] Foroblique is the trademark registered by the American Cystoscope Makers, Inc. to designate their obliquely forward telescopic optical system.

lumen of the vesical orifice is at this edge of the field. When the dorsal area of the bladder neck is in view, the portion closest to the orifice of the vesical neck is in the upper edge of the field (Fig. 8A). When the instrument is rotated to bring the ventral area of the bladder neck into view, the portion closest to the orifice of the vesical neck is in the lower edge of the field (Fig. 8B). Similarly when the instrument faces lateralward, the edge of the bladder neck is on the side of the field of vision — in the left of the field when the objective lens is directed toward the left of the patient, and in the right of the field when directed toward the right of the patient.

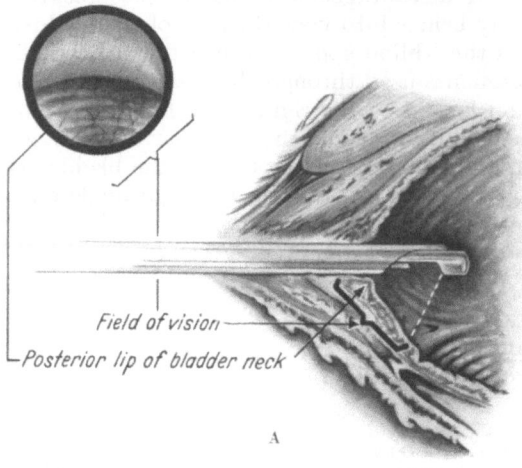

Field of vision

Posterior lip of bladder neck

A

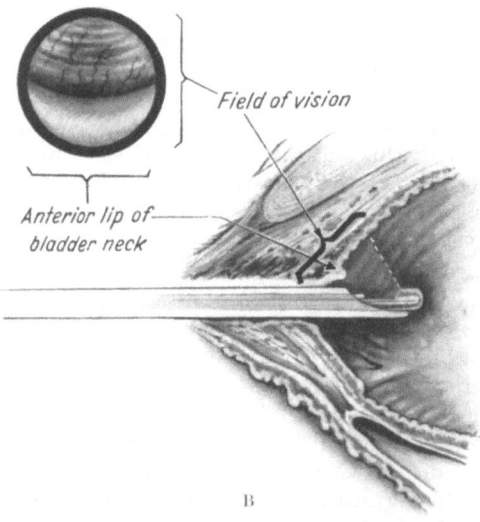

Field of vision

Anterior lip of bladder neck

B

Fig. 8 A and B. Orientation with retrograde lens. A View of dorsal lip of bladder neck. B View of ventral lip of bladder neck

d) **Directly forward.** The *directly forward vision* optical system is used principally for examination, although endoscopic instruments may be used with this type of telescope. It is most useful in the urethra, and is standard equipment for the Lowsley and the Buerger Universal urethroscopes. The Vest telescope is used through the resectoscope sheath to examine the prostatic urethra from its distal margin. The resectoscope made by RICHARD WOLF (Germany) has a telescope which is only slightly deflected from the direct forward view. The comprehensive view thus obtained enables the operator to discover encroaching prostatic tissue which might otherwise be missed (Vest). The field of vision encompasses about 70 degrees (Fig. 7D).

e) **Adjustable.** A visual system which combines all these views except the direct forward, and which is *adjustable*, features a small rear surface mirror mounted on a pivot in front of the objective lens (Mc CARTHY 1947). A small knob raises or lowers the mirror, changing the angle of vision. Thus a total field of about 170 degrees can be brought into view without changing the telescope. The McCarthy "periscopic" cystoscope is the one manufactured by the American Cystoscope Makers. A similar instrument called the Retrograde Cystoscope is manufactured in Japan and is used widely in that country. This instrument has a prism which is attached to the tip of the telescope and about a 50° right angle vision is obtained. By moving the telescope forward, about 45° of retrograde vision is also obtained. This instrument thus provides right angle and retrograde views, but not forward and foroblique views.

3. Telescopes

The telescope of lens endoscopes may contain other essential elements in addition to the optical system. Whatever space is required for these other elements reduces the space for the optical system, thus making the field of vision smaller. Therefore, a telescope designed for examining only gives a larger and brighter field of vision than one made to be used with catheters and instruments. Both an examining and a catheterizing model are available in most telescopes utilizing the various optical systems.

a) **Wiring circuit.** The *wiring circuit* is contained within the McCARTHY foroblique telescope and also in the right angle telescope used in the Stern-McCarthy electrotome. A separate tube which carries the wiring circuit and the light bulb is attached to most retrograde and direct vision telescopes and also to the right angle telescope used in the McCarthy routine cystoscope. Telescopes which have an attached light can be removed and the bulb changed easily but when the light is incorporated in the tip of the sheath, the entire instrument must be removed from the urethra whenever the light fails to function during the examination.

b) **Catheter guides and deflectors.** Catheterizing and operating telescopes are equipped with *guides* for the instruments which are passed through the cystoscope. The field of vision is necessarily reduced by the space used for the passing of catheters and operating instruments and of their guides. It is therefore more difficult to visualize the interior of the bladder through a catheterizing or operating telescope than through an examining lens system used in the same sheath. The *catheterizing* telescope is equipped with two catheter guides, and the *operating* instrument has one larger centrally placed guide through which operating instruments and larger catheters can be passed. Many catheterizing and operating telescopes are now combined into a *convertible* telescope which can be used for both catheterizing and operating. The advantages of the convertible over the separate operating and catheterizing telescopes are: 1. the initial cost of the single combined instrument is less than that of two; 2. it is often desirable to use three catheters or instruments simultaneously, which can be done with the convertible telescope. *Multiple channels* are present in the Brown-Buerger convertible telescope (Fig. 9) which has three catheter guides, the central one for insertion of large instruments or catheters and two lateral ones for smaller catheters. A removable fin separates the compartment within the sheath into two catheter tunnels. The Shulte panendoscope adapters are made with one, two, three, or four catheter channels; they are used with the 28 Fr. panendoscope sheath. The telescope of the Vest dilating cystoscope has one large catheter channel for passage of large instruments (Vest). Two small catheter tunnels, each with a deflector spring roller mechanism at its objective end, is used in the McCarthy *ejaculatory duct catheterizing* instrument.

A catheter *deflector* is attached near the objective end of right angle operating, catheterizing and convertible telescopes, and is controlled by a knob near the ocular end to which it is attached by wires. Some of the obliquely forward vision telescopes such as the McCarthy convertible close vision cystourethroscope and the McCarthy deflecting foroblique panendoscope have a similar deflector. The end of a catheter or instrument being used through a cystoscope can be manipulated more easily and widely with the deflector than is possible without it. Rigid cystoscopic instruments cannot, however, be used with it, but they can be used through an instrument such as the McCarthy panendoscope which has a straight sheath and no movable deflector on the telescope. In these a fin or guide attaches to the foroblique telescope. The tip of the guide is deflected away from the objective end of the telescope and directs the catheter toward the bladder wall (Fig. 10). The single guide is used for operating; the double guide divides

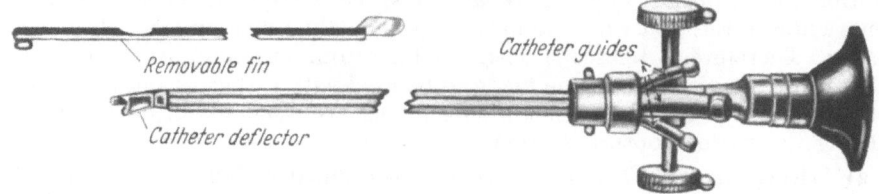

Fig. 9. Convertible telescope for Brown-Buerger cystoscope. Center catheter guide is used for large instruments and two lateral ones for smaller ureteral catheters. A removable fin may be attached to keep catheters separate as they are passed along telescope through sheath

the space within the sheath into two catheter tunnels (KIEFER). In some cystoscopes such as the Braasch-Bumpus and the Wolf (Germany) operating cystoscope the catheter or instrument guide is in the sheath rather than in the telescope. In the Wolf the catheter channel extends entirely through to the tip of the curved beak. The operating instrument or catheter used through this channel

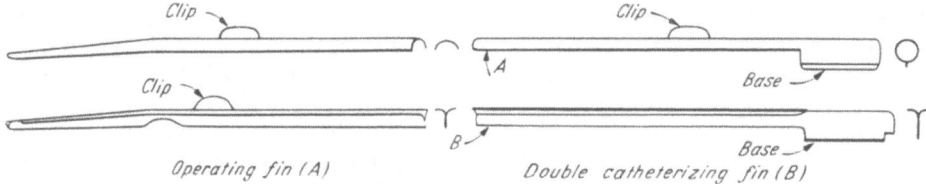

Fig. 10. Catheter deflectors for McCarthy panendoscope

is deflected by the curved beak and the right angle telescope projects beyond the beak, thus keeping the tip of the instrument within the field of vision.

c) Protection of catheters. Increased *protection* of the urethral catheters from contamination is afforded by the *longer telescope* and a wide *disc* near the eyepiece in the Young instrument (YOUNG).

d) Carriage for telescopes. The *carriage*, including the locking device, is an integral part of most telescopes. In a few, however, such as the McCarthy foroblique, the telescope can be removed from the carriage.

III. Endoscope sheaths

1. Illumination. Types of sheaths

An essential part of the modern endoscope is the sheath which is passed through the urethra into the bladder and through which the telescope is inserted in lens endoscopes. The simplest sheath is the Kelly (Fig. 2) which is a tube with no attachments, the light being reflected through it from a head mirror. The sheaths of open tube urethroscopes are also simple. They are described on page 12. The sheath of the Braasch cystoscope (Fig. 1) is more complicated. The light bulb is contained within its vesical end and the wiring circuit within its walls. The post for connecting the light cord and the stopcock for the water inlet is affixed to its side near the ocular end. A plain glass or a magnifying lens window covers the ocular end, and guides for operating instruments and catheters are located near it. A valve for clearing air bubbles from the sheath is optional (TYVAND).

Sheaths for lens endoscopes may be simple, such as the McCarthy, or may be more complicated such as the Brown-Buerger, the Wolf, and the Gentile. The

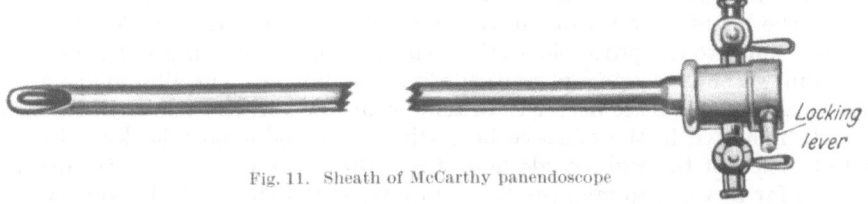

Fig. 11. Sheath of McCarthy panendoscope

sheath for the McCarthy panendoscope is a *straight* tube with an oblique opening in the objective end and a water inlet stopcock and a locking lever near its ocular end (Fig. 11). The various telescopes used in this sheath carry their own light. There are several advantages to sheaths of this type: (1) various sizes can be

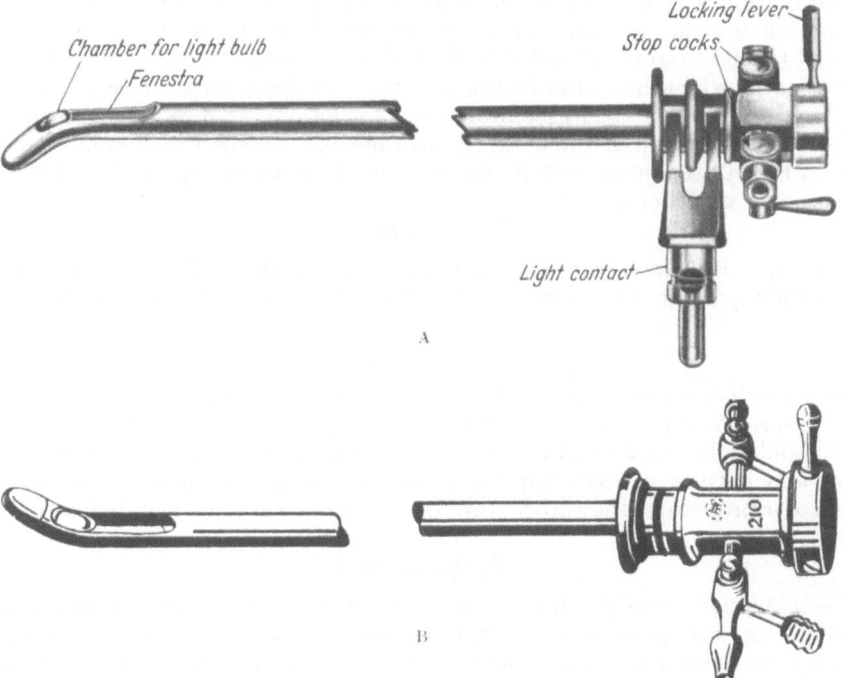

Fig. 12 A and B. A Brown-Buerger convex sheath. B Wolf concave sheath

acquired with little expense; (2) the simple construction, without a light or other attachments makes them inexpensive; (3) the same telescope and bridge assembly can be used in all sizes from 16 Fr. to 28 Fr.; (4) rigid instruments can be used through them. The most commonly used cystoscope sheaths and a few urethroscope sheaths carry the light in the objective end and the wiring circuit extends through the sheath (Fig. 12).

2*

2. Beaks and fenestrae of sheaths

The beak of most light carrying sheaths is curved or angulated for passing through the posterior urethra more easily. In the *concave* sheath the light and fenestra are on the concave side of the sheath (Fig. 12 B). When the curve is away from the light and fenestra, the sheath is called *convex* (Fig. 12 A). The advantages of the convex sheath are: 1. The objective lens can be placed very close to the bladder wall, thus magnifying the object in view; 2. This sheath can be drawn out into the prostatic urethra where a good view can usually be obtained while the water is flowing in and dilating the lumen. The illumination in the concave sheath is a little better than that from the convex because in the former, the bulb mounted in the concave beak directs the light more backward into the field of vision of the right angle lens; the bulb in the convex sheath directs the light too far forward to give the best illumination of the field of vision when the right angle lens is being used. When it is desired to leave ureteral catheters in place, the concave sheath is removed more easily over them than is the convex sheath. The beaks of some endoscope sheaths including the Buerger and the McCarthy cystourethroscopes, the McCarthy routine and the Ravich convertible cystoscopes, are only *slightly convex*, they are suitable for visualization of the prostatic urethra. The beak of the Kirwin cystoscope is jointed or *flexible* so as to function as either a convex or a concave sheath; it also has a fenestra on each side and the light bulb is mounted in the end between the two fenestrae (KIRWIN). Most sheaths which have the fenestra in the side have an *inclined plane* at the distal end of the fenestra to facilitate the passing of catheters; when there is no inclined plane, the tip of the catheter or other instrument which is being passed through the cystoscope may strike against the distal right angle end of the sheath and become obstructed.

3. Light posts

Sheaths which have the light bulb mounted in the distal end are equipped with a light post near the ocular end. Most of these rotate around the sheath.

4. Stopcocks

In some instruments such as the Young, the water inflow and outflow stopcocks also *rotate* around the sheath; the sheath can be rotated without the rubber tubing and light cord winding around it. *Large* size stopcocks permit rapid inflow and outflow of irrigating fluid, thus reducing the time required for the cystoscopic procedure; they are featured on the Young and the Kirwin instruments.

5. Obturators

The *obturator* (mandrin) fills the fenestra of the sheath and is inserted into it before passing the instrument into the bladder or urethra. It removes the hazard of injury from the edges of the sheath as the instrument is being passed. The obturator is reinserted before the sheath is removed unless it is an open end urethroscope; this may be withdrawn gradually through the urethra as the examination is in progress. The obturator used with some open end endoscopes is jointed or flexible (WEBSTER) to allow deflection during introduction; the curve of the posterior urethra can be more easily followed with this type of obturator (Fig. 13).

6. Locks

The lock for holding the telescopes and obturator in the sheath is near the external end of the sheath. Most lens endoscopes used in adults have a locking device; but many used in children and infants depend upon a snug fit to hold telescopes and obturator in position. The lock used on most endoscopes is a bayonet lock mechanism consisting of a sleeve which rotates on the outside of the ocular end of the sheath. It is manipulated by means of an attached arm

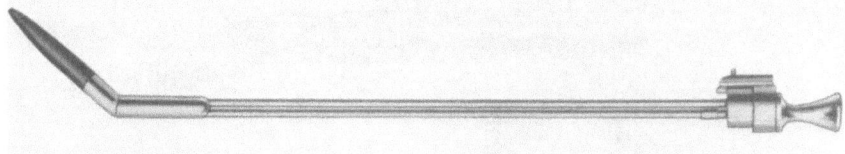

Fig. 13. Timberlake-Alcock obturator. (Courtesy American Cystoscope Makers)

which when rotated through approximately 45 degrees locks the telescope or obturator firmly in place. The older type spring catch lock is used on most endoscopes for transurethral surgery; this type allows more rapid disengagement of the telescope.

IV. Sizes of endoscopes

A variety of sizes of endoscopic sheaths is available in many instruments. Adult endoscopes vary from No. 16 Fr. to No. 30 Fr., and will accommodate ureteral catheters up to one No. 16 Fr. A few instruments are made for use in the adult female and are shorter than the ones designed for routine use. An extra long sheath is available for the McCarthy panendoscope.

Endoscopes designed for use in infants and children are shorter than those for use in adults, and vary in diameter from No. 7 Fr. to No. 16. Fr. The smaller sizes are used for examining only; the larger ones will accommodate two No. 5 Fr. ureteral catheters or one No. 8 Fr. Except for their size, they are in general similar to the adult models.

V. Instruments designed for endoscopic surgery

In general, these are similar to the corresponding instrument used for routine cystoscopy.

The Braasch-Bumpus resectoscope is a direct vision instrument similar to the Braasch-Bumpus cystoscope (Fig. 14). The Thompson resectoscope is a modification of the Braasch-Bumpus instrument, and is the most widely used cold punch resectoscope. The line of vision is through a window, either plain glass or one-half diopter magnifying lens, at the ocular end of the sheath. Additional modifications have been made by other urologists (HUTCHINS; TYVAND). Endoscopic resection of tissue is accomplished by means of a cold tubular blade which is manipulated on the inside of the sheath and punches out successive pieces of tissue. An electrode is passed through a small tunnel along the upper edge of the sheath and is used for spot coagulation of bleeding points.

1. Stern McCarthy visual prostatic electrotome

This is more commonly called the McCarthy resectoscope; it is one of the first instruments designed for endoscopic surgery (Fig. 15). The foroblique optical system is in the telescope of the working element. The sheath is nonconductive bakelite or plastic and the active electrode loop is manipulated by a control handle through a rack and pinion mechanism. The electric cutting and coagulating currents are delivered from an electrosurgical unit to the loop through a wire connection inserted into the active terminal post. The irrigating fluid inlet is a tube on the sheath to which the stopcock is connected. Right angle, retrograde, and direct forward view telescopes are available for examination purposes to be used

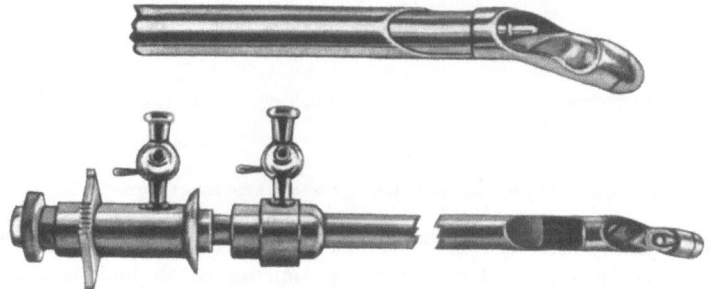

Fig. 14. Braasch-Bumpus resectoscope

through the resectoscope sheath. Sheaths of sizes 24, 26, and 28 Fr. are available for the McCarthy electrotome; the 28 Fr. is the most widely used. The size 16 Fr. McCarthy electrotome has a smaller telescope and working element, but is the same length as the larger instrument. The 12 Fr. size especially designed for use in infants is shorter and the telescope is an extremely small foroblique optical system.

2. Resectoscope made by Wolf (Germany)

This is somewhat similar to the McCarthy instrument. The chief differences are that the sheath has a slightly convex beak and the fenestra is on the side of

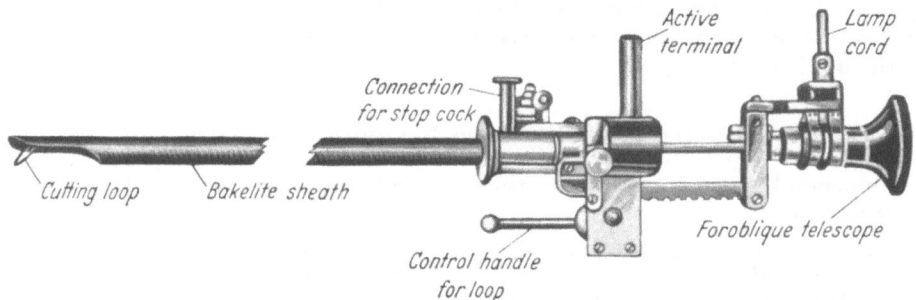

Fig. 15. The Stern-McCarthy visual prostatic electrotome (McCarthy resectoscope)

the sheath rather than at the end; the rack and pinion of the working element is on the opposite side of the sheath from the fenestra. A needle for injecting fluid into the prostate and which is on a carriage for insertion through the sheath is available with this instrument.

3. Modifications of the McCarthy electrotome

These have been designed by a number of urologists. Most of them are operated with one hand. The *Nesbit* (Fig. 16) modification is that most commonly used. The fingers of one hand hold a handle which is rigidly attached to the sheath, while the thumb manipulates the cutting loop. The other hand is left free so that the index finger can be inserted into the rectum and the prostate palpated; three dimensional perception can thus be obtained while the resection is in progress. The Nesbit resectoscope is available in different sizes similar to the McCarthy electrotome. The Nesbit perineal resectoscope is size 33 Fr. and is shorter than the one used routinely. The *Creevy* modification provides control of motion of the loop by means of two finger grips; one attached to each side of the working element. The *Miller* and the *Foley* modifications provide a circular ring for manipulation of the loop. The *Iglesias* has a leaf type spring to return the loop to the retracted position and there are two finger rests on the sheath. The *Baumrucher*, the *Gibson* and the *Scott* modifications all provide for control of the loop

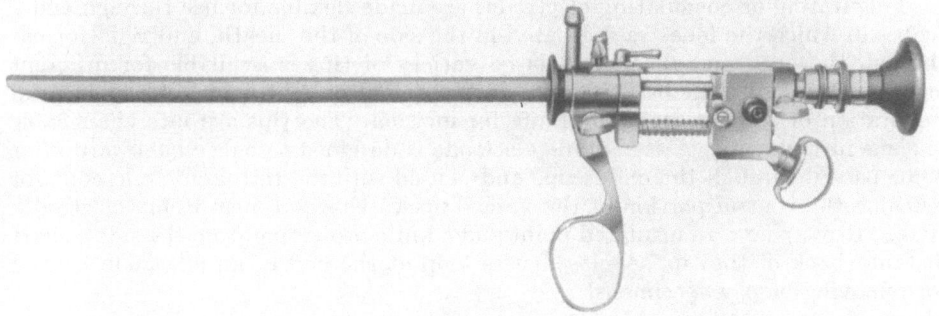

Fig. 16. Nesbit resectoscope. (Courtesy American Cystoscope Makers)

with the finger rather than with the thumb. A pistol grip which provides more firm manipulation is rigidly attached to the working element of the Gibson, the Scott and the *Hosel Wolf* instruments. The sheath, telescope and cutting loop of the Scott and the Hosel can be rotated a full 360 degrees, while the pistol grip, trigger, water inlet and light post remain stationary. The Foley modification is also rotatable; the water inlet, lamp terminal and high frequency terminal remain stationary while the sheath, the telescope and the working element are rotated.

4. Visual lithotrites

These are endoscopic instruments for crushing bladder stones under vision. The optical systems are similar to those in cystoscopes. By their use small stones can be seen and grasped with the jaws of the lithotrite. Stones located in cellules or diverticula or behind an enlarged middle lobe of the prostate can be more easily located by vision than by feeling with the blind lithotrite. A disadvantage of the visual lithotrite is the inability to construct it as strong as the blind instrument. Another is the rapid clouding of the field of vision with finely crushed fragments of stone during the lithotripsy; considerable time is required to clear the visual field by irrigating the bladder. By skillful manipulation of the blind lithotrite most stones can be crushed atraumatically and more rapidly than with the visual instrument.

Telescope. The *telescopes* of all visual lithotrites and rongeurs carry the light bulb. A right angle optical system is used in the Alcock and Wolf lithotrites and in the Deming, the Kirwin, the Young and the Wolf rongeurs or forceps. The foroblique telescope is used in the Hendrickson lithotrite and in the Ravich lithotriptoscope. All these instruments except the Alcock have forceps like handles for opening and closing the jaws. A rack and pinion operated by a wheel is the motive power for the Alcock and the Wolf lithotrite. The jaws open and close parallel to the shaft of the instrument in all but the Young and the Kirwin instruments; in them the jaw motion is at right angles. The Wolf combination bladder stone and foreign body forceps has interchangeable sharp-edged and serrated jaws; they and the forceps type handle open at right angles to the sheath.

B. Instruments used through endoscopes

I. Electrodes

Fulgurating or coagulating electrodes are made flexible for use through endoscopes in which the fenestra is located in the side of the sheath, and rigid for use through the open end sheath. A large variety of tips is available for different uses (McDONALD). The ball tip is for deep coagulation; the conical for superficial fulguration of a small area; the knife for incision. The tips are interchangeable in some instruments. A retrograde electrode is designed to curve backward after being passed through the endoscope and is used with the retrograde telescope for reaching the ventral portion of the vesical neck. Ureteral meatotomy electrodes (RIBA; ROWE) have an insulated point and a knife projecting from the side a short distance back of the tip. A closed wire loop at the end of an electrode is used for removing biopsy specimens.

II. Forceps, rongeurs, and scissors

These are made with either flexible or rigid stems. They are necessarily small in order to pass through operating cystoscopes. The rongeurs are cupped and sharp edged, and are sometimes used for removal of tissue for biopsy. The specimen obtained, however, is so small and superficial that a satisfactory examination of the tissue is seldom possible — it is better to use a larger endoscopic rongeur or the loop of a resectoscope. Forceps are satisfactory for removal of small foreign bodies. Larger jaws are possible on the rigid stemmed instrument. Scissors are used mostly for ureteral meatotomy. Some have a long bougie-like tip on one blade for insertion into the ureter to facilitate meatotomy. A forceps type handle motivates the jaws or blades. One model by American Cystoscope Makers, Inc., has interchangeable rigid or flexible stems for use with the same handle. A forceps for grasping and removing small bladder stones is made with removable handle and locking jaws to allow the cystoscope to be removed, leaving the flexible stem passing through the urethra; the forceps with the stone in its grasp is then pulled out (I. SIMONS).

III. Infiltration needles

These are used through endoscopes to inject medication into the bladder wall, prostate or urethra. Either flexible needles, for use in endoscopes with a side fenestra, or rigid for the open end instrument, are available. Radon seeds may be implanted into the bladder or urethral wall through a needle equipped with a special plunger for forcing the seeds out into the tissue (RINKER).

IV. Ureteral catheters

for routine endoscopy are described in Chapter II.

V. Special ureteral catheters

Catheters made from plastic material such as polyethylene are chemically inert and nonirritating to living tissue (BROWN and HARRISON). They are non-radiopaque and nongraduated, and are used when it is desirable to leave the catheter in the ureter for continuous drainage. Sizes 3 to 12 Fr., either round or whistle tip, are available. Double lumen polyethylene ureteral catheters (McCarthy) permit continuous irrigation. A two-wing olive tip self retaining catheter (DAVIS) and the inflatable bag ureteral catheter (LIVERMORE) will remain in place without tying unless forcibly pulled out.

VI. Ureteral instruments

1. Bougies

These are similar to catheters in size and in shape of the tip. They are used to dilate the ureter and are stiffer than catheters. Their disadvantage is that they cannot be left in for continuous drainage after having passed by a stone or other obstruction in the ureter.

2. Calculus dislodgers

Most ureteral instruments other than catheters used through cystoscopes are for removal of calculi from the ureter. There is danger of injury from any of these, especially when the calculus is high in the ureter. When gentleness and discretion are exercised while using them, the danger is no greater than that from the average surgical procedure.

a) Wire basket. The *wire basket* ureteral stone remover consists of from three to five or six strands of spring wire into which the stone is manipulated (Fig. 20, p. 40). The wires are parallel to the shaft of the instrument and converge both at the distal and proximal ends of the basket. A filiform tip (NATION) guides the basket into the ureter and the shaft is a tight spiral of semi-flexible metal. An inflatable balloon may be located below the basket to dilate the ureter as the instrument is withdrawn (ROOME). In some, such as the JOHNSON, the spring wires are kept apart to receive the stone by their arched contour, and in others, such as the Councill, the wires are spread apart by approximating the two ends of the basket — a center wire is attached to the distal end of the basket and runs through the shaft to a ringed handle on the outside; a pull on the handle spreads apart the wires of the basket.

b) Looped ureteral catheter. A *looped ureteral catheter* is sometimes used to dislodge a calculus (ELLIK). One end of a nylon cord is fastened securely to the tip of a No. 5 Fr. olive tip catheter. The other end is passed into the lumen through a small hole drilled in the side of the catheter 5 cm. from the tip and is threaded down to emerge at the proximal end of the catheter. The loop is formed by pulling on the cord. The tip is passed into the renal pelvis before forming the loop; as it is drawn down through the ureter, the calculus may be caught in the loop.

c) Forceps. *Forceps* which are passed up the ureter to grasp a stone may injure the ureteral mucosa by pinching and tearing. They are available with or without a filiform tip. The Ferguson instrument has two flexible metal bands which are made to arch out from a cut away section near the end of the shaft.

3. Transilluminator

A ureteral *transilluminator* (TATUM) may be passed up the ureter to help locate stones, strictures or other lesions during open surgery upon the ureter.

A small light bulb is in the distal end of a ureteral catheter stem and is connected to a battery with a wire running through the lumen of the catheter. The light shines through the ureteral wall when the ureter is exposed.

The detection of ureteral calculi may be accomplished by using the wax bulb on a catheter or bougie (KEYES). The bulb is formed, then the catheter is passed up the ureter using care not to allow the wax to be scratched while passing it through the cystoscope. After removing the catheter the wax is examined with a magnifying lens; scratches or grooves are diagnostic of ureteral calculus.

A teleprobe (COUNCILL) may aid in detecting stones. The probe, a ureteral catheter stem with a piezo-electric crystal on its tip, is passed up the ureter through a panendoscope. When the tip contacts a stone or other hard object a sound is produced through an audio amplifier and loud speaker system.

C. Cystoscopic attachments

I. Cystoscope holders[1]

These are useful when it is desired to leave the cystoscope in place during completion of a retrograde kidney study and when radium is applied at the distal end of an endoscope. For demonstration purposes the holder is of special value; it maintains the particular field of vision for study by several individuals.

II. Teaching attachment (KAHN)

Such an attachment with one straight and one angulated eyepiece screws on to the ocular end of the telescope after removal of the bakelite eye shield disc. The vision directed by a prism through the angulated eyepiece to the observer is not as bright as that going through the straight eyepiece to the operator-instructor. The image to the observer is often dark and unsatisfactory unless there is exceptionally brilliant illumination of the field of vision in the bladder.

III. Photographic attachments

Equipment for endoscopic photography has been designed by McCREA (McCREA) and is made by American Cystoscope Makers. Similar equipment is made by RICHARD WOLF in Germany and GENTILE in France. Special telescopes which provide a wider and brighter field of vision and extra brilliant lamps are used with a robot camera attached to the optical end of the telescope. The field of vision to be photographed is found through a movable prism which is raised out of the optical path when the exposure is made. The light bulb is overloaded to intensify the illumination during the time of exposure. Black and white film is exposed for one-fourth to one-half second; color requires 1 to 3 seconds. Satisfactory photographs can be obtained by an expert technician, but on the whole are not as suitable for teaching and demonstration as drawings because the field of vision is necessarily limited; a composite of several visual fields is usually shown in a drawing. Satisfactory photography including motion pictures (cinema) can be made through endoscopes whose source of illumination is through a quartz tube (see p. 27). The brightness of the field of vision gives adequate illumination for short exposures through the lens system (FOURESTIER et al.) (JAUPITRE).

---

[1] COUNCILL (1930). BLASUCCI, RIABOFF, ROSENTHAL, MALLARD, SEGAL

D. Sources of light for endoscopes

I. Bulbs

American made light bulbs used in most urological endoscopes of adult size are three volt capacity. The smaller endoscopes for use in children and infants are equipped with bulbs of smaller capacity. Very few bulbs are interchangeable among different instruments. It is therefore important to specify the exact bulb desired by catalogue number or by the name and model of the telescope, sheath, or light carrier when ordering replenishments.

II. Quartz tube

A recent development by Fourestier, Gladu, Vulmiere and the Centre National de la Recherche Scientifique uses a quartz tube for transmitting light to the field of vision through endoscopes. The instruments are manufactured by GENTILE. It is possible to transmit a very bright light; the illumination from this source is much brighter than is possible from previous methods of illumination.

III. Batteries

Batteries are the safest and most accurate and dependable source of current for light bulbs of endoscopes (BINKLEY). Two batteries of $1^1/_2$ volts each connected in series without a rheostat provide the proper current for the commonly used 3 volt bulb. When smaller capacity bulbs are used, one battery may give the proper amount of current, but sometimes a rheostat is required for adjusting the voltage. Battery containers equipped with built-in rheostats and some with a voltmeter are convenient for use with different size bulbs; the container can be hung on the cystoscopic table. A small Mercury battery may be attached directly to the light post of the endoscope. It is of such small size that manipulation of the instrument is not hampered by its presence (BODNER et al.; SEID and COHLER).

IV. Electric house current

This may be brought from a wall plug through a *rheostat* to the endoscope light bulb. From this source of current there is danger of electric shock to the patient and of a short circuit while using an electrosurgical unit during endoscopic surgery. The danger is minimized by the use of an isolation transformer and by grounding the cystoscope (DAVIS and WHITEHEAD).

E. Care and maintenance of endoscopes

I. Routine care

Many of the component parts of endoscopes are necessarily of delicate construction. The small size telescopes, the minute light bulbs, the delicate wiring circuits and the tiny lenses in the complicated optical systems cannot be constructed to withstand rough handling.

1. Basic precautions to prevent breakage

The observance of a few *basic precautions* in the handling of endoscopes will reduce the amount of breakage.

1. Put the case containing the endoscope on the table before opening it. Even the most clever juggler may let the case slip from his hands on to the floor when he opens it in midair.

2. Hold the instrument over a table covered by a pad or thick towel while cleaning, drying, repairing or adjusting it. The most careful handler may let it slip; a short fall to a padded table is much less damaging than a long fall on to a tiled floor.

3. Pick up and transport instruments with the hands not with clamps or forceps. The delicate walls of the sheaths and telescopes may be damaged by clamping them. The instrument may slip from the grasp of the forceps.

4. Handle only one or at the most two parts of an instrument at a time. Grasping a handful might bend or break the thin telescopes or tiny light bulbs.

5. Hold one hand under the instrument while transporting it or handing it to someone else; if it does slip there is a secondary line of defense before it hits the floor.

2. Disinfection

Disinfection of endoscopes is best accomplished by immersion in an efficient antiseptic solution at room temperature. A quaternary ammonium compound produced under the trade name of Urolocide or Detergicide has been found to be satisfactory for disinfection of endoscopes under ordinary conditions (LOWSLEY, KIRWIN and DAVALOS). Immersion of the instrument in a 1:1000 aqueous solution of this antiseptic for twenty minutes is sufficient to kill the bacteria most commonly found in the urinary tract. The endoscope can be taken from the solution and inserted into the urethra without rinsing, for the material is nonirritating to tissue. It is also indefinitely stable, clear, colorless and odorless. LATTIMER and his coworkers, however, have found it to be ineffective in destroying all tubercle bacilli on instruments when they have been seeded with large numbers of these organisms. They also report that mercuric oxycyanide, even as strong as 1:100 does not kill M. tuberculosis in one hour; formalin vapor is not lethal to this organism, nor will it kill B. proteus after one hour's application. Ten percent formalin solution is, however, lethal to all organisms including M. tuberculosis. After five minutes' immersion of the cystoscope in this solution it was found to be sterile. After disinfecting with 10 per cent formalin it is necessary to rinse the instrument in sterile water or a nonirritating antiseptic solution such as 1:1000 quaternary ammonium compound before introducing it into the urethra, as formalin is irritating to tissue.

Endoscopes are easily ruined by improper methods of disinfection. Mercury bichloride corrodes the metal. Alcohol and alcoholic solutions such as phenol dissolve the cement used to fasten the lenses in place. The heat from boiling and autoclaving distorts the delicate telescopes, melts the wax and cement and may crack the prisms, lenses and lamps.

To insure thorough disinfection, endoscopes are thoroughly cleansed of blood, pus, mucous and debris, preferably by washing with soapy water, before immersion in the disinfecting solution. All parts of the instrument are separated: the telescopes and obturators are removed from the sheath, and the loop and telescope are taken out of the working element of resectoscopes. All stopcocks are opened.

After use, and before being put away, endoscopes are cleansed and disinfected, then dried. When compressed air is available, the moisture is easily blown from crevices, cracks and openings after separating the component parts. Thorough wiping with a cloth is necessary when there is no compressed air. The inside of sheaths is wiped by pushing pledgets of cotton through them with a cleaning rod which is supplied with the instrument. The window covering the light bulb at the end of the sheath is easily broken by the rod unless care is used; the rod must not be rapidly pushed through the sheath.

II. Minor repairs and adjustments

These can be made by the urologist or an assistant.

1. Light failure

This often requires correction. When it occurs, the light cord connection is changed to the light post of a different telescope or sheath. If this also fails, the trouble is probably in the battery, the cord or its connections. If the second light burns, the trouble in the first one is more likely in the telescope or sheath, in the bulb, or in their connections. By thus eliminating one group of points to be checked, the search for the trouble is narrowed. Following are the points to be examined (Fig. 17).

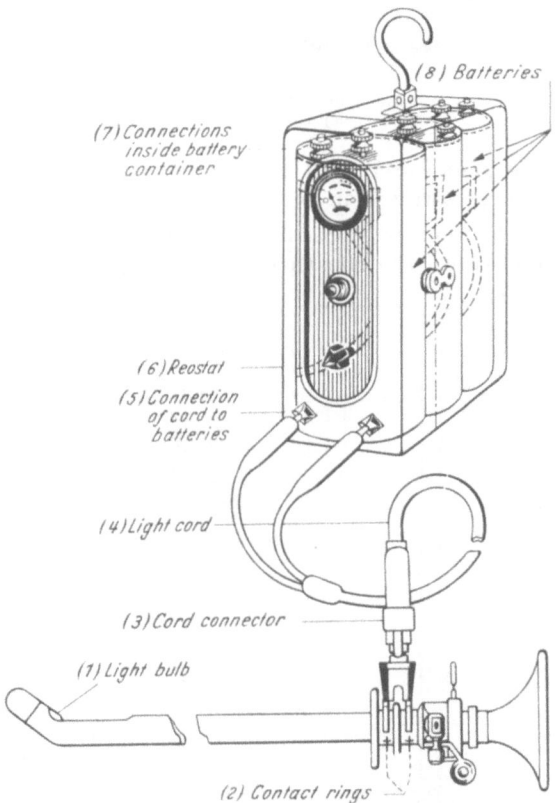

a) Light bulb. Remove it without the use of forceps for they may damage the delicate threads. If a cap on the end of the sheath covers the bulb, slip a large dry rubber tip over the cap to unscrew it (Fig. 18). The bulb can then be removed with the fingernail or a blade of a knife. Do not use forceps. If the bulb screws directly onto the telescope, a small rubber tip can be slipped over it for a better grip while unscrewing it. Care must be taken that any pressure be exerted at the base of the bulb over the metal cuff rather than at the tip over the thin glass of the bulb itself. The contact wire at the base of the bulb is cleaned and elevated slightly with the

Fig. 17. Points for checking to find cause of light failure: *1* light bulb; *2* contact rings; *3* contacts between cord and lamp post; *4* light cord; *5* connection of cord to battery terminals; *6* rheostat; *7* connections inside battery container; *8* batteries

point of a common pin. The disc contact on the cap is cleaned with the blade of a knife or a small piece of emery paper; the disc, which is fastened to a spring behind it, is pulled out slightly to provide firmer contact with the bulb connection. When there is no cap and the bulb screws into the end of the telescope, the contact inside the socket is gently scratched with the point of a pin; or a pointed rubber pencil eraser is inserted into the socket and rotated. The light bulb may be tested in two ways: (a) by grasping the metal part of the bulb gently with thumb forceps, then making contact between the two poles of the battery or battery box by holding the wire of the lamp against one pole and the shaft of the forceps against the other (Fig. 19). If the bulb is good it will light when these contacts are made; (b) by holding the bulb against the two prongs of the rotating light contact which has been removed from the instrument (Fig. 20). The wire of the bulb is held against one prong and the metal cuff against the other.

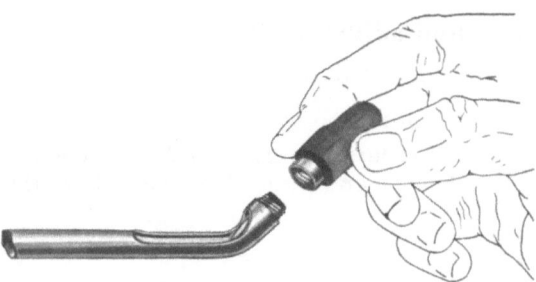

Whenever the light globe is replaced into the sheath or telescope, wax, which is supplied with the instrument, is placed on the threads of the cap or of the base of the bulb before screwing it into place. A toothpick is used for application of the wax to the threads. It makes

Fig. 18. To remove light bulb from sheath, slip a large dry rubber tip over cap covering bulb to unscrew cap

the joint water tight and helps to prevent corrosion of the contacts. The wax is kept away from the contacts to prevent insulating them.

b) **Contact rings of lamp post.** Clean the inside of the contact rings and the groove on the sheath or telescope into which the rings fit; wipe them firmly

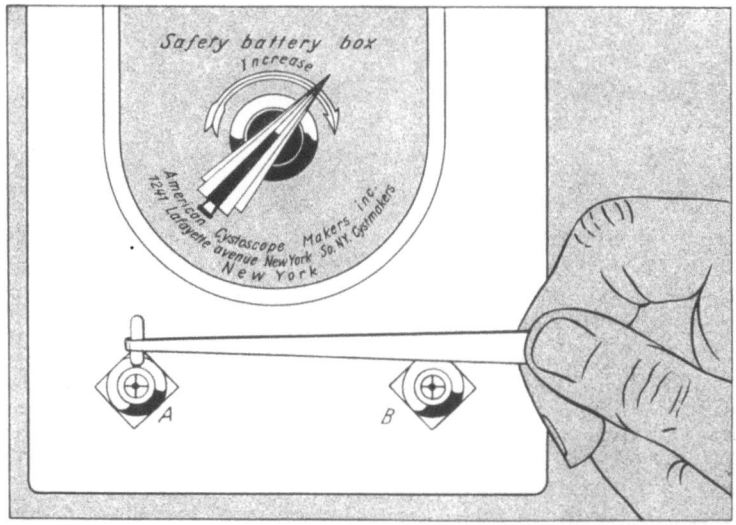

Fig 19. Testing light bulb. Hold metal cuff of bulb with thumb forceps. Contact wire of bulb to one pole of battery and shaft of forceps to other pole. If bulb is good, it will light when these contacts are made

with a cloth or use fine emery paper. Squeeze the contact rings slightly toward each other so they will grasp the sheath or telescope more firmly.

c) **Contacts between cord and lamp post.** Clean the post, the connecting arm of the cord and the socket into which the post fits. Be sure the connecting arm is contacting the metal of the post and is not on one of the black spots which are insulated areas made for turning the light off without disconnecting the cord.

d) **Light cord.** Move the light cord back and forth while testing light. If the light flickers there is probably a broken wire or connection in the cord. Try another cord.

e) Connection of cord to battery terminals. Clean the cord tips and the sockets of the battery terminals into which they fit.

f) Rheostat. Eliminate rheostat by making direct contact of cord tips to battery terminals. If bulb lights, the fault is probably in the rheostat.

g) Connections inside battery container. Check contacts and connections inside battery container.

h) Batteries. Try new batteries. Old ones give bright light when the instrument is first connected, which rapidly grows dim.

If the bulb fails to light after all these points have been checked, the trouble is probably in the wiring circuit of the sheath or telescope; this necessitates shipping the instrument back to the manufacturer for repairs. Before doing this, however, be certain the trouble is not elsewhere. Recheck by trying the same bulb, cord, batteries, and connections on another sheath or telescope.

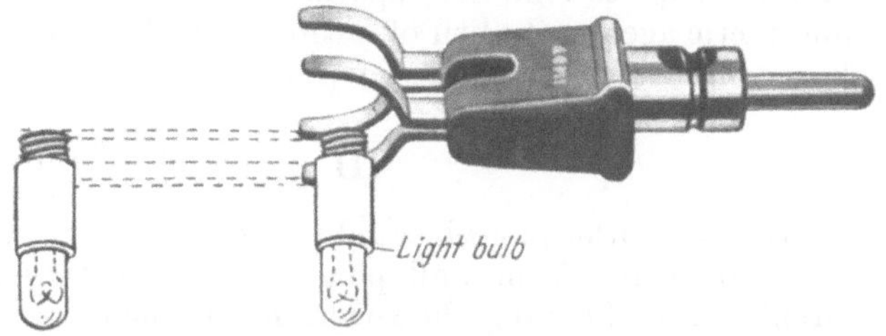

2. Blurred vision

Blurred vision through the optical system is usually due to foreign material,

Fig. 20. Testing light bulb. Hold metal cuff of bulb against one prong of rotating contact and wire of bulb against other prong

such as dried lubricating jelly, on the lenses. They are checked for clarity by looking through the telescope at a window or electric light in the ceiling. Foreign material can usually be seen on the ocular lens; it is cleaned by wiping at first with wet gauze or cotton then with dry material A very small bit of foreign material on the objective lens may blur vision through it, but may be so small that it cannot be seen. The objective lens is best cleansed with point of a toothpick which may be wet with water but never with alcohol. The point of a common pin may be used to clean the small angle between the lens and the adjacent telescope; care must be used to avoid force which might scratch the lens or loosen it.

Blurred vision may be due to loosening of the lenses in the optical system, to leakage of moisture into the tube of the telescope or to other causes which require major repairs by the manufacturer.

When vision is blurred during the endoscopic examination, but is clear when the telescope is checked outside the sheath by looking through it at a window or light, the difficulty is probably due to faulty technique of procedure (see Chap. II).

F. The cystoscopic room (theatre)

Satisfactory endoscopy is much more easily accomplished in a *special room* built and equipped for the purpose than in a regular operating room (theatre) used for open surgery. The cystoscopic table and most of the other equipment used in the cystoscopic room is entirely different from that required for open surgery.

I. Aseptic technique, cleanliness and decorum

Aseptic technique

Careful observance of these features is as important in the cystoscopic room as it is in the operating room (SINCLAIR). Although all the elaborate precautions against contamination of an open wound are not necessary during endoscopy, strict aseptic technique to prevent infection in the bladder — especially while performing endoscopic surgery — must be carried out. Adequate *drapes* are necessary for maintaining aseptic technique. Waterproof drapes are desirable to prevent infection (GARSKE) but are clumsy and more difficult to sterilize then cloth drapes.

II. Floor

A considerable quantity of water is often spilled on the floor of the cystoscopic room even though the urologists and their assistants take care to prevent it. It is therefore advisable to have the floor slightly inclined toward a *drain* located on the floor under the cystoscopic table. A floor which is *conductive* to electric current helps to eliminate sparks which might ignite inflammable gases used as anesthetic agents. In lieu of a conductive floor, the cystoscopic table is grounded by a wire attached to a water pipe.

III. Electric switches

Electric outlet plugs located in the floor are convenient because they eliminate extension cords from wall plugs to electrical equipment; the danger of short circuits from flooding, however, make them hazardous. The safest place for electric outlets in any surgical theatre is four or five feet above the floor — above the dangerous area for explosion of heavy gases which might escape from an anesthetic machine.

IV. Darkened room

Excessive light from windows in a cystoscopic room is undesirable. It is not necessary to make the room entirely *dark* but there should be no direct beam of light shining into the operator's eyes while he is looking through an endoscope. Shades for darkening the windows should be available, but a complete blackout is not necessary.

V. Anesthetic equipment

This is the same as supplied for the room where major surgery is done. Oxygen and suction must be immediately available whenever a patient is given an anesthetic.

G. Cystoscopic room equipment

The well equipped cystoscopic room can be used for all types of endoscopic work including endoscopic surgery. All equipment and supplies are kept in the room to avoid delay in obtaining it after a procedure has been started. Each cystoscopist has his own peculiar likes and dislikes; a detailed list is not necessary, for it would not be generally applicable.

I. Cystoscopic table

The table which is best suited for all purposes has a Bucky diaphragm incorporated in the table and a tube stand with tube attached to facilitate the taking of x-rays during the endoscopic procedure. It is easily adjustable, and the surface can be tipped into a vertical position to allow exposures with the patient in the erect posture. The knee rests, the pan under the table and the foot extension are easily and quickly moved. It is advantageous to be able to elevate and lower the level of the table, but mechanisms to accomplish this are expensive and cumbersome; a more simple arrangement for the same purpose is to use a rapidly adjustable cystoscopic stool. The ideal table is built so it can be used for endoscopic surgery as well as for routine diagnostic endoscopy. The knee rests can be easily spread apart, the pan is deep to prevent splashing and has a wire screen to collect the pieces of tissue as they are resected. The table is high enough to allow the surgeon to manipulate the resectoscope without continually bending down.

An arm projecting over the center of the cystoscopic table is useful for suspension of the water irrigating tube and electrical connections to the endoscope. The dragging of these connections when they are not suspended impedes the free manipulation of the instrument and dulls the delicate sense of touch which every cystoscopist should have and use while manipulating endoscopes. The cross piece of the stand which holds the x-ray tube above the table makes a convenient suspension bar. An elastic connection, such as a long thin rubber band, extends from the crosspiece to a hook or clamp which holds and suspends the supply lines to the endoscope. Pulleys and counterweights are preferred by some (TRATTNER). However, even the small amount of weight required to pull the supply lines through the pulleys is sufficient to drag on the endoscope and disturb its balance.

A special leg holder is available for use during cystoscopy in children; it fastens to the cross bar of the adult knee rest (CAMPBELL).

Tables are available which incorporate all these requirements and yet are simple in construction. The more simply a table can be built the longer it will function without need of costly and time consuming repairs.

II. Cystoscopic stools

A stool which can be raised and lowered rapidly by the cystoscopist himself facilitates manipulation of the endoscope, especially during endoscopic surgery. The resectoscope must be moved into different positions to visualize the interior of the bladder, and it is necessary to change its position frequently while performing endoscopic prostatectomy. The ability of the operator himself rapidly to raise and lower his position saves him from stooping to visualize the dome of the bladder or the ventral portion of the prostatic urethra, and from standing up to see into the depth of a cystocele or the bottom of an excavated prostatic urethra. Two types of rapidly adjustable stools designed by the authors are available. One is raised by a spring and lowered by the operator's weight; the other is powered by reversible electric motors and hydraulically operated. Each one can be rapidly adjusted to any convenient height (Fig. 127, p. 156, 157).

III. Irrigating fluid supply

1. Flask system

The supply of sterile irrigating fluid for distending the bladder during endoscopic examination or surgery must be adequate. There are several practical systems of supply. The "flask system" provides numerous 2000 cc. stoppered flasks containing water or glycene or glucose solutions which are autoclaved and then stored in a warming cabinet in the cystoscopic room or adjacent workroom. As irrigating fluid is needed during the endoscopic procedure, the contents of one or more flasks is poured into a large glass bell jar or other container hung at the proper height above the cystoscopic table. Rubber tubing conducts the fluid from the container to the endoscope. This system is flexible; the type of irrigating fluid can be changed easily and quickly, and the fluid pressure can be adjusted by

raising or lowering the suspended container. It is aseptic: the flasks and contents are autoclaved; the bell jar is filled with disinfectant solution after each
day's use and is emptied before its next use. Irrigating fluid which is kept warm
is less irritating to the bladder mucosa than when it is cold. The disadvantage
is the necessity of continually watching the supply in the bell jar and replenish-
ing it.

2. Sterilizer near ceiling

A convenient system of irrigating fluid supply is a *sterilizing tank* installed
next to the ceiling in the cystoscopic room or an adjacent workroom. The sterile
water is run by gravity through pipes and rubber tubing direct to the endoscope
or to a bell jar suspended from the ceiling. Additives such as glycene or glucose
can be mixed with the water in the jar. The conducting pipes can be sterilized
by forcing live steam through them. This system is labor saving but expensive
to install.

3. Pressurized from container on floor

Apparatus for pumping fluid or forcing it by compressed air from a container
on the floor into the endoscope is satisfactory as long as it functions properly;
the fluid pressure is indicated on a meter and is adjusted by a valve. Mechanical
failure is a hazard of this more complicated system.

4. Water sterilizer — pitcher — jar

The most simple system of irrigating fluid supply is from a water sterilizer
to the bell jar by means of a large pitcher. A sterilized enema can may be used
in lieu of the bell jar. Repeated fillings and numerous trips between the sterilizer
and the container are necessary. Additives for use during endoscopic surgery may
be mixed with the water before it is poured into the suspended container.

5. Control of water by foot switch

Some urologists prefer to control the flow of irrigating fluid to the endoscope
by means of a foot switch (CHANNESON and MURPHY). The stopcocks on endoscopes are, however, so convenient and easily manipulated that the use of a foot
control seems superfluous.

H. Endoscopic armamentarium in the armed forces

Endoscopic armamentarium in advance units must be easily mobile, compact
and light (SPENCE). Only the necessities are allowed, and the urologist must
often improvise for apparatus which is not available.

Chapter II

The cystoscopic procedure

A. Value of properly performed cystoscopy

Cystoscopic examination is one of the most valuable diagnostic procedures in
medicine. Without it, the urinary tract is a system of sealed dark passageways
and reservoirs, and the true nature of its maladies is often only a conjecture. But
the cystoscope, handled by an expert, and aided by x-ray, opens the system and
transforms conjectures into definite diagnoses. Failure to obtain a cystoscopic
diagnosis often leads to "confusion and deadly errors" (PELOUZE).

The cystoscopist

Mistaken diagnoses may be made when the cystoscopic procedure is carelessly or unskillfully performed, or when the cystoscopist has had inadequate training or experience to enable him to interpret correctly the findings. The cystoscopic appearance of different lesions may be similar (KAUFMAN). In these cases, especially when urography has been imperfectly performed, an error in diagnosis easily becomes possible.

1. Training

Proficiency in diagnostic and operative cystoscopy requires intensive training for a period of three years. The medical graduate who is to become a urologist should have at least three years' training on an active urological service in addition to a year in general surgery and any other time spent in the basic subjects — surgical anatomy, pathology and physiology.

There is a need for diagnostic cystoscopic training for the physician in general practice who is located in a rural area where no urologist is available. During a course of four to six weeks he will learn to catheterize ureters and to recognize uropathology. Although he may not be able to make a definite diagnosis or to treat the lesion, he can refer the case to a larger center with confidence that the patient is not making the trip for naught. The physician in general practice who has had this short training will discover many existing urological diseases in his patients which otherwise would pass unnoticed.

2. Dexterity

Manual dexterity, gentleness and the ability to place himself figuratively at the inner end of his cystoscope are valuable qualifications of the cystoscopist. He must make a composite picture of the many fields of vision which he sees through the lenses of his instrument as he manipulates it to encompass the whole of the interior of the bladder.

B. Indications and contraindications for cystoscopy (BUMPUS)

I. Indications

With but few exceptions every patient who has a suspected disease of the urinary tract which cannot be diagnosed or excluded otherwise, should undergo a cystoscopic examination. One of the more specific indications for cystoscopy is blood in the urine. The examination should be made at the time of active bleeding; the source of the blood can then be more accurately located. The list of other indications for diagnostic cystoscopy is long, (Low), and there is little difference of opinion among urologists as to the necessity for such an examination in order to make an accurate diagnosis or to exclude the urinary tract when it is suspected of being involved in a morbid process (HEIDER). There is on the other hand considerable difference in view-points regarding indications for endoscopic treatments and surgical procedures. These are discussed in Part II of this volume.

II. Contraindications

A few contraindications are obvious. Urethral stricture of small calibre and other obstructions in the urethra will not permit the passage of a cystoscope. Patients who are debilitated or extremely ill should be given supportive and palliative treatment before cystoscopy is done, unless it is the only means whereby

3*

the best immediate therapy can be determined or given. It is usually better to give anti-bacterial medication to a patient with acute urethritis or prostatitis or other acute urinary infection and high fever than to cystoscope him immediately. Conversely, uremia due to bilateral ureteral obstruction requires immediate cystoscopy and catheterization of the ureters for both diagnosis and treatment. Cystoscopy is not indicated in most urological diseases in which a reasonably sure diagnosis can be made without this examination. In an elderly man presenting symptoms and findings typical of prostatic hypertrophy nothing can be gained by cystoscopy. Catheterization of the ureters is definitely contraindicated in the presence of an enlarged prostate and residual urine. An upper urinary tract disease which has been diagnosed by excretory urograms does not need a cystoscopic examination. It is only in the absence of a definite diagnosis or in a questionable diagnosis that cystoscopy is indicated. Medical diseases of the kidney can usually be diagnosed without cystoscopy. The symptoms and findings by physical examination and urinalysis in patients who have glomerular nephritis, nephrosclerosis, lipoid nephrosis and other nonsurgical renal diseases are usually sufficient to make a definite diagnosis without recourse to cystoscopy.

C. Routine supplies for cystoscopy

I. Sterile set-up

All supplies which come into contact with cystoscopes and other instruments which are to be inserted into the bladder or kidney, must be sterile. Although aseptic technique does not need to be as elaborate as for open surgery, contamination of cystoscopes and catheters with unsterile hands and drapes and portions of the table is a source of infection and must be avoided. If the cystoscopist is careful, he can perform cystoscopy aseptically by wearing sterile gloves, and will not need a sterile gown or mask. He must avoid breathing on the cystoscopes or catheters and must prevent their touching his face. When a mask is worn the operator's breath is deflected upward onto the ocular lens of the cystoscope, resulting in fogging and decreased vision (HEIDER). The normal bladder is difficult to infect but this fact does not give the cystoscopist license to be careless about his technique. The majority of the urinary tracts which he examines are not normal and are more susceptible to infection.

A small table or tray with sterile cover is placed within easy reach of the cystoscopist. The contemplated procedure and the preferences of the operator determine the sterile supplies to be placed upon the table. There are syringes for instilling local anesthetic and for injecting radiopaque fluid into the renal pelves; containers for urine from the bladder und from the kidneys; sounds or bougies up to the size of the cystoscope being used; material for renal function tests; lubricant and other items.

II. Lubrication

Oil for lubrication of instruments is usually preferable to a water soluble lubricant; the latter is washed away soon after introduction of the cystoscope or catheter. Sterile mineral oil, castor oil, peanut oil or preferably one with an antiseptic contained within it such as Metaphen (R) in oil are satisfactory.

III. Drapes

A one-piece sterile drape, large enough to cover the patient from the nipples to the feet, is preferable to a smaller one (CRABTREE).

IV. Media for distending the bladder

1. Water

Sterile tap water is the usual medium and is satisfactory for diagnostic cystoscopy. Physiological saline solution may be slightly less irritating. Warming the medium to body temperature helps prevent bladder spasm. The addition of a small amount of antiseptic solution for bacteriostasis is recommended when there is danger of contamination. When endoscopic surgery is to be performed, an isotonic solution which is not an electrolyte is preferable (Chapter XV).

2. Urine

Clear urine may be used as a medium through which visualization of the interior of the bladder is accomplished. When the bladder is already distended, the urine can be kept in when the cystoscope is passed. There may be less danger of trauma and of infection by using this natural medium.

3. Oil

Sterile mineral oil as a medium for distending the bladder has its advantages (GREENBERG). Blood and pus do not mix with it easily, providing clearer vision, and it is soothing to the bladder mucosa.

4. Air

This is the distending medium when the KELLY cystoscope and its modifications are used (p. 12) (RIDLEY).

D. Preparation of the patient

I. Prophylactic antibiosis

The administration of antibacterial medication before cystoscopy may be indicated in selected cases (FELBER). When the patient has had fever following previous similar instrumentation and when there is a probability of carrying bacteria to the kidneys from an infected bladder, it is wise to give prophylactic therapy. As a routine procedure, however, it is not indicated. The nonspecific administration of these drugs is poor medical practice. They may mask a urinary infection which needs diagnosis. Careful aseptic technique of cystoscopy precludes the danger of introducing infection in most cases.

II. Bowel preparation

This is necessary previous to cystoscopy if the patient has had no evacuation for 24 hours, especially if x-rays are to be taken in conjunction with the procedure. A saline enema is usually adequate. When upper urinary tract x-rays are to be taken and the preliminary scout film shows excessive intestinal gas, an intramuscular injection of pitressin or similar smooth muscle stimulant will usually clear the gas sufficiently to provide urograms without confusing shadows. When there is a fecal impaction, abnormal intrusion of the dorsal bladder wall occurs, giving the impression of a retrovesical tumor at the time of cystoscopy. In most instances, however, when cystoscopy for examination of the bladder only is done, preliminary bowel preparation is not necessary.

III. Analgesia

Patients who undergo cystoscopy for examination of the bladder only, usually do not need precystoscopic analgesia. The procedure is usually not unbearably painful if a small cystoscope such as the Brown-Buerger No. 16 Fr. is used and is manipulated with care and gentleness. However, when it is necessary to use a larger instrument, and when a more extensive procedure such as ureteral catheterization is contemplated, a preliminary analgesic should be administered. Morphine sulphate gr. $^1/_4$ and hyoscine gr. $^1/_{150}$ given 30 minutes before cystoscopy is usually adequate to relax the patient and to prevent excessive pain (LAIBE). The dose is reduced for children and for most individuals over 65 years of age. Nisentil[R] given intravenously for analgesia is effective (GRAVES; MATHIAS and BUNTS). Morphine and hyoscine do not alter results obtained from renal function tests.

IV. General or spinal anesthesia

One of these may be necessary when the patient has a low pain threshold or when there is a disease which causes increased bladder sensitivity. Cystoscopy is very painful in the presence of severe tuberculous cystitis, acute infection, vesical calculus or tumor, or any malady which causes inflammation of the bladder. The majority of cystoscopic examinations, however, can be done without general or spinal anesthesia (BAURYS).

Triethylene (trilene) (HORNE) may be self administered by the patient. It is comparatively safe and there is no post anesthesia drowsiness. It is especially useful in out patient practice, for the patient can leave the office or clinic without assistance within 30 minutes of completion of the procedure. Other general anesthetics such as *pentothal sodium* (BLUSGER and DIXON) and *Evipal*, (MERRITT) and *spinal anesthesia*, are useful in hospital practice or when the patient may remain in the clinic for two or three hours following their administration.

V. Local anesthesia

The male posterior urethra is more sensitive than the anterior urethra, and probably also more than the female urethra. The bladder mucosa itself is sensitive to distention when it is inflamed for any reason; when it is normal, there is no discomfort in the bladder from routine cystoscopy. Nevertheless, it is advisable to instill the local anesthetic into the bladder while the urethra is being anesthetized. The use of local anesthesia applied 10 minutes before passing the cystoscope is recommended in most cases.

1. Anesthetic agents

The multiplicity of local anesthetic agents which are recommended for use in the urethra is evidence that none is entirely satisfactory. None of them completely anesthetizes the urethra but most of them benumb the area sufficiently to be worth using. A *tablet of cocaine* implanted into the posterior urethra with a special instrument is claimed by some to be the most effective method of local anesthetisation (LEWIS; THOMAS). *Novocaine* (procaine hydrochloride) 5 cc of 5 per cent solution instilled into the anterior urethra followed by a waiting period of 15 minutes has been used. Some urologists believe that passing the cystoscope into a partially filled bladder produces less pain than into an empty one (PARKER). *Cocaine onehalf per cent solution*, nupercaine (a-butyloycinchoninic acid-y-diethylenediamide hydrochloride) 1:500 solution or even as concentrated as $1^1/_2$ per cent (DOUGLAS) solution may be either instilled in the urethra or injected into the ureter to relieve spasm and reduce pain (BRIGGS). *Intracaine* (BELT and BAVETTA; CORBUS) (beta-diethylaminoethyl-p-ethoxy benzoate) hydrochloride, *metycaine* (LA GUETTE) (benzoyl-y-(2methylpiperidino) propanolhydrochloride), *anesthesin* (ethylaminobenzoate), *anethain* (aminothocain hydrochloride) 2 per cent solution and numerous others are available, (MARSHALL; PERSKY and DAVIS) and each is extolled by the detail man as being the best. Some of them are incorporated into a water soluble lubricant which is instilled into the urethra and held there for 10 minutes before the cystoscope is passed (MUSCHAT). Preliminary administration of pyridium has been reported as producing adequate local analgesia (MORRISSEY and SPINELLI).

2. Application

Anesthetic applied to the *female urethra* is best accomplished by inserting into it an applicator which has been dipped into the anesthetic solution. A more concentrated solution of the anesthetic can be used on the applicator than is safe to inject into the bladder or into the male urethra. Preparatory instillation of five to 10 cc of the solution into the empty bladder through a catheter benumbs the vesical mucosa. A small amount of the solution is injected into the urethra as the catheter is being withdrawn.

Application of the anesthetic to the *male urethra* is most often accomplished by instillation with a blunt tip syringe pressed against the meatus. The patient is instructed to relax the urethral sphincter as if he were attempting to void; relaxation allows the solution to pass into the posterior urethra. Extreme care must be exercised to avoid forcing the fluid; overdistention of the urethra injures the mucosa and rapid absorption of the solution may result in a severe reaction. The solution is held in the anterior urethra by a strip of gauze tied around the phallus, or preferably by a penis clamp applied near the end of the penis. More adequate anesthesia of the prostatic urethra may be obtained by carrying the anesthetic agent to it. The solution is injected through a tube such as the Keyes instillator or the Guyon catheter, (PELOUZE; KEYES and FERGUSON) or a tablet of cocaine may be implanted with a special instrument (LEWIS and THOMAS). An applicator dipped in the anesthetic solution and inserted into the urethral meatus helps when the meatus is small. At least 10 minutes should elapse between the application of the anesthetic agent and the beginning of the cystoscopic procedure; otherwise the full benumbing effect of the anesthetic will not be obtained.

3. Untoward reactions

Untoward reactions from a local anesthetic after instillation into the urethra are rare, but may occur from any of the agents used. The cystoscopist who has never seen an unfavorable reaction among his patients has either given very few local anesthetics or else he is an extremely fortunate individual. It is definitely true, however, that gentleness of technique and the avoidance of forceful instillation reduces these reactions to a minimum.

The usual untoward reaction is perspiration, faintness, sometimes rigor, rapid weak pulse and a sudden lowering of blood pressure (ALEXANDER). More severe symptoms are twitching or spasm of the muscles, coma, absence of pulse and blood pressure, and in extreme cases death. These reactions are due to sensitivity to the anesthetic agent. The cause of the sensitivity is not definitely known, although it is probably similar to anaphylactic shock (SADOVE et al.). *Treatment* consists of administration of an antihystaminic such as Benadryl, (beta dimethylamino-ethyl benzhydryl ether hydrochloride) lowering the patient's head and loosening all tight clothing. Inhalation of oxygen and artificial respiration may be necessary.

E. Position of the patient

For cystoscopy this is usually the relaxed lithotomy position. The thighs are only mildly flexed and are abducted, and the knees rest in crutches. An extreme lithotomy position with the thighs flexed more sharply over the abdomen increases the curve of the urethra, causing increased difficulty in passing the cystoscope. A few urologists prefer the patient's legs to be straight out and abducted; the examination can then be done on a stretcher and time consumed in placing him on the cystoscopic table is saved (LANDES and RANSOM). It is well for the cystoscopist to have some practice with the patient in this position for he may some day need to perform the examination with the patient lying in bed, such as when he is encased in a hip or leg cast. The knee chest position is used when air is the distending medium.

F. Checking of equipment

I. Instruments

All instruments used during the cystoscopic procedure are made ready and checked before the cystoscope is introduced. Delay occasioned by minor adjustments and arrangements during the examination prolongs the procedure unduly

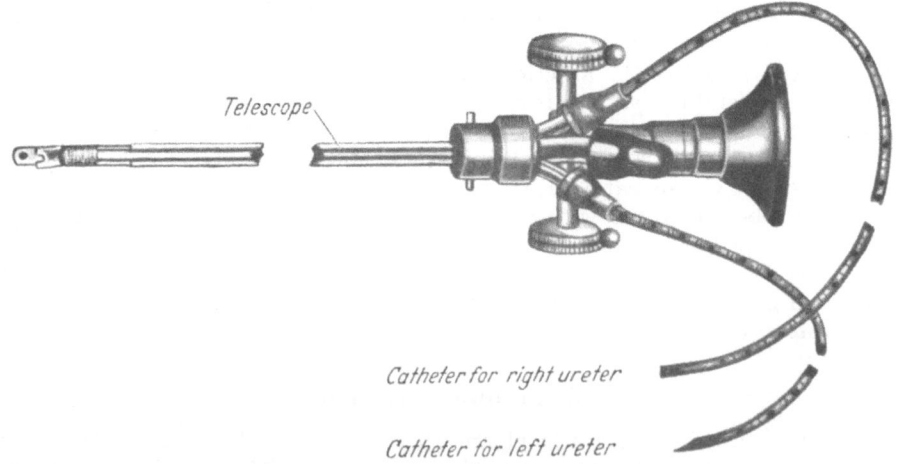

Telescope

Catheter for right ureter

Catheter for left ureter

Fig. 21. Insert ureteral catheters into guides in telescope before inserting cystoscope. Outside end of catheter in right ureter is cut at right angles, left one at an acute angle

and increases trauma. If the ureters are to be catheterized the inner ends of the catheters are dipped into lubricating oil and inserted into the catheter guides of the telescope and left ready for use on the table. The outside end of the one going to the right kidney is cut at a right angle; the one for the left side is cut at an acute angle (Fig. 21). The obturator is inserted into the sheath, the cystoscope is lubricated, then the light cord is connected to the post on the sheath and the rheostat, if one is used, is adjusted by slowly turning it up until the yellow glow which first appears in the filament turns to a white glow.

II. Light bulbs

These are tested before introducing the cystoscope, because it is traumatic to the urethra and sometimes embarrassing to the physician to have to remove the sheath immediately after it has been passed because there is no light.

G. Introduction of the cystoscope

This is accomplished most easily by putting the phallus on a stretch over the pubis and inserting the instrument straight toward the perineum. When the tip enters the bulbous urethra there is obstruction to its passage caused by the beginning of the posterior urethral curve. The ocular end of the instrument is then gently depressed while the tip is elevated. Inasmuch as the inner end of the cystoscope has a short coudé tip or is straight, it does not follow the curve of the posterior urethra as easily as a sound. The tip must be guided around the curve by elevating it away from the floor and toward the roof of the urethra. "Rebound obstruction", or the sensation of elasticity encountered when an obstruction is met, is an indication that the tip is not in line with the urethral lumen but is against the urethral wall. When the cystoscope meets an obstruction, gentle manipulation of the tip by elevating it against the roof, then pointing it slightly to one side or the other, then depressing it toward the floor will help to line it up with the urethral lumen. The inward pressure on the instrument is alternately relaxed, then gently increased as the manipulation is continued. When it passes inward a little and does not rebound, but is rather held by the urethra, it is an indication that the tip is passing through the urethral lumen. Spasm of the external urethral sphincter causing obstruction to the passage of the cystoscope may be overcome by maintaining gentle pressure with the tip of the instrument against the sphincter; the muscle becomes fatigued and relaxes. The passing of one or more sounds or bougies preparatory to inserting the cystoscope may help to open the sphincter, to smooth out irregularities or to dilate a stricture.

When there is elevation of the dorsal lip of the bladder neck it is necessary to depress the occular end of the cystoscope more than usual, sometimes even below the level of the buttock, in order to pass the instrument into the bladder; the tip is thus elevated to ride over the high bladder neck. It sometimes helps for the cystoscopist to insert a finger into the rectum and press forward on the sheath as it is passed through the prostatic urethra.

I. Information gained from passing the cystoscope

Some information is gained from mere passage of the cystoscope.

1. Stricture

A stricture may be discovered if the urethra has not been previously calibrated. When the point of the instrument enters the stricture there is a tight sensation rather than a rebound such as occurs when the point is against the lateral urethral wall.

2. Elevated posterior lip

If it is necessary to depress the ocular end of the cystoscope to get the beak into the bladder, an elevated posterior lip of the bladder neck is suspected; it might be a *median bar*, a *collar contracture* or an *intravesical middle lobe prostatic hypertrophy*. Sometimes an enlarged middle lobe elevates the bladder neck to such an extent that it is impossible to pass the cystoscope over it.

3. Elongated prostatic urethra

When the prostatic urethra is exceptionally long due to hypertrophy, it may be impossible for the cystoscope to reach entirely into the bladder.

4. Residual urine

A measure of the amount is obtained through the cystoscope when the patient has voided immediately previous to the examination. *Bladder tone.* The force with which the urine or cystoscopic irrigating fluid runs out through the cystoscope gives an estimate of bladder tone.

II. The causes of difficulties encountered during passage of the cystoscope

These may be summarized as follows: (1) Failure sufficiently to elevate the beak of the cystoscope; (2) failure sufficiently to depress the ocular end of the cystoscope; (3) urethral stricture; (4) spasm of the urethral sphincter; (5) large middle lobe prostatic hypertrophy; (6) extensive elongation of the prostatic urethra by huge hypertrophy; (7) exceptionally long phallus; (8) erection of the penis.

H. Procedures for obtaining clear visualization

I. Adequate intensity of illumination of the interior of the bladder

This is obtained by maintaining good contact between the light source and the cystoscope bulb (p. 30), by proper adjustment of the rheostat (p. 29), or by using the correct voltage battery for the light source (p. 27).

II. Distention of the bladder

When the bladder is empty the objective lens is in contact with the mucosa of the collapsed bladder wall. In this position there is no clear vision because very little or no light reaches the area which is against the lens; this area is also so small that structures cannot be identified. Distention moves the mucosa away from the ocular lens and smooths out the folds of bladder wall. Observation through the telescope is continued while the bladder is being distended; thus a closer view of the fundus and dome is obtained. A lesion located immediately above the ventral lip of the bladder neck may be within view while the bladder is partially collapsed but hidden behind the rim of the bladder neck when the bladder is full. Overdistention of the bladder is to be avoided. Ecchymotic areas appear and sometimes bleeding occurs from overdistention. Bladder spasm may be initiated and become severe enough to force all of the distending fluid out around the cystoscope, making further visualization impossible. Spasm of the ureteral orifices precludes passing of catheters through them. The cystoscopist often becomes so engrossed in looking through the cystoscope that he forgets the inlet stopcock is open until the patient can tolerate the pain of overdistention no longer. The cystitis caused by overdistention may persist for many days.

III. Washing debris from the bladder

When there is blood, pus or other opaque material in the bladder, it is washed out with the fluid used for distention before the telescope is introduced and before visualization is attempted.

IV. Manipulation of the inflow of fluid through the sheath

Such manipulation aids visualization. When some opaque material remains in the bladder, the inflow of fluid keeps the area between the lens and the object clear unless there is too much space between them. However, when there is a small amount of bleeding from a spot in the bladder mucosa, the bleeding point may be identified more easily by temporarily shutting off the inflow stopcock; otherwise the slow stream of blood oozing from the lesion may be washed away by the constantly inflowing fluid and may not be noticed.

V. Proper manipulation of the objective lens

Magnification or reduction in size of the image seen through the optical system depends upon the distance of the object from the objective lens (Fig. 22 A

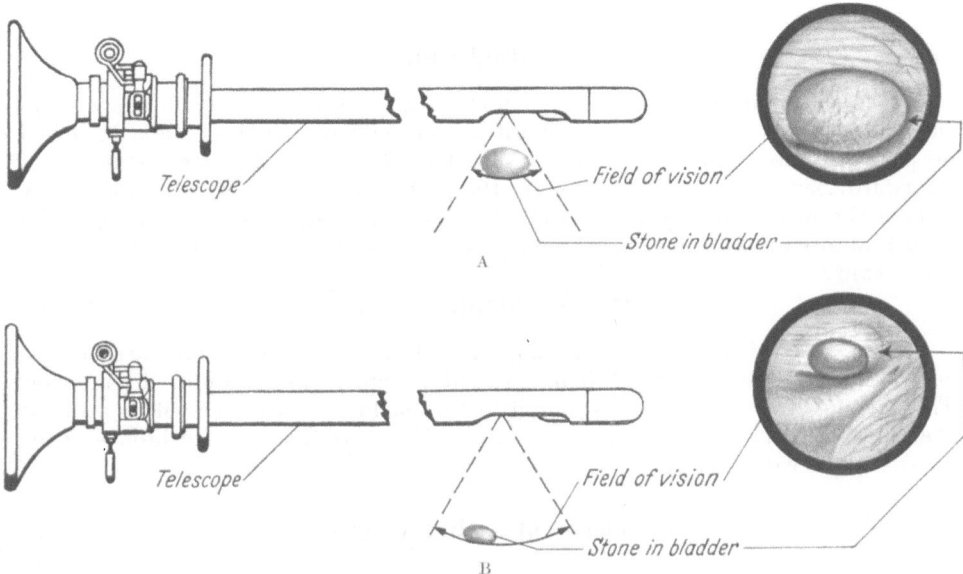

Fig. 22 A and B. A When objective lens of telescope is close to object, image is enlarged or magnified.
B When lens is far away, image is reduced

and B). When they are only a few millimeters apart the object is magnified, but when several centimeters separate them, the object is reduced in size and may appear smaller than it actually is. The size of the field of vision also depends upon the proximity of the lens to the object. When they are close, the field of vision is small; the farther apart they become, the wider area the field of vision encompasses. A good routine is first to view the object from a distance, then to move the objective lens closer to it while visualization is continued. The object gradually becomes larger. It is the same technique used by the cinema cameraman.

If it is difficult to identify a small object, the objective lens is manipulated close to it; a small red area may appear from a distance to be ecchymosis, but when the lens is close to it, the projections of a papilloma may be seen. A large object may, however, be more readily identified from a distance; a vesical calculus may cover the entire field of vision when close to the lens, but when farther away its edges and contours may be discernible.

I. Orientation with different lenses,
see Chapter I

J. Routine bladder examination

The entire surface of the interior of the bladder is inspected in an orderly manner during cystoscopy. The floor of the bladder is first surveyed by advancing the cystoscope gradually farther into the bladder, starting from the bladder neck. When the upper portion of the fundus is visualized, the instrument is rotated about 30 degrees and drawn back outward again to the bladder neck. The maneuver is repeated until the entire inner surface of the bladder has been covered. During this survey the objective lens is kept one to two cm. away from the bladder mucosa. However, if a small abnormality is seen from this distance, the lens is manipulated closer to it for a more detailed view.

I. Blind spot

When the right angle optical system is being used, there is a "blind spot" on the bladder wall opposite the vesical orifice. Visualization of this area is described on page 4, Chap. I. When the convex sheath of the Brown-Buerger cystoscope or a forward view telescope is being used, the posterior urethra can be examined by pulling the end of the cystoscope out beyond the bladder neck. The inflow of fluid is allowed to continue for the purpose of dilatation which separates the urethral walls.

II. Diverticular cavity

When a diverticular orifice is found, it is usually possible to insert the objective end of the cystoscope through it into the diverticulum and examine its interior. This maneuver is more easily accomplished by using a cystoscope with a straight sheath such as the McCarthy panendoscope. Stone, tumor or other lesions in the diverticulum can be thus diagnosed.

K. Ureteral catheterization

I. Ureteral catheters

1. Tips

a) **Whistle.** The *whistle* tip catheter (Fig. 23 A) with an opening in the end is the one most commonly used for routine diagnosis of upper urinary tract disease. Urine flows more freely through this type than through those in which the openings are all on the side.

b) **Olive.** The *olive* tip catheter (Fig. 23 B) passes into the ureteral orifice more easily than the whistle tip, and is more suitable to dilate the ureter and to pass when the ureteral lumen is constricted.

c) **Coudé.** The *coudé* or elbow tip (Fig. 23 C) is useful to pass through an angulated portion of the ureter. In some patients the ureter enters the bladder almost transversely and it is not possible to manipulate the catheter to be parallel with the axis of the ureter. The coudé tip catheter can, however, be rotated so that the tip goes straight into the meatus.

d) **Filiform.** *Filiform* tips are used for dilating tight ureteral strictures. The spiral filiform (Fig. 23 D) is more useful than the straight one, for it passes through angulations in the ureter similarly to the coudé tip. Each of these catheters is rotated as it is passed to allow the off center tip to pass around irregularities and angulations in the ureteral lumen.

e) Conical or Garceau and Braasch bulb. The *conical* or *Garceau* tip (Fig. 23 E) and the *Braasch bulb* (Fig. 23 F) are designed for dilating the ureter. The former is more useful when the ureteral stricture is tight and difficult to dilate. The bulb, located near the end of the catheter, dilates the entire ureter as far as the renal pelvis as it is passed upward.

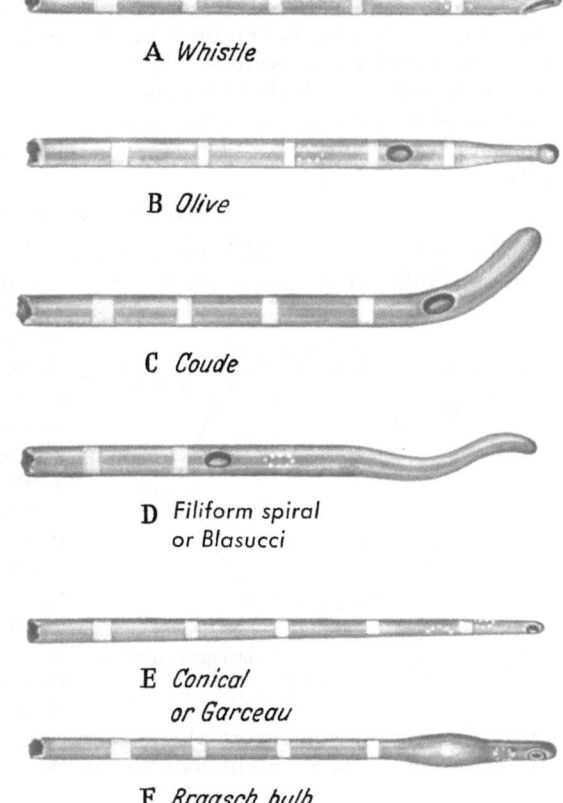

A *Whistle*

B *Olive*

C *Coude*

D *Filiform spiral or Blasucci*

E *Conical or Garceau*

F *Braasch bulb*

Fig. 23 A—F. Ureteral catheter tips. A Whistle. B Olive. C Coudé. D Filiform spiral or Blasucci. E Conical or Garceau. F Braasch bulb

2. Size

The size of the ureteral catheter used for any purpose should be the smallest that will accomplish the desired result. Immediate trauma and subsequent reactions and sequelae are increased when larger catheters are used. For routine kidney study a size 5 Fr. is usually large enough to collect an adequate specimen of urine from the renal pelvis. This size passes up most normal adult ureters with little trauma. A size 4 Fr. or even sometimes a No. 3 is indicated when the ureter is abnormally small and in children under four years of age. A small catheter can also be used when the purpose of ureteral catheterization is for retrograde pyelography only. Larger catheters are indicated for dilatation of the ureteral lumen, for draining a hydronephrosis or pyonephrosis, and for plugging the ureteral orifice to obtain the retrograde ureterogram.

3. Flexibility

The maneuverability of a catheter is partially dependent upon its flexibility. When it is too stiff, the tip tends to stick into the ureteral wall instead of following the lumen. Stiff catheters are therefore more traumatic than soft ones. They may be softened by warming the solution in which they are sterilized immediately before use or by immersion in warm water after sterilization. When too limber, the catheter buckles while being passed; the thrust applied to it at the outside of the cystoscope does not reach the tip; it is like a string which folds up in the cystoscope, in the bladder, or in the ureter. In countries which are warm and humid throughout the year, most silkwoven catheters and bougies become excessively limber. In order to maneuver soft ureteral catheters, the fine wire stilette which is supplied is left in place while the catheter is being passed. After

the tip has entered the ureteral orifice, the wire is pulled out about one centimeter; the tip then becomes flexible and follows the ureteral lumen more easily,
but the rest of the catheter remains rigid enough to transmit the thrust. Catheters
which are too limber may be stiffened by keeping them in a cold dry place or
by adding ice to the solution in which they are sterilized.

4. Opacity

Most ureteral catheters contain sufficient radiopaque material to be visible
by x-ray; the course of the ureter is outlined on the radiograph and its relation
to other opacities determined. Excessive radiopaque material in the catheter
may cast so dense a shadow that the opacity of a small stone is completely covered
by that of the catheter. It is preferable to utilize catheters which are only slightly
opaque; these may be called "modified shadowgraph" ureteral catheters
(HERMAN).

5. Graduation markings

Ureteral catheters which are graduated by marks one centimeter apart are
more desirable than those which are not graduated. The marks are watched

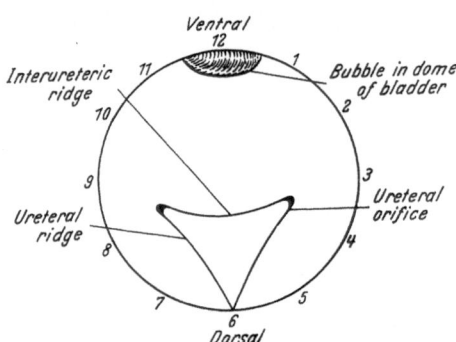

Fig. 24. Orientation of positions in bladder in relation
to numbers on a clock dial

through the lens of the cystoscope as
the catheter is passed up the ureter.
When an obstruction is met, its distance
above the bladder can thus be measured
and recorded. The renal pelvis is 30 cm.
above the ureteral orifice in the average
adult. The cystoscopist knows that
the tip of the catheter has entered the
pelvis when he sees the 30 cm. mark at
the ureteral orifice. If the catheter is
passed more than 30 cm. there is danger
of pushing it into the renal parenchyma.
In adults who are smaller in stature
than average, and in children, the length
of the ureter can be estimated by
measuring the distance from the suprasternal notch to the pubis. The length
of the ureter is equal to one-half this distance (COFFETT and CHARNOCK).

II. Technique of ureteral catheterization

To locate and to identify the ureteral orifices is at times difficult even for
the experienced cystoscopist. Their normal position is between one-half cm. and
$1^1/_2$ cm. above the bladder neck at the 5 and 7 o'clock positions (Fig. 24). To
locate them with the right angle and obliquely forward view lenses, the cystoscope is pulled out until the bladder neck at the 6 o'clock position comes into
the lower edge of the field of vision. The cystoscope is then slowly advanced
into the bladder with a slightly rotary motion toward the patient's left (for the
left orifice); the ocular end is moved toward the patient's right and elevated.
By this maneuver, the ureteral ridge can be followed closely with the objective
lens. The small slitlike ureteral orifice soon comes into view (Fig. 30, p. 57,
Chap. IV); it is usually on the crest of the ureteral ridge. Sometimes this ridge
is not sufficiently developed to be identified; then it is necessary to scan a larger
area by a close view with the objective lens.

III. Manipulations to facilitate ureteral catheterization

When the ureteral orifice is not in the normal position a wider close search of the entire bladder floor, the bladder neck, and sometimes even the posterior portion of the urethra becomes necessary. When the orifice is not seen, the tip of a ureteral catheter may be gently moved inward against the bladder mucosa in the region where the orifice should be located. When the ureteral ridge is evident but the orifice cannot be seen, the tip of the catheter is gently moved

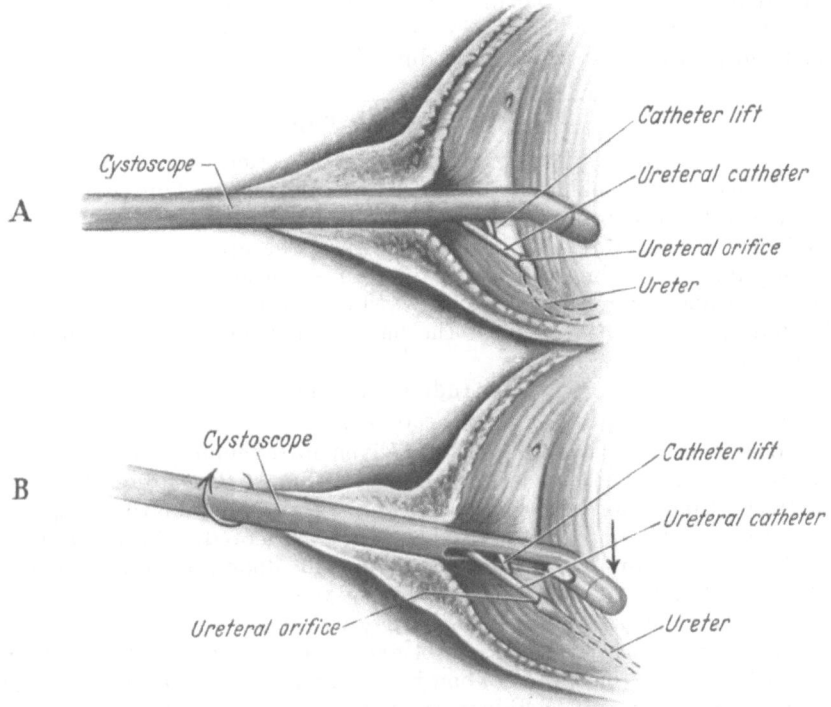

Fig. 25 A and B. Maneuver for passing a catheter up a ureter which enters bladder transversely. A Tip of catheter is inserted into ureteral orifice. B Inner end of cystoscope is moved lateralward. Tip of catheter in meatus displaces it lateralward thus straightening lower end of ureter

inward along the crest of the ridge. By this maneuver the tip of the catheter may enter the meatus even though the orifice cannot be seen. Intravenous injection of dye such as indigocarmine aids in locating the ureteral orifice when the kidney has sufficient function to excrete the dye; the spurt of deep blue liquid is seen and followed to its source. When edema, ulceration, tumor or other lesions distort or change the bladder mucosa it may be impossible to locate the ureteral orifices; gentle probing of the most probable area with the tip of the catheter may be rewarded by seeing it pass into the ureter.

The ureter which enters the bladder transversely instead of obliquely is diffi-cult to catheterize; the catheter must enter the meatus at an angle instead of in a line parallel with it. There is a maneuver which is helpful in these cases (Fig. 25). The tip of the catheter is inserted into the meatus, then the inner end of the cysto-scope is moved lateralward; the catheter, with its tip still in the meatus, displaces the orifice lateralward thus straightening the intramural portion of the ureter so that it enters the bladder in a more oblique direction. The catheter can then be more easily passed up the straightened ureter. If this maneuver fails, a coudé tip or a spiral filiform tip catheter may be passed.

L. Differential renal function

An estimation of the function of each kidney is desirable in most cases of suspected renal disease. In nonsurgical maladies such as glomerular nephritis and nephrosclerosis, a combined renal function estimation is adequate; but a differential function is essential in all cases of suspected surgical disease of the kidneys. The most accurate determination of the function of each kidney can be obtained during the cystoscopic procedure.

I. Chromocystoscopy

Chromocystoscopy has largely replaced the collection of excreted substances for the estimation of renal function (DIETZ; FRIEDRICH). This method utilizes observation of the ureteral orifices through the cystoscope for appearance of the dye; it is not necessary to leave the catheters in the ureters to collect the dye. The time required for the cystoscopic procedure considerably shortens chromo-cystoscopy. Unless ureteral obstruction exists, a more accurate appearance time can be determined, for there is not the hazard of inadequate flow through the ureteral catheters.

1. Indigocarmine
(Secretan)

This dye is commonly used; 5 cc. are given intravenously either just before or immediately after the cystoscope has been passed. The ureteral orifices are watched; the time from injection to the first appearance of the dye, and the concentration of the spurt through the orifice are noted. Normal appearance time is three to five minutes. Concentration depends upon the dilution of the urine as well as renal function. A poorly functioning kidney will be both slow in excreting and deficient in concentrating it. The kidney which has good function may show a prolonged appearance time because of excessive urine concentration and slow passage of it through the ureter; when it does appear, however, concentration is excellent. Conversely a good kidney may show poor concentration when there is excessive excretion of urine, even though the appearance time is normal. Therefore, two coexisting findings, — delayed appearance of the dye and poor concentration are necessary to make a diagnosis of poor renal function (RITT-MANSBERGER).

2. Trypan red

This has been injected simultaneously with indigocarmine to give more accurate indication of renal function (JASIENSKI). This dye is retained in the circulation for a long time when the filtration apparatus is intact, but is eliminated readily when there is impairment of filtration. Its excretion with indigocarmine indicates some renal damage. When trypan red is excreted and indigocarmine is withheld, renal damage is severe.

3. Neoprontosil
(disodium 4-sulfamido-phenyl-20z0-7-acetylamino-1-hydroxy-naphthylene, 3,6 disulfonate)

This has been used for chromocystoscopy when other material was not available (DE LA PENA). Its antiseptic action may help to reduce the incidence of infection following cystoscopy.

II. Phenolsulphonaphthalein (P.S.P.)

Determination of the amount of P.S.P. excreted in the urine obtained from each kidney is a common method of estimating the separate function of each. One cubic centimeter (0.6 gram) of the dye is given intravenously after the ureteral catheters have been passed. The time of appearance from each side is noted by allowing the urine to drip into a weak alkali such as sodium bicarbonate or soap solution. The urine containing the dye is then collected from each catheter for a ten or a fifteen minute period, and the amount of the dye in each specimen is measured by comparing with the color standard. P.S.P. for differential renal function is not used as widely now as it was two decades ago. Its use consumes time which is precious to the patient; the quicker he can be rid of the ureteral catheters the better for him. (Catheters remaining in place for over a half hour predispose to ureteral edema and postcystoscopic pain.) Sometimes urine flows around the ureteral catheters into the bladder; this is collected and measured, but the side from which it came is not known. Occasionally the ureteral catheter becomes plugged and the urine containing the excreted dye is retained in the kidney pelvis.

III. Urea clearance

The collection of urine from each kidney for estimation of urea clearance is probably a more accurate method of determining renal function, but is more complicated (VAN KEERBERGEN).

M. Kidney study (retrograde cystoscopy)

This is the term applied to the cystoscopic procedure which includes examination of the upper urinary tract. It consists of the following steps: (1) Passing the cystoscope and examining the interior of the bladder; (2) differential renal function test by chromocystoscopy; (3) catheterization of both ureters; (4) collection of a specimen of urine from each kidney; (5) differential renal function test if not already done by chromocystoscopy; (6) x-ray of the urinary tract with ureteral catheters in place; (7) bilateral retrograde pyelogram in the recumbent position; (8) bilateral retrograde pyeloureterogram in the erect position while catheters are being pulled down to the lower end of the ureters.

Usually. almost every urinary tract disease can be diagnosed by a kidney study. The procedure, however, may be painful and sometimes the postcystoscopic reaction disables the patient for several days or may rarely lead to serious complications (Chap. III). There is a trend away from routine complete retrograde kidney studies for diagnosis of upper urinary tract diseases, and toward excretory urography combined with cystoscopy and partial kidney study when necessary. Excretory urography is done first, followed by cystoscopic examination of the bladder. When a survey of the urographic films leaves any doubt regarding the condition of the upper urinary tract, the ureter on the questionable side — or both ureters if indicated — is catheterized and a retrograde pyeloureterogram made. Thus, complete information regarding the entire urinary tract can be secured by a procedure which is less painful and less likely to cause reaction than complete kidney study.

N. Removal of the cystoscope

When the cystoscopic examination has been completed the cystoscope is removed in the reverse manner to which it was introduced. When a convex or a concave sheath is used, the concavity is kept toward the pubis. When catheters have been passed up the ureters, they may be left in place for collection of urine and for urograms. The catheterizing telescope is first withdrawn, leaving the catheters in the sheath. Then the sheath is removed as the catheters are pushed inward forming a loop in the bladder: this avoids dislodging them from the ureters. When the catheters have been passed only a few centimeters up the ureters it may be necessary to leave the cystoscope in place during the remainder of the procedure for fear of dislodging the catheters.

O. Cystoscopy hipogastrica

Cystoscopy may be performed through a suprapubic cystostomy opening (GORO). A direct or an obliquely forward telescope provides the best view unless the cystostomy passage extends obliquely downward instead of directly dorsal-ward; in these a right angle view is needed to visualize the fundus of the bladder. The vesical orifice is well seen. Stones, tumors and foreign bodies can be identified. An attempt can be made to pass a ureteral catheter outward through the urethra, in cases of impassable stricture. The ureteral orifices can be visualized but cannot be catheterized through this approach. Foreign bodies and small calculi can be removed.

P. Experimental and practice cystoscopy

I. Female dogs

Experimental cystoscopy can be done on female dogs (BARRINGER). The growth of induced bladder tumors can be watched and the results of cystoscopic procedures observed. However, animals are not practical to use for learning cystoscopy.

II. Phantom bladder

When this apparatus is available, it provides the best means of learning how to manipulate the cystoscope and ureteral catheters. Otherwise supervised practice on patients is the best method of learning cystoscopy.

Chapter III

Postendoscopic care, reactions and complications

A. Postendoscopy care

Routine orders following endoscopic examinations anticipate mild reactions which may occur. Administration of a sedative when the precystoscopic anal-gesic is wearing off helps to prevent spasm with its resulting frequency and urgency. Forcing fluids, 8 oz each hour for 12 hours following the procedure, aids in holding down infection and inflammation. When nausea prevents ingestion of fluids by mouth, 5 per cent glucose in saline is given intravenously. Application of heat such as short wave diathermy through the bladder immediately following cysto-scopy is helpful in preventing spasm. Routine administration of an antibacteria agent has been recommended; but when aseptic technique is used and when there is no evidence of latent infection, this prophylactic measure is not necessary. Administration of antibacterial drugs is reserved for the treatment of frank and latent infections and should not be used indiscriminately; the chances of benefiting the patient are not worth the risk of obtaining an unfavorable reaction to the medication, nor of developing an immune strain of bacteria.

B. Reactions and complications

Immediate mild symptoms almost always occur and are expected. These are burning on urination, frequency and urgency, a small amount of blood in the urine or a slight bloody urethral discharge; they usually disappear within six to 12 hours. Severe reactions arising from endoscopic examination, and complications following the procedure, are more often due to faulty technique than to any other cause (WEHRBEIN). When the immediate symptoms are severe, and when they occur consistently after each examination, the cystoscopist should examine his technique minutely to discover his faults and to correct them. Any endoscopic procedure is traumatic, but excessive trauma is almost always due to improper handling of the instruments or of the patient, or both. In exceptional cases it may be advisable for the cystoscopist to assume a calculated risk and proceed with the examination even though excessive trauma is unavoidable for its completion. Mercurial poisoning following use of mercury oxycyanide solution in cystoscopies has been reported (PAGE and WILSON).

C. Prophylaxis of complications

I. Gentleness

In the handling of endoscopic instruments gentleness minimizes trauma. A "sensitive touch" is necessary for the expert cystoscopist. He "feels" with the instrument which he has in his hands; the least resistance while passing the cystoscope is transmitted through the instrument to his finger tips. When an obstruction is encountered he feels first one way then another with the tip of the instrument until a way is found around the obstacle. The "light touch" which he has prevents his thrusting through, tearing and perforating; these are only for the man with the "heavy touch", for the rough handler of instruments. Rapid jerky motions are more likely to be traumatic than slow deliberate ones; but every motion must count. When the entire interior of the bladder is surveyed, slow systematic movements are made first inward and then outward as the cystoscope is slowly rotated. The posterior urethra is examined only once; to move the instrument in and out of the bladder neck several times is traumatic. The ureteral catheter is slowly and gently inserted into the meatus and passed upward (STOLZ); jabbing it into the mucosa produces pain and starts spasm.

II. Alertness

The skilled cystoscopist is alert. He has evaluated the patient and has ordered adequate sedation both before and after the examination. He remembers to connect the light cord and check the bulb before introducing the cystoscope; unnecessary trauma is caused by removing the instrument to adjust the light or replace the bulb. Most important of all, when fluid is running into the bladder, he is aware of it and he shuts it off before the bladder becomes overdistended (LEWIS). Edema and petechial hemorrhages occur from mild overdistention. When the pressure of the irrigating fluid is high, severe trauma to the bladder mucosa and massive hemorrhage may result because the cystoscopist is not alert to the flow of water into the bladder.

4*

III. Carefulness

Care in observing aseptic technique reduces contamination of instruments to a minimum and helps to prevent postcystoscopic infection. The wearing of sterile gloves is a quick way of making the hands aseptic. Changing gloves between cases prevents carrying infection from one patient to another. The ureteral catheter, as it is passed through the cystoscope, is held away from the face of the operator, and the ends are kept within the sterile drapes and not allowed to flap about.

IV. Good judgment

The use of good judgment at all times will obviate many complications. Errors in judgment and miscalculations will continue, however, as long as the human element is involved. For the beginner, it is encouraging to know that improved judgment comes with experience. A patient may not have appeared nervous or hyperirritable before cystoscopy, but when the cystoscope is passed, he becomes tense and rigid; better judgment would have suggested more sedation. A man with an enlarged prostate is cystoscoped, then reacts with complete urinary retention and fever due to pyelonephritis. It would have been more judicious to have obtained the information by a less traumatic procedure such as a cystogram or excretory urogram; these would probably have given sufficient information without causing an undesirable reaction.

V. Avoidance of overeagerness

The astute cystoscopist is not overeager. Sometimes efficiency and the urge to accomplish the task outweighs good judgment. The urethra may be small or strictured; it should not be excessively dilated, for divulsion is traumatic and may result in immediate hemorrhage or rupture and subsequent stricture formation. Persistence in trying to catheterize the ureters or to accomplish some other procedure may lengthen the examination unduly; it is better to give up trying than to keep the cystoscope in place for longer than one-half hour, especially when only local anesthesia is being used.

VI. Definite prophylaxis

A few definitive measures for the prevention of complications are considered. Adequate precystoscopic sedation reduces spasm and tension; a relaxed patient is much less likely to be injured than one who is straining and moving about. The administration of profenil (KAHN), a powerful smooth muscle depressant, helps to prevent spasm and lowers blood pressure by its peripheral vasodilatory action. When the diagnosis of infection is not the major reason for examination, an antibacterial drug administered before cystoscopy may prevent acute infection following the procedure. This is especially true when the patient has a chronic or latent prostatitis which might be exacerbated into an acute stage by the manipulation. Overdistention of the renal pelvis for retrograde pyelography is a frequent cause of renal pain and sometimes of anuria, following cystoscopy (WOLF). When the tip of the catheter does not reach to the pelvis, the ureter may be ruptured by forceful injection (HENLINE). The gravity method of filling the renal pelves or ureters for retrograde urograms seldom if ever causes injury. This method is, however, slow and cumbersome. Injection with a small glass piston syringe (10 cc. Luer) is quicker and easier, and is as safe if the cystoscopist is alert to the pressure required to inject the fluid. If he has the "light touch" he can feel the increased resistance to the plunger when the pelvis has become filled, and will stop the injection at that point. Repeating the urogram to obtain more adequate filling is better than overdistending the pelves the first time. The ureteral catheters should, however, be left in place as short a time as possible; their presence in the ureter over a period of a half hour predisposes to edema, trauma and untoward reactions.

D. Unavoidable reactions and complications

I. Sensitivity to drugs

This cannot always be predetermined by present methods of testing. Reactions to local and general anesthetics have caused deaths. Sometimes spasms occur as a reaction to the anesthetic at which time severe trauma may be inflicted during the cystoscopic procedure (WILHELM). Mucosal edema subsequent to the examination may be so severe as to obstruct the urethra or the ureters. Combined anuria and uremia is an exceedingly rare complication and is due either to severe edema of the mucosa of both ureters or to renal trauma or drug hypersensitivity. Other manifestations of uremia such as uremic colitis (D'ABREU), coma and even death have been reported. Cystoscopy in cases of genitourinary tuberculosis may be followed by an acute flare-up of the disease. A safe procedure is to give streptomycin and isoniazide for a few days before the examination.

II. Presence of disease

The presence of severe urinary tract disease is often an unavoidable cause of reaction or complication following cystoscopy. It is sometimes necessary to take a calculated risk in order to make a definite diagnosis or to administer treatment by cystoscopy. A diagnosis of the cause of anuria may be urgent, and the only way of telling whether or not it is due to bilateral ureteral obstruction is to catheterize the ureters. If no obstruction is found, the procedure would have done more harm than good, but this risk must sometimes be taken. A necrotic tumor filling a small bladder may allow the cystoscope to perforate the bladder wall without resistance. The majority of deaths following cystoscopy are due to an aggravation of the disease for which the examination was made.

E. Diagnosis and treatment of reactions and complications

Most reactions and complications are easily recognized and their treatment is simple; the more severe complications, however, may present problems.

I. Fever, spasm and pain

Febrile reactions are common and are treated by a wide spectrum antibiotic; spasm and pain require sedation and application of heat such as diathermy or hot sitz baths.

II. Sensitivity to the local anesthetic

These reactions are discussed on above.

III. Urethral bleeding

Excessive bleeding from the urethra is best treated by indwelling hemostatic bag urethral catheter. Slight tension on the catheter helps to control hemorrhage into the prostatic urethra. Narrow thick pads pressed against the perineum and the penile urethra and held in place by a T binder drawn tight helps to suppress bleeding when it is external to the membranous urethra.

IV. Perforation

Perforation of the urethral or bladder wall with the tip of the cystoscope is seldom immediately serious unless the false passage is deep. Periurethral and perivesical hemorrhage is almost impossible to control by open operation; conservative treatment is much safer and just as effective.

V. Extravasation

Occasionally urinary extravasation occurs following perforation of the urethra or bladder. In recognized cases of perforation a retention catheter kept in place and kept open for five to six days prevents extravasation. A cystogram or urethrogram is sometimes enlightening when extravasation is suspected.

VI. Anuria

Anuria which persists more than two days following ureteral catheterization should be treated by passing catheters up the ureters and leaving them in place for continuous drainage. Even though the anuria is not due entirely to constriction from ureteral edema or blood clots, there may be sufficient obstruction to hold back the urine when excretory renal pressure is low. Other routine measures such as the administration of 50 cc of 50 per cent glucose solution in water intravenously twice daily and limitation of fluids to within 1000 cc. of the urinary output are worthwile carrying out.

Chapter IV

The normal bladder and prostatic urethra

A thorough knowledge of the normal bladder and posterior urethra is essential before abnormalities and disease can be recognized and identified. The appearance of structures is not the same through different optical systems. The descriptions which follow are those seen through the *right angle vision telescope* unless otherwise specified.

A. Divisions of the bladder

For convenience of description and for location of structures and of lesions the bladder is divided into the following areas (Fig. 26):

1. *The bladder neck or vesical orifice* is the narrow rim which encircles the outlet from the bladder.

2. *The trigone* is the triangular area bounded above by the interureteric ridge or a line extending between the ureteral orifices, and on each side by the ureteral ridge or a line from each ureteral orifice to the dorsal midline of the bladder neck.

3. *The fundus* is the area extending from the upper margin of the trigone to the region opposite the vesical orifice; it includes the lateral walls of the bladder.

4. The *base or floor* is the trigone and that portion of the fundus which constitutes the dorsal bladder wall.

5. The *dome* includes the rounded portion of the upper and ventral bladder wall.

6. The *ventral wall* extends from the anterior lip of the bladder neck upward to the dome; it includes the ventral portion of the dome.

The boundaries of these areas are not fixed and definite. They may overlap to some extent, but are specific enough to localize lesions satisfactorily.

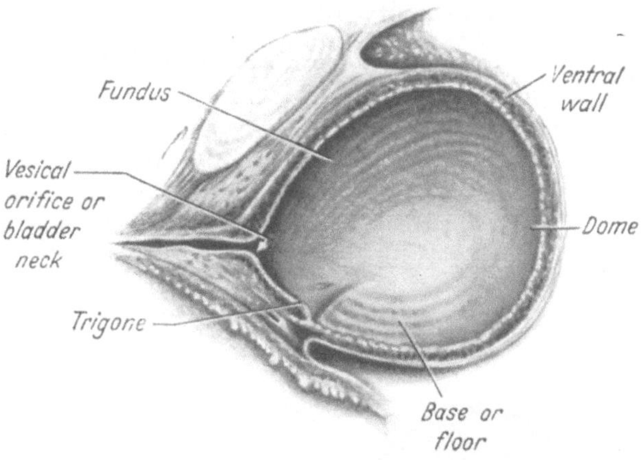

B. Vascular pattern

The vascular pattern of the bladder mucosa is observed minutely during cystoscopy (HOYT and MESSIER). Early disease may be indicated

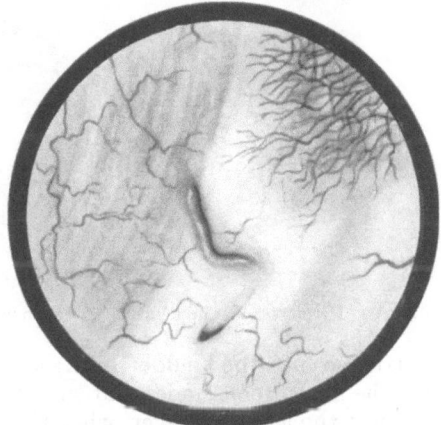

Fig. 26. Divisions of bladder for use in locating structures or lesions

by changes in the blood vessels before there is any indication of abnormality elsewhere. The mucosa covering the trigone is more vascular than that in the remainder of the bladder. Numerous small vessels are commonly found on the

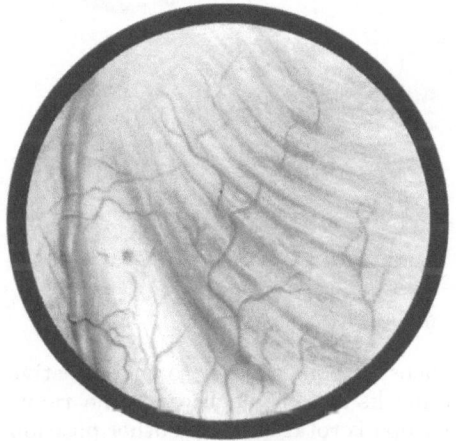

Fig. 27. Normal vascular pattern of bladder mucosa in region of right ureteral orifice. Large vessel lateral to meatus is a vein larger than average but not abnormal

Fig. 28. Varicose vein and network of small vessels on apex of trigone

apex of the trigone. Frequently larger vessels are near the ureteral orifices and often originate in that area (Fig. 27). A network of tiny vessels is commonly found near the ureteral orifices, usually below or lateral to them. When the meatus is difficult to identify, this vascular pattern may give a clue to its location. Dilated veins of various sizes (Fig. 28) are sometimes seen on or adjacent to the trigone, and are occasionally found in the fundus and dome of the bladder. They are not considered abnormal, nor do they cause symptoms unless they are exceptionally large or unless they rupture.

C. Bladder neck

The normal bladder neck is smooth, regular and convex (Fig. 29). It is elevated from the adjacent bladder wall except in the region from the 5 o'clock to the 7 o'clock positions where it is on a level with the floor of the bladder. When the field of vision through the telescope is directed dorsalward and the cystoscope is slowly drawn outward into the bladder neck, there is a gradual transition from

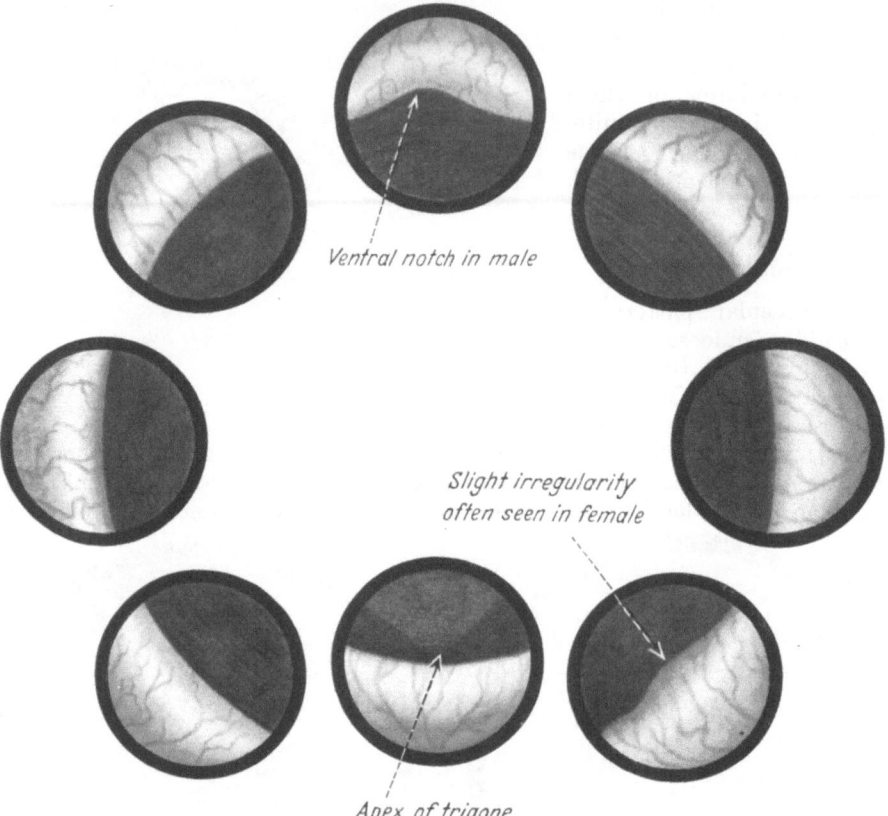

Ventral notch in male

Slight irregularity often seen in female

Apex of trigone

Fig. 29. Normal bladder neck outline as seen in multiple fields of vision while cystoscope is being rotated

the trigone into the bladder neck; as the mucosa becomes closer to the objective lens, the sharp focus fades and then no landmarks are visible because the tissue is against the lens. However, when the cystoscope is rotated to any other position on the bladder neck and is gradually withdrawn, the outline of the internal vesical orifice comes into the field of vision suddenly; there is no gradual transition of the bladder wall into the bladder neck. There is usually a shallow notch at the 12 o'clock position on the bladder neck in the *male*; it is formed by the junction of the lateral lobes of the prostate. The outline of the vesical orifice in the *female* is usually a little irregular in places, especially when the patient is middle aged or older.

D. Trigone and ureteral orifices

The trigone is sometimes, but not always, outlined by the interureteric ridge which connects the ureteral orifices, forming the base of the trigone, and by the

ureteral ridge on each side extending from the orifice to the 6 o'clock position on
the bladder neck. The absence of these markings does not constitute an abnormality. When the ureteral ridge is present the ureteral orifice is on its crest near
the upper end. The distance of the ureteral orifices above the bladder neck varies
from 1 to 3 cm., averaging 2 cm. The orifices are usually equidistant from the
bladder neck, although it is not abnormal to find one lower than the other. The

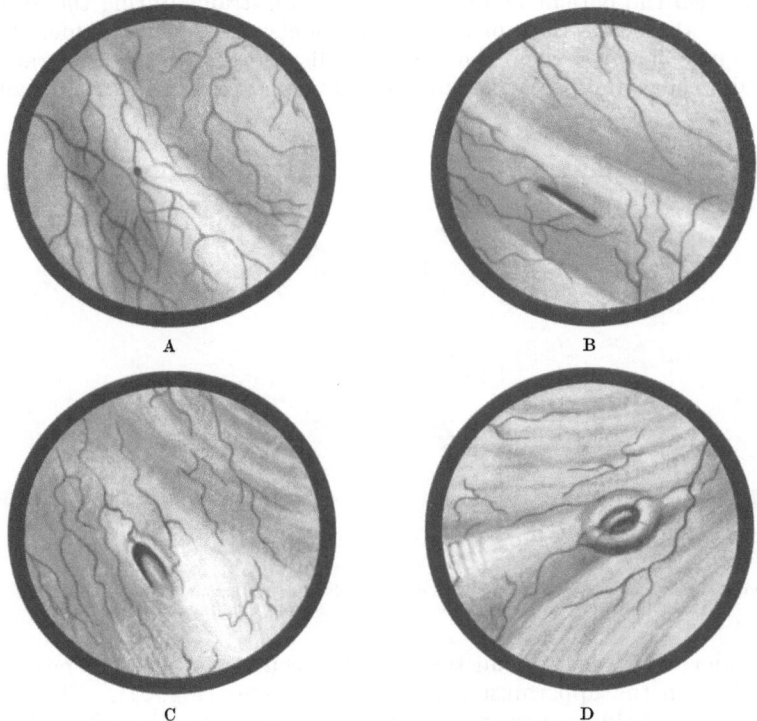

Fig. 30 A—D. Different forms of normal ureteral orifices. A Small almost invisible slit. B Large slit.
C Crescent shaped. D Circular

most common type of ureteral orifice is a small slit 1 to 2 mm. long; when closed
it may be almost or entirely invisible (Fig. 30A), but usually the rounded edges
of the slit are distinct (Fig. 30B). Sometimes the meatus is crescent shaped
(Fig. 30C), and occasionally a circular orifice (Fig. 30D) is seen. When a circular
orifice remains open it is evidence of fibrosis and is not normal. Each ureteral
meatus opens every three to ten seconds while urine which is propelled downward
from the kidney by a peristaltic wave is ejected from it. Some ureteral orifices
are visible only during this diastolic phase; in order to identify them, the cystoscopist watches the general area of the meatus until he sees it open. Congenital
variations of the ureteral orifices are described in Chapter V.

E. Distending the bladder

When the bladder is collapsed, the folds of bladder wall intrude into its cavity
and cover the trigone and all other structures. As the bladder is distended by
allowing fluid to run into it, these folds gradually smooth out; a view of the entire
interior of the bladder can then be obtained. Observation through the cystoscope should begin when the bladder is empty and continue as it is being distended.

The folds of the bladder wall are watched as they become flattened. Small lesions in the fundus and dome are more easily observed when the bladder is partially collapsed.

Sometimes each lateral portion of the fundus of the bladder is recessed upward for several centimeters; a broad partial septum separates the recesses. This contour usually occurs in women and is due to the normal uterus pressing on the bladder. When the patient is unable to relax but strains during the cystoscopic examination, the entire floor of the bladder is elevated. The bladder floor may also be slightly elevated when the rectum is distended with feces. Straining and reflex bladder spasm often cause marked coarse wrinkling of the fundus and dome

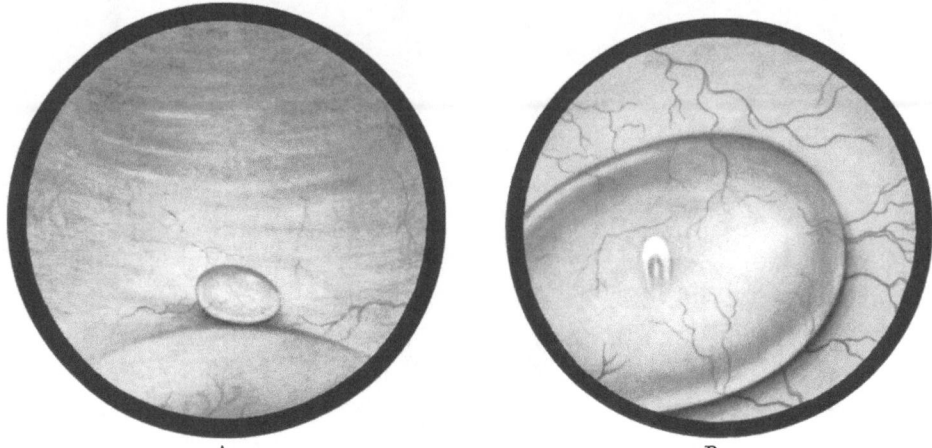

A   B

Fig. 31 A and B. Cystoscopic view of air bubble in dome of bladder. A Distant view. A fold of bladder wall of partially distended bladder is seen in lower edge of visual field. B Close view

of the bladder, which may be mistaken for trabeculation. There is nearly always an air bubble in the uppermost portion of the dome (Fig. 31). The air enters from the lumen of the cystoscope or rubber tubing as the distending fluid is run in at the beginning of the examination.

F. Bladder tone

Bladder tone can be estimated during the cystoscopic examination. Urine and fluid used for distending the bladder normally flow out through the cystoscope in a good stream without manual pressure over the suprapubic region. When the patient is relaxed, especially if the head is low, suprapubic pressure may be necessary to express the last few ounces from the bladder. Abnormal *hypotonia* can be diagnosed when suprapubic pressure is necessary to empty more than a few ounces of fluid out through the cystoscope sheath. Increased intravesical pressure and bladder spasm often occur in the normal bladder during cystoscopy, especially when the patient is tense and is insufficiently sedated. Abnormal hypotonia is difficult to evaluate by cystoscopic examination because the normal bladder can become very *hypertonic* during this procedure.

G. Capacity

The capacity of the normal bladder varies within wide limits. Two hundred cubic centimeters is about the minimum normal capacity. The maximum may be 2000 cc. or even more; this amount is normal unless it is not all passed each time the individual voids. When there is residual urine, a large capacity is abnormal.

H. Variations of the normal bladder

I. During pregnancy

Normal variations of the bladder occur during pregnancy. As the pregnant uterus enlarges, there is a gradual increasing protrusion of the upper portion of the fundus and the dome of the bladder in the midline. The lateral recesses of the fundus are narrow and apparently deeper. During the last month of pregnancy the protrusion sometimes includes the upper portion of the trigone; the ureteral orifices are on the edges of the rounded mass. When the cystoscope is introduced the beak strikes against the mass before it has extended more than 2 cm. into the bladder. It is necessary to deflect the beak into one of the lateral recesses and to rotate it medially in order to visualize or catheterize the ureteral orifice. First the recess on one side is examined, then the cystoscope is withdrawn to the vesical neck, the beak pointed toward the other recess and passed inward again to survey the bladder on the other side of the protrusion.

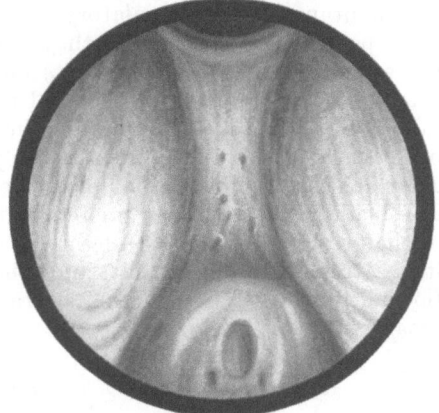

Fig. 32. Prostatic urethra as seen through foroblique telescope

II. In the aged

In elderly persons there is usually some fine trabeculation in the upper portion of the fundus and dome of the bladder, even though no lower urinary tract obstruction is present. The mucosa is also somewhat paler and the blood vessels are less prominent.

I. The prostatic urethra

This is frequently surveyed during cystoscopic examination. When a convex or straight sheath is used, it can be withdrawn into the prostatic urethra; even with the right angle lens a good view can usually be obtained while distending fluid is flowing into the bladder, thus distending the urethra. The prostatic urethra is distended more easily by hydrostatic pressure than is the vesical orifice. When the foroblique telescope is used a good view of the bladder neck can be obtained from the prostatic urethra with the verumontanum (colliculus seminalis of the urethral crest) in the foreground (Fig. 32). It may be necessary to move the cystoscope in or out as much as two centimeters in order to obtain a view of the complete prostatic urethra including the vesical neck. The length of the normal prostatic urethra varies from 25 to 38 millimeters (BARNES). The verumontanum is about 1 cm. distant from the membranous urethra but may vary from seven to 17 mm. (Table 2). The floor of the prostatic urethra rises gradually to the posterior lip of the bladder neck. The verumontanum is in the lower edge

Table 2. *Length of Prostatic Urethra in millimeters, 10 normal specimens measured* (BARNES).

Points of measurement								
External sphincter to verumontanum			Verumontanum to bladder neck			External sphincter to bladder neck		
greatest	least	average	greatest	least	average	greatest	least	average
17	7	10.7 (1 cm)	21	10	17.1 (1.7 cm)	38	25	29.5 (3 cm)

of the field of vision. The sinus pocularis or prostatic utricle (homologue of the uterine cavity) can be seen as a shallow depression on the distal declivity of the verumontanum. The ejaculatory duct orifices are located one on each side of the sinus pocularis. Occasionally they can be identified as tiny slitlike depressions, but are usually visible only when seminal fluid is being discharged through them. There is usually a slightly elevated ridge, extending inward from the verumontanum and losing itself by fanning out into the bladder neck. The prostatic duct orifices are located in the floor of the prostatic urethra on each side of the ridge. They are usually not visible unless there has been infection and resulting fibrosis. The bladder neck and verumontanum are deep red in color, and the remainder of the prostatic urethra is a lighter red. When the prostatic urethra is collapsed it is a transverse curved slit, and when distended by the endoscope and by inflowing fluid it is circular; the lateral lobes of the prostate do not normally encroach into its lumen.

Chapter V

Abnormal ureteral orifices

A. Congenital anomalies

I. Agenesis

1. Unilateral

Agenesis of a ureteral orifice may accompany renal agenesis on the same side. More often, however, when there is congenital absence of a kidney the upper end of the ureter is a blind pouch. When there is absence of a ureteral orifice in the bladder, the normal markings of the trigone on that side are usually absent as well: there is no ureteral ridge, and half or more of the interureteric ridge is absent.

2. Bilateral

This rarely occurs (GRIM; BENJAMIN and TOBIN).

II. Imperforate

A ureter may be imperforate, in which case the orifice which has failed to open is usually ectopic (GIBSON; KREUTZMANN).

III. Ectopic location

1. Below normal

When a ureteral orifice is in an abnormal position, it is usually located lower than normal on the ureteral ridge. It may be at the *bladder neck* or in the floor of the *urethra*. In the female an ectopic orifice is sometimes located *between the urethra and the vagina*, and may be any place along the anterior vaginal wall. In the male an ectopic orifice is rarely located in a *seminal vesicle*.

2. Above normal

When a ureteral orifice is found higher than normal, the malposition is occasionally congenital in origin but is more frequently due to retraction from fibrosis or to surgery.

Ureteral orifices located in abnormal positions are usually difficult to identify. They are more likely to be crescent shaped or rounded, and are often covered by a ridge of bladder mucosa. They may be atresic.

IV. Duplication

Duplication of the ureteral orifice is common. The incidence of unilateral duplication exceeds by six times that of bilateral. Autopsy reveals about equal distribution between the sexes (CAMPBELL).

1. Unilateral

The upper orifice is usually in the normal position at the upper end of the ureteral ridge, and the lower one is close below it. Sometimes they are farther apart; the lower one may be in any one of the abnormal positions given above for a single ectopic ureteral orifice.

2. Bilateral and multiple

Bilateral duplication is often complete on one side and incomplete on the other (CAMPBELL). The occurrence of three orifices on one side has been reported (WOODRUFF).

V. Abnormal shape and size

1. Atresic

Congenital atresia of the ureteral meatus is often seen.

2. Constricted

The orifice may be only pinpoint in size; when it is very small dilation of the intramural portion of the ureter occurs, forming a ureterocele (see below).

3. Dilated

Congenital incompetence of the ureterovesical valve results in a wide open ureteral orifice surrounded by normal bladder mucosa. It may be circular, crescent shaped or oblong. It is usually accompained by megaloureter.

4. Unusual shape

Congenital variations in the shape of the ureteral orifice are common, and are not considered abnormal unless there is a marked difference from the average slit or slightly crescent shaped meatus. A large crescent with a fold of mucosa along its upper margin is usually indicative of an incompetent ureterovesical valve.

B. Acquired abnormalities of size, shape and position

Disease or surgical interference often causes abnormalities in the size, shape and position of the ureteral orifices.

I. Dilated

Chronic infection sometimes results in dilation and fibrotic hardening.

1. Golf hole

The so-called "golf hole" ureteral orifice (Fig. 33) remains open, is circular in shape and fibrotic. It is usually due to chronic tuberculous ureteritis and cystitis but may occur from any severe chronic infection.

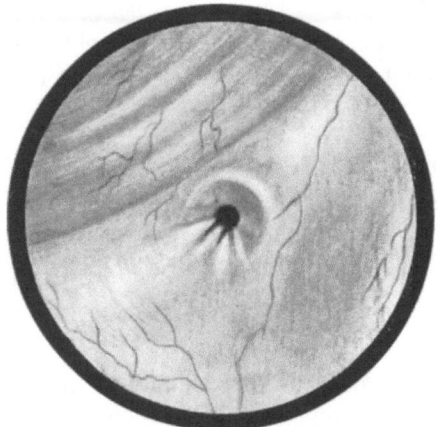

2. Impacted calculus

The ureteral meatus may remain dilated for several days following the passage of a calculus which has been impacted in the meatus.

3. Incompetent ureterovesical valve

When there is complete incompetency of this valve, the ureteral meatus remains open whenever the bladder is distended. An incompetent valve may be due to a central nerve lesion, to congenital or acquired neuromuscular dysfunction or to urinary back pressure from lower urinary tract obstruction.

Fig. 33. Dilated fibrosed ("golf hole") ureteral orifice due to chronic infection, usually tuberculous

II. Position higher than normal

1. Retracted

Retraction of the ureteral orifice is due to a shortened fibrotic ureter; tuberculous ureteritis is the most common cause. The orifice is located higher than normal and there is a funnel shaped invagination of the bladder mucosa leading up to the retracted meatus. The orifice may be entirely covered by the infolded mucosa, so that all the cystoscopist sees is a depression with a fold of mucosa covering it. It is difficult to pass the tip of a catheter through a retracted meatus; a spiral filiform tip is most likely to find the orifice when it is passed into the funnel shaped depression of the mucosa.

2. Surgical reimplantation

When the ureter has been reimplanted into the bladder the position of its new orifice is nearly always higher than normal. In some cases the only clue to its whereabouts is the presence of one or more ridges and some fibrosis at the site of the implant. By manipulating the inner end of the cystoscope to the extreme side of the bladder and rotating the lens toward it the orifice can be visualized and identified and sometimes catheterized.

3. Following ureteral meatotomy

As a result of this procedure the orifice is abnormally high, is usually crescent shaped, and a groove representing the floor of the intramural portion of the ureter leads up to it.

4. Following resection of bladder tumors

After a portion of all of the intramural ureter has been removed during surgery for excision of a bladder tumor, the orifice is higher than normal. It usually remains patent if the ureter has been dilated before the surgery; but when the preoperative condition of the ureter was normal, a fibrotic constricted meatus is more likely to occur.

III. Constricted

1. Following surgery

When a ureteral meatus is inadvertently removed during endoscopic surgery, the result is a constricted fibrotic orifice surrounded by light colored scar tissue which extends down into the prostatic urethra.

2. Following infection

Healed severe infection may also cause fibrous constriction of the ureteral orifice.

C. Edema

I. Calculus

Edema which is localized at the ureteral meatus is most often due to the presence of an intramural ureteral calculus (Fig. 34). The orifice may be entirely covered by the edema or it may remain slightly dilated. Occasionally the stone projects from the orifice. There is usually a small moundlike elevation of the area which is due both to edema and to the stone situated above the meatus.

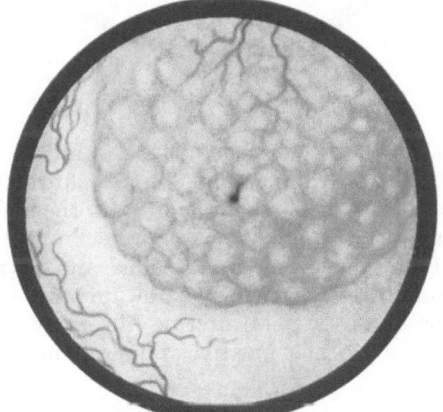

II. Catheterization

Localized edema of the meatus is frequently a result of trauma due to ureteral catheterization or attempted catheterization. There is usually also ecchymosis of the mucosa adjacent to the orifice. The more persistent and forceful the attempts at catheterization, the greater the amount of edema and ecchymosis.

Fig. 34. Localized edema at ureteral meatus due to calculus just above orifice

III. Tumor

Tumor involving the ureteral orifice may cause local edema.

IV. Infection

Severe chronic inflammation of the ureter due to infection may also cause edema which is localized about the meatus. More frequently, however, upper urinary tract infection creates more generalized changes in the bladder mucosa.

D. Protrusion of the ureteral meatus

I. Calculus

Ureteral calculus located just above the meatus causes protrusion of the intra-mural portion of the ureter in addition to edema of the mucosa.

II. Ureterocele

A smooth spherical protrusion of the ureteral meatus covered by normal bladder mucosa is usually due to ureterocele. This arises from a congenital con-striction of the ureteral orifice resulting in dilation of the intramural ureter above it. Each time a ureteral peristaltic wave forces urine into the intramural ureter, the thin walled ureterocele becomes distended and protrudes into the bladder (Fig. 35A). A thin stream of urine spurts out through the constricted orifice,

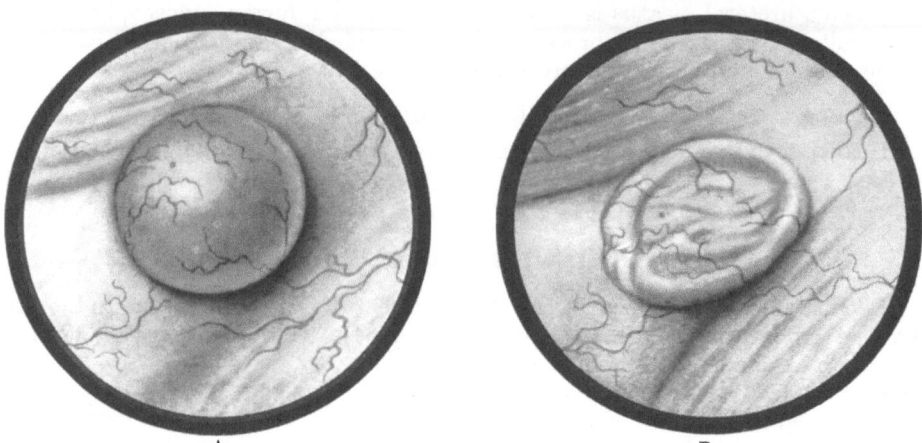

A   B

Fig. 35 A and B. Ureterocele. Note pinpoint uretera lorifice. A Distended with urine. B Collapsed

then the ureterocele collapses and becomes flattened (Fig. 35 B). Most uretero-celes are from one-half to one centimeter in diameter. They are sometimes larger and may be large enough to fill the entire bladder. Rarely they are pedunculated, and in the female may project out through the urethra (ORR and GLANTON). Large ones remain distended and do not collapse at regular intervals as smaller ones do. Protrusion of the ureteral meatus simulating a ureterocele may rarely be due to a small uterine fibroid or other extravesical mass located immediately dorsal to the meatus and pressing it forward.

III. Tumor

A tumor involving the intramural portion of the ureter may cause an elevation of the meatus. When the tumor is invasive there is inflammation and edema of the mucosa surrounding the orifice.

E. Ulceration

I. Tuberculous

Ulceration of the ureteral meatus is usually due to tuberculosis of the ureter and kidney on the side involved. It is more common for the ulceration to be adjacent to the meatus than for it to involve the edges of the orifice (Fig. 74, p. 92).

II. Nontuberculous

Occasionally nontuberculous infection causes similar ulceration.

F. Projections from the ureteral orifice

I. Blood clot

When there is bleeding from the kidney or ureter a blood clot may project from the ureteral orifice (Fig. 36). The clot may be due to trauma inflicted by

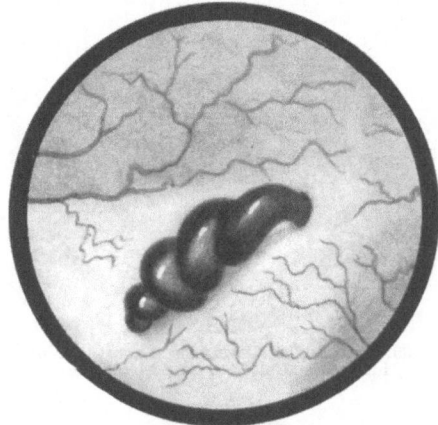

Fig. 36. Blood clot projecting from ureteral orifice   Fig. 37. Calculus projecting from ureteral orifice

a ureteral catheter; when this is the cause there is also edema and ecchymosis of the meatus. A fresh clot is bright red in color; one which has come down from the kidney and is old is dark red.

II. Calculus

A calculus sometimes becomes lodged in the ureteral orifice and a portion projects out through it (Fig. 37). There is accompanying elevation of the meatus and localized edema.

III. Pus

In severe pyogenic infection of the kidney a collection of mucus and pus may temporarily project from the ureteral orifice; it can usually be washed away easily by water flowing in through the cystoscope.

IV. Tumor

A small portion of a papillary tumor involving the intramural ureter may project from the ureteral orifice (Fig. 38) (SORRENTINO). Sometimes it retracts up into the ureter except while urine is being ejected through the orifice.

V. Prolapse of ureteral mucosa

This is of rare occurrence. The ureteral mucosa protrudes as a tube-like structure. The ureteral meatus ring may be partially or completely obliterated

G. Propulsions through the ureteral orifice

I. Bloody jet

Fluid material other than urine may be ejected from the ureter and observed during cystoscopy. Gross blood appears as a cloudy jet which is small and concentrated at the orifice, then fans out and disappears into the clear medium in the

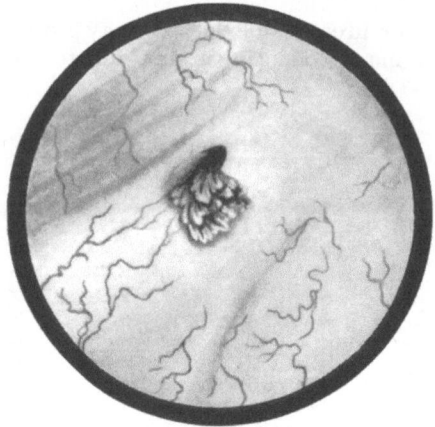

Fig. 38. Papillary tumor projecting from ureteral orifice

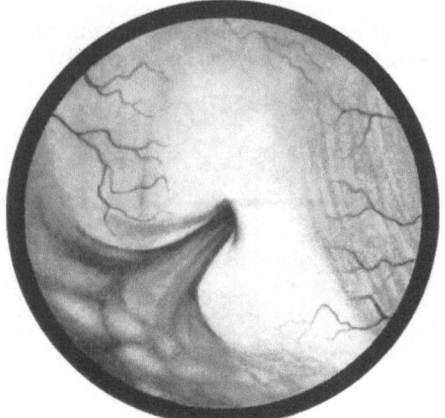

Fig. 39. Dark reddish brown bloody spurt from ureter indicates old blood passing from kidney

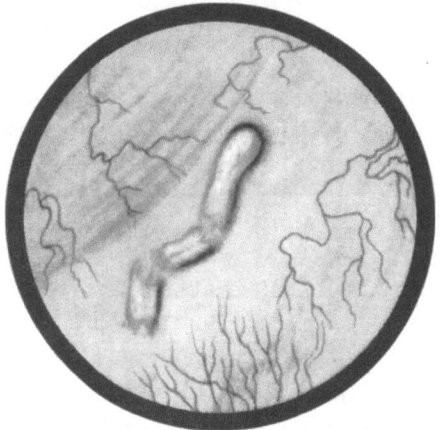

Fig. 40. Pus exuding from ureteral orifice

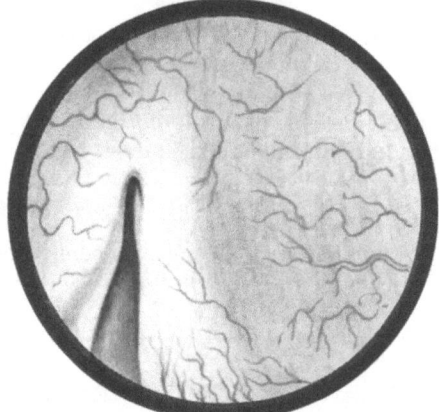

Fig. 41. Indigocarmine being ejected from ureteral orifice following intravenous injections for determining differential renal function

bladder. A small amount of blood may cause only a slight degree of cloudiness, and close observation is necessary to detect it. The spurt is a dark reddish brown cloud when there is much bleeding from the kidney (Fig. 39).

II. Pus

A small amount of pus in the urine is not discernible by cystoscopy as it is ejected through the ureteral orifice, but when a large amount is present there is a white or light cream jet through the orifice. More frequently, however, when pus is observed coming from the ureter, it is thick and pasty; a creamy cylindrical mass extrudes slowly from the orifice, then breaks to pieces as it slithers down onto the trigone (Fig. 40). It has the appearance of tooth paste being squeezed from a tube.

III. Dye

Indigocarmine, after intravenous injection for differential renal function determination, normally appears as a dark blue jet from the ureteral orifice (Fig. 41) in three to five minutes. When the urine is highly concentrated and peristaltic action of the ureter is decreased by sedation, the deep blue fluid may flow slowly and intermittently from the orifice. When renal function is poor the colored jet is slower in appearing and lighter in color. If ureteral peristalsis is poor due to a dilated ureter, to sedation, or to some other cause, the light blue color oozes only from the ureteral orifice.

Chapter VI

Abnormal appearance of mucosal blood vessels in the bladder and posterior urethra

The cystoscopic appearance of certain lesions involving the bladder may vary within wide limits. The same disease may show change in bladder contour with no change in color of the mucosa in one patient, and change in color or vascular pattern without change in contour in another. Moreover, subsequent examinations of the same patient may show an appearance markedly different from those of preceding observations. The same lesion may at times be irregular, flat and ulcerated, and at other times protrude into the bladder and be covered with sloughing tissue or calcareous deposit. It is therefore not possible to make a definite classification of bladder lesions according to their appearance. Diagnosis by endoscopy can, however, be fairly accurate, and conclusions are based upon the appearance of the lesion as seen through the endoscope. The classification which is being used here is therefore based upon the endoscopic appearance of the lesion. It is obvious that many diseases will appear under more than one heading in this classification.

Changes in the appearance of the mucosal blood vessels accompany many diseases of the bladder and posterior urethra. Occasionally the vascular abnormality is the only indication of the presence of pathological changes.

A. Abnormal grouping of blood vessels

Abnormal grouping or arrangement of blood vessels may be due either to an actual increase or to a decrease in the number of vessels in a limited area, or to either more marked or less marked prominence of the vessels in that area.

I. Acute hemorrhagic cystitis

This occurs following overdistention of the bladder or other mild trauma to the mucosa, and is visualized as numerous small groups of closely packed capillaries which are distended, many of which are ruptured. They may be so numerous and close together that they are visualized as a hemorrhagic area rather than as individual tiny vessels (Fig. 42).

II. Hunner ulcer

This localized area of interstitial cystitis is characterized by a radial or stellate arrangement of the blood vessels about a central ischemic hub (Fig. 43). There may be also a red velvety appearance surrounding the hub due to prominence of

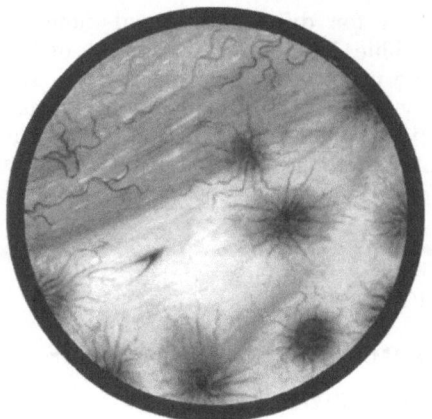

Fig. 42. Patchy areas in acute hemorrhagic cystitis due to groups of prominent and engorged capillaries

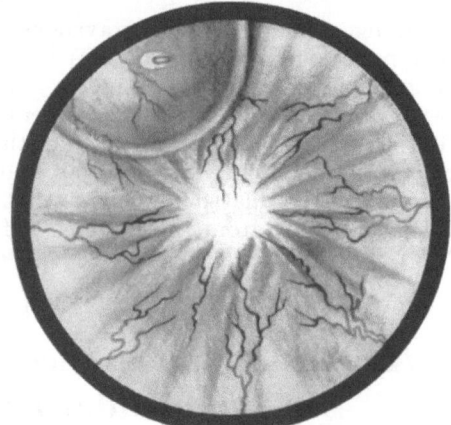

Fig. 43. Radial or stellate arrangement of mucosa vessels in Hunner ulcer (localized interstitial cystitis) Note air bubble at edge of ulcer

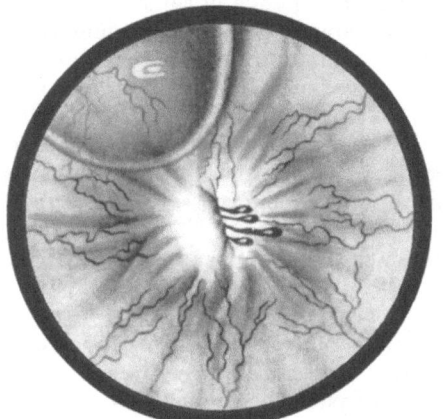

Fig. 44. Slight bleeding from center of a Hunner ulcer following bladder overdistention

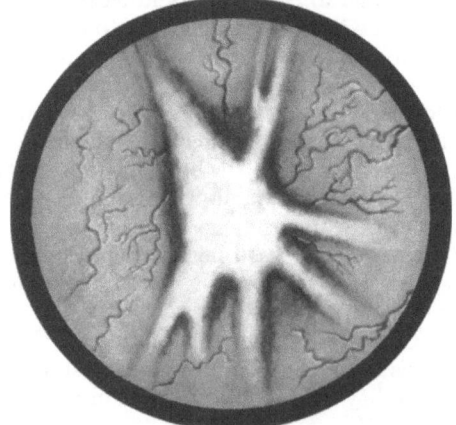

Fig. 45. Scar in bladder mucosa which forms after transurethral resection of bladder tumor or other bladder surgery or injury. Note radiating vascular pattern

the capillary network. Slight oozing of blood occurs from the lighter central area when the bladder is overdistended (Fig. 44).

III. Scars

A scar such as that occurring after transurethral resection of a bladder tumor may display a radial grouping of vessels pointing toward the central ischemic area (Fig. 45). There is no red velvety appearance of the surrounding mucosa; the central light colored area of fibrosis is more glistening than that in Hunner ulcer, and it does not crack and bleed when the bladder is overdistended.

B. Decrease in number and size of blood vessels

I. Chronic cystitis

The number and size of blood vessels in the bladder mucosa gradually reduces with advancing age. Mild fibrotic changes are commonly found in elderly individuals.

1. Herpes vetularum

This term is sometimes applied to the chronic aplastic cystitis of elderly women. The mucosal blood vessels are decreased in size and number.

2. Fibrosis

Fibrosis resulting from healed tuberculosis or other infection of the bladder may be extensive if the infection has been severe; the bladder mucosa appears bloodless, there are ridges and depressions, and the capacity is almost nil. The scar from a healed localized ulcer has fewer blood vessels than normal mucosa, but is not as devoid of vascular pattern as the scar following electrosurgical removal of a bladder tumor.

II. Anemia

In anemia the bladder mucosa is pale. Vessels are few and indistinct. The scarcity of blood vessels in the bladder is as characteristic of anemia as is their absence in the sclerae.

C. Increase in number and size of blood vessels

I. Subacute cystitis

1. Infection, trauma, chemical irritation (Vassallo)

An increase in number and size of blood vessels occurs in subacute inflammation. The arterioles and venules enlarge, and many become visible which normally are too small to be seen. There is an apparent increase in the number of vessels but actually only more of them are visible. Subacute inflammation may be due to one of many causes. A mild infection from any source, generalized trauma as from overdistention, and irritating fluids instilled into the bladder, result in an enlargement of the vessels and an apparent increase in their number.

2. Allergy

The mucosal change resulting from an allergic reaction is most frequently an engorgement of the vessels and an apparent increase in their number (Fig. 46); sometimes there are also patchy areas of increased redness.

3. Endocrine imbalance

Especially at the time of the menopause, this may produce an appearance similar to allergy.

II. Bladder tumor

The mucosal vessels surrounding a bladder tumor are usually more numerous and larger than those in the remainder of the bladder.

III. Prostatic adenoma

The mucosa covering a large benign prostatic adenoma contains more numerous and larger vessels than does the mucosa of the normal or slightly enlarged gland (Fig. 41). These engorged vessels rupture easily and may result in profuse hemorrhage.

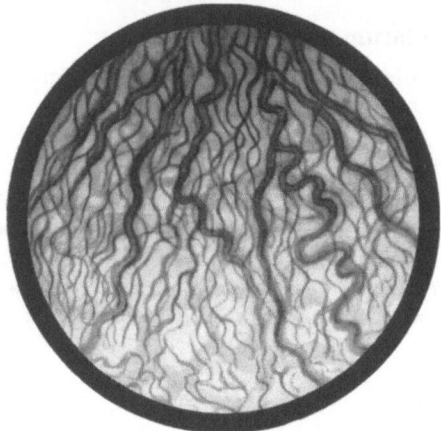

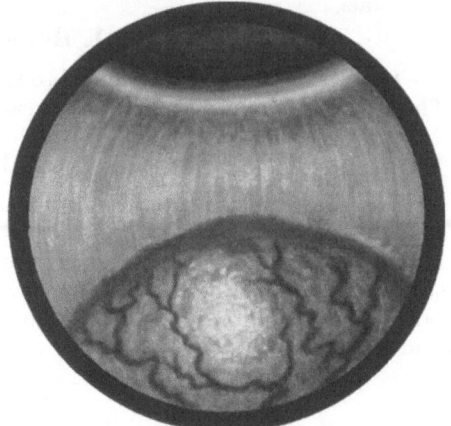

Fig. 46. Increase in size and number of mucosal vessels due to allergic reaction

Fig. 47. Enlarged tortuous blood vessels covering recurrent adenoma of prostate appearing as lobular outgrowths in prostatic urethra. Note granulation tissue at base of nodule

D. Prominent blood vessels

I. Bladder neoplasm

Prominent or enlarged blood vessels without increase in number are sometimes seen in the bladder mucosa which overlies a neoplasm.

II. Large prostatic adenoma

Prominence, as well as increased number, of blood vessels in the mucosa overlying a large prostatic adenoma is a constant finding.

III. Recurrent prostatic adenoma

Benign adenoma of the prostate may recur in the prostatic urethra as lobular outgrowths with enlarged tortuous vessels in the mucosa. These are usually distinguishable from the nodules of prostatic carcinoma by increased tortuosity of the vessels and accompanying granulation tissue (Fig. 47). The lobular outgrowths of recurrent adenoma are also larger and more distinct than the nodules of prostatic carcinoma. The vessels covering a recurrent adenoma rupture and bleed more easily than those overlying a carcinoma of the prostate.

IV. Sclerosis of blood vessels of the bladder mucosa

This occurs in patients with generalized arteriosclerosis. The appearance of the mucosal vessels is similar to that seen in the fundus of the eye. They are like crooked pipe stems and where arteries and veins cross, the vein is obliterated for a fraction of a millimeter on each side of the artery, forming a fine light colored line crossing the vein on each side of the artery (Fig. 48) (HOYT and MESSIER).

V. Varicosities of the bladder

These are considered to be normal (Fig. 28, p. 55, Chap. IV) unless the dilated veins are large and numerous. Their presence is not significant of any pathological

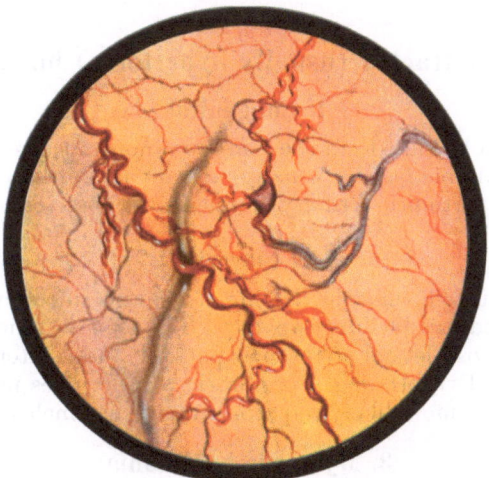

Fig. 48. Sclerosis of blood vessels of bladder in generalized arteriosclerosis. Note obliteration of vein where an artery crosses it

condition. Varicosities are blue, tortuous, tubelike structures lying immediately beneath the bladder mucosa and easily visible through it. Most often they are located at the base of the bladder, near the vesical neck more frequently than elsewhere. They may easily be ruptured from mild trauma such as bladder over-distention, and produce profuse hemorrhage.

Chapter VII

Bladder contour abnormalities associated with normal mucosa

The normal bladder may be irregular in contour (Chap. IV). The lateral recesses are sometimes deep, or the trigone may be elevated in one individual more than in others. When the bladder is incompletely distended, the folds of its wall in the fundus and dome are irregular in contour; these flatten out as the bladder is filled. During the latter months of pregnancy the dome and fundus protrude inward. These contour irregularities due to normal conditions may at times be more noticeable than those arising from abnormal ones.

Some contour abnormalities are rarely accompanied by abnormal changes in color of the bladder mucosa; others are usually accompanied by changes in color. In this chapter, the contour changes are described which are usually not accompanied by changes in color. Some will also be described in Chapter IX.

A. Abnormalities in bladder size and tone

Bladder capacity and tone are often estimated during cystoscopy. The amount of fluid the bladder will hold while the examination is in progress is usually less than its normal capacity of urine because instrumentation often causes spasm. Bladder tone is estimated during cystoscopic examination by noting the force with which fluid is expelled from the bladder through the cystoscope. If it is necessary to press on the suprapubic region to empty the bladder, the tone is obviously poor.

I. Contracted (usually hypertonic) bladder

1. Congenital

This is one which has less than the minimum normal capacity of about 200 cc. A congenitally small bladder is usually normal in appearance by cystoscopic view, and is not hypertonic. Urinary frequency in these cases dates back to infancy.

2. Fibrosis

Extensive fibrosis following the healing of a severe infection or after extensive surgery markedly reduces bladder capacity. The bladder mucosa is light colored inasmuch as it is almost devoid of vascular pattern, and is often irregular, showing ridges and hollows. Even though the fibrotic bladder does not distend easily, it is not hypertonic; it is more likely hypotonic due to the inelasticity of the fibrosis.

3. Myogenic hypertonia

Irritation of the bladder mucosa from trauma, pain, infection, neoplasm or other causes usually results in bladder spasm and reduced capacity; the bladder muscle is hypertonic and there are frequently changes in appearance of the bladder mucosa. *Hunner ulcer* (interstitial cystitis) is an irritative painful lesion which causes reduced bladder capacity; the mucosa is normal except for the small areas of submucous inflammation.

4. Neurogenic hypertonia

Pathological processes causing facilitation of the sacral parasympathetics causes hyperactivity of the bladder resulting in reduced capacity (BORS).

II. Enlarged (usually hypotonic) bladder

An enlarged bladder is of no clinical significance unless it fails to empty completely during micturition. In the normal adult, the average maximum bladder capacity is about 600 cc., but capacity up to 1500 cc. is not abnormal if all the urine is passed when the individual voids.

1. Congenital

The presence of residual urine in a congenitally large bladder is nearly always due to an associated congenital lesion (STEINHARDT). A congenital loss of muscle tone in these bladders may be due to a defective myoneural junction.

2. Myogenic

Bladder neck or urethral obstruction may be the cause of distended hypotonic bladder. The bladder muscle becomes stretched and looses its tone. Myogenic hypotonia may also be due to debility or other systemic causes. In early obstructive disease, the ridges are fine and the depressions shallow; later, coarse, heavy trabeculation is usually found in the fundus and dome.

3. Neurogenic

Most large hypotonic bladders which are not secondary to urinary obstruction, are due to certain lesions of the central nervous system. Among neurogenic bladders, hyper- and hypotonic states of the bladder are essentially due to a matter of facilitation and inhibition. This, of course, is dependent on reflex activity. For example, pathological processes causing destruction of the conus medullaris (micturition center) have no reflex activity and, consequently, there is hypoactivity and hypercapacity of the bladder. Lesions above the conus will yield in most instances hyperactivity and hypocapacity of the bladder (COMARR).

B. Abnormal contour of ureteral orifices

(see Chapter V)

C. Abnormal orifices in the bladder wall

I. Cellules

Cellules frequently occur in a bladder wall which is heavily trabeculated (Fig. 49). The orifice of the cellule is circular or oval and the bottom of its shallow cavity can be seen through the cystoscope. The cavitation of a cellule becomes wider beyond its orifice, whereas a deep trabeculation is a depression without a definite orifice. When a cellule is deeper and larger it becomes a diverticulum. Cellules may retain a small amount of residual urine and be the cause of persistent infection following prostatectomy.

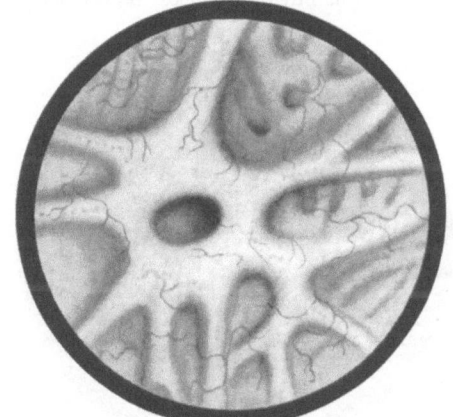

Fig. 49. A cellule. Its bottom can be seen through the cystoscope

II. Diverticular orifice

A diverticular orifice appears through the cystoscope lens as a circular ring surrounding a block cavity (Fig. 50 A) (LLANOS). The ring is often slightly puckered and is sometimes lighter in color than the surrounding bladder mucosa. When the orifice is small and spastic it may be difficult to identify (Fig. 50 B). The most frequent location of a diverticular orifice is on the floor of the bladder lateral to and above the ureteral orifices.

Appearance of interior of diverticulum

When the cystoscopic examination is being performed with a straight sheath, it is frequently possible to pass the objective end of the instrument through the orifice into the diverticulum. The interior of the cavity can then be examined and a stone, tumor or other pathology found if it is present. The mucosa lining the interior of the diverticulum is smoother, somewhat lighter in color and less vascular than the mucosa in the bladder.

III. Fistulous orifice

A fistulous orifice in the bladder may be so small or surrounded by so much edema that it is not visible through the cystoscope.

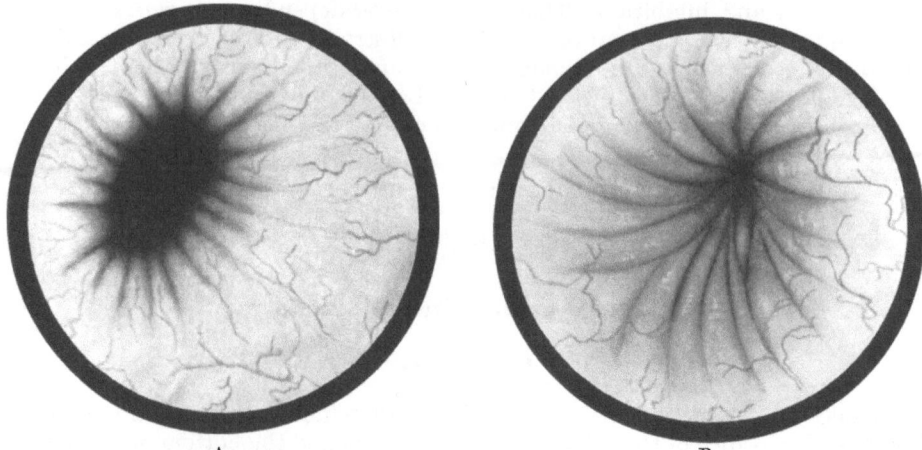

A   B

Fig. 50 A and B. Diverticular orifice. Distant view. A Usual size. B Small spastic orifice sometimes difficult to identify

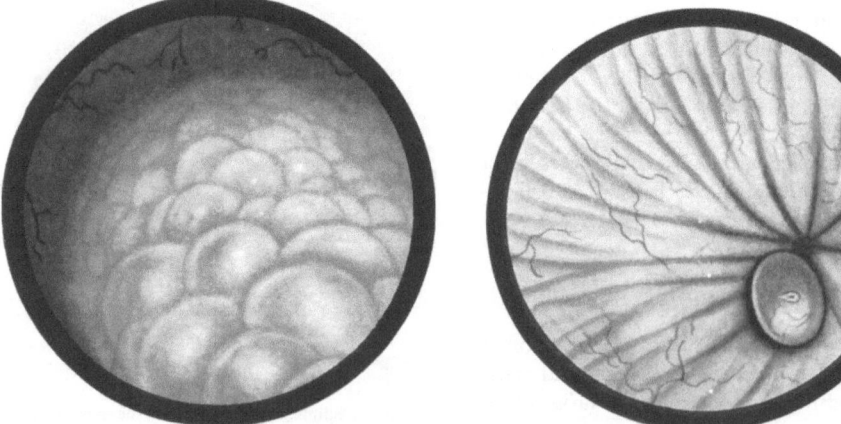

Fig. 51. Intestinovesical fistula surrounded and covered by bullous edema. The orifice itself is not visible. Close view

Fig. 52. Vesicodermal fistula in dome of bladder following suprapubic prostatectomy. Fistula persisted because of hypertrophy of interureteric ridge

1. Congenital

The fistula from a patent urachus is similar in appearance to that following suprapubic surgery (see Vesicodermal).

2. Intestinovesical or from abscess

An intestinovesical fistula is nearly always surrounded and covered by bullous edema (Fig. 51). The fistulous opening itself is usually not visible unless it has been present for a long period of time and all inflammatory reaction has subsided. Whenever a localized area of bullous edema is seen above the trigone or higher in the fundus of the bladder, an intestinovesical fistula should be suspected. A fistula into a perivesical abscess is similar in appearance but the opening can

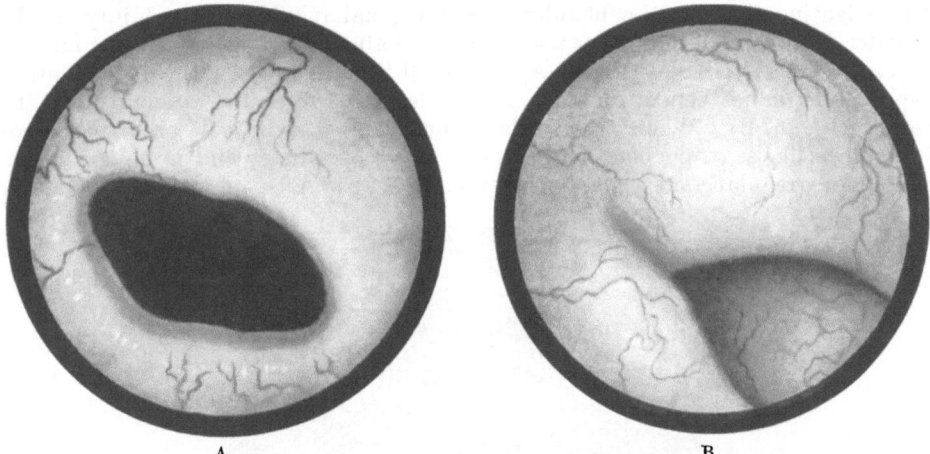

A   B

Fig. 53 A and B. Vesicovaginal fistula due to operative injury. A Medium sized fistula; distant view. B Small fistula covered by overhanging ledge of bladder wall; close view

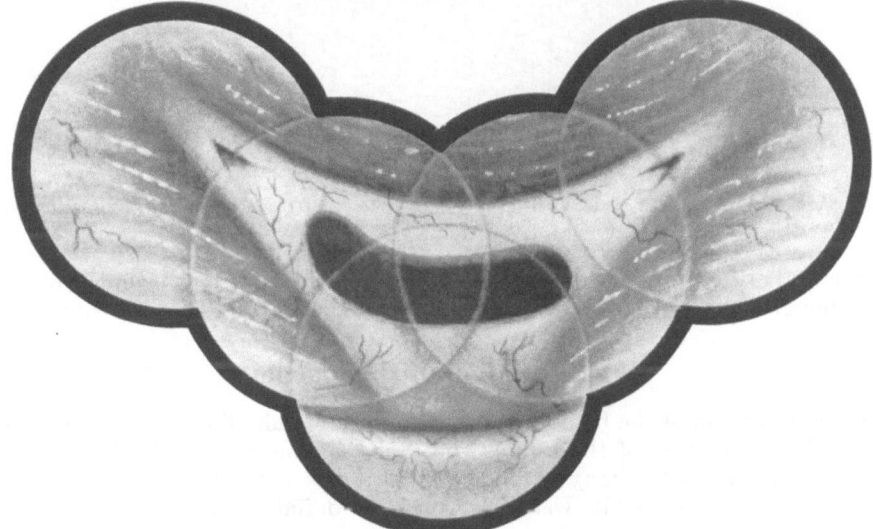

Fig 54. Vesicovaginal fistula due to prolonged difficult labor. Composite of five visual fields

usually be identified; it is shaggy and there may be light colored sloughing tissue at the edge of the fistula or protruding through it.

3. Vesicodermal fistula

This is common following any suprapubic operation on the bladder; it persists when there is obstruction to the outflow of urine from the bladder. There is a tentlike retraction in the dome and the bubble of air usually in the center of the invagination (Fig. 52). The fistulous opening itself is usually so small that it cannot be identified at the extreme upper end of the depression.

4. Vesicovaginal fistula

This is usually easily seen through the cystoscope lens. When it is large, there may be difficulty in distending the bladder sufficiently to identify clearly the edges; one or more fingers inserted into the vagina over the orifice helps to keep the irrigating fluid in the bladder. Vesicovaginal fistula due to injury of the bladder during pelvic surgery is usually located above the trigone near the midline and is irregularly circular (Fig. 53 A). A small fistula may be difficult to identify, especially if it lies beneath a ledge of bladder wall which has resulted from the injury (Fig. 53 B). When the fistula is the result of a difficult prolonged labor it is more likely to be placed below the ureteral orifices, and is oblong transversely or irregularly crescent shaped (Fig. 54). It may be impossible to identify

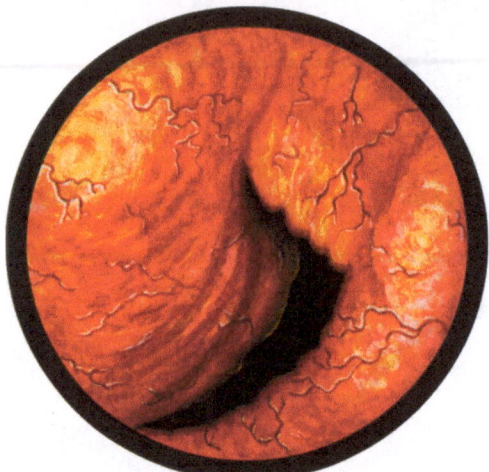

Fig. 55. Perforation of posterior bladder wall with beak of cystoscope. Bladder was contracted due to chronic tuberculous infection

the ureteral orifices when the fistula is very large. The openings from the ureters, even though damaged, are more likely to be in the bladder mucosa than in the vagina.

IV. Herniation of the bladder

When a portion of the bladder is drawn into an inguinal or other hernia, there is a tentlike retraction of the fundus and dome on the side involved (SCANDINO and UPSON). An orifice may rarely be seen at the point of exit of the hernia through the abdominal wall. Usually there is no definite orifice, only an invagination.

V. Rupture through the bladder wall

A ragged opening may be visible at the site of a rupture of the bladder wall. More frequently there is an irregular lacerated area and dark red discoloration from hemorrhage into the bladder mucosa. Rupture or perforation may be due to instrumentation; the beak of a cystoscope may easily perforate the bladder wall when passed into a small contracted bladder (Fig. 55). When the objective lens is through the perforation and in the extravesical space, the view is similar to that of a large perforation through the prostatic capsule (Fig. 15, p. 200). There are spiderweb like strands of tissue, and shiny yellow perivesical fat.

D. Depressions in the bladder wall

I. Cystocele

Cystocele is the term commonly applied to a depression of the floor of the bladder into the vagina. In order to obtain a close view of the mucosa of the depressed area through the cystoscope it is necessary to raise the ocular end abnormally high. When the cystocele protrudes out through the vagina, it may not be possible to visualize the depth of the depressed area through a right angle or foroblique telescope. The bottom of the depression will, however, be seen in the visual field of the retrograde lens. An estimation of the amount of damage to the trigonal muscle (BELL's muscle) can be obtained during cystoscopy by having the patient strain. If the depression of the cystocele is mostly above the interureteric ridge (MERCIER's bar) it is evidence that the trigonal muscle has not been damaged. However, if the depression includes the trigone also, it is inferred that this muscle has been damaged (VAN DUZEN). Following repair of the cystocele the result is more likely to be good if the trigonal muscle has shown no evidence of damage. If it has, difficulty of urination may persist postoperatively due to inability of the muscle to pull open the vesical orifice.

II. Following surgical removal of the rectum

Depression of the floor of the bladder following this procedure is, by cystoscopic examination, similar to the deformity caused by mild cystocele. Hypotonia and an enlarged dome of the bladder persist for several months following radical excision of the rectum.

III. Sacculation

1. From chronic overdistention

Sacculation or increased size of one portion of the bladder is usually due to chronic overdistention. When there is retention due to lower urinary tract obstruction, the dome of the bladder is elongated in the form of an inverted cone. When overdistention is the result of paralysis due to a central nerve lesion, there is an increase in all diameters of the bladder; the entire upper and anterior portions of the bladder cavity form a large sacculation which is usually in the midline but may extend upward and forward on either side.

2. Following surgical procedures on the bladder

The bladder wall may become adherent to the anterior abdominal wall or to other structures and result in a sacculation toward the scar.

E. Elevation of the bladder floor

This is due to abnormalities dorsal to it.

I. From anteflexed or anteverted uterus

The uterine fundus, when enlarged or displaced anteriorly, is a common cause of elevation of the bladder floor. There may or may not be protrusion of the dome of the bladder also.

II. From cervix

Sometimes the cervix presses against the posterior bladder wall when the fundus of the uterus is retroverted. The cervical stump, if it is large and if the broad ligaments and other tissues have been sutured to it following subtotal hysterectomy, may press upon and elevate the bladder floor.

III. From miscellaneous masses posterior to the bladder

Neoplasms of the cervix or rectum or of the vesicorectal septum, fecal impaction, retrovesical abscess and inflammatory lesions dorsal to the bladder may displace the floor of the bladder forward, resulting in its elevation as seen through the cystoscope.

F. Protrusions of the fundus and dome

Protrusions of the fundus and dome of the bladder may be due to the same pathological conditions that cause elevations of the floor if the mass is very large

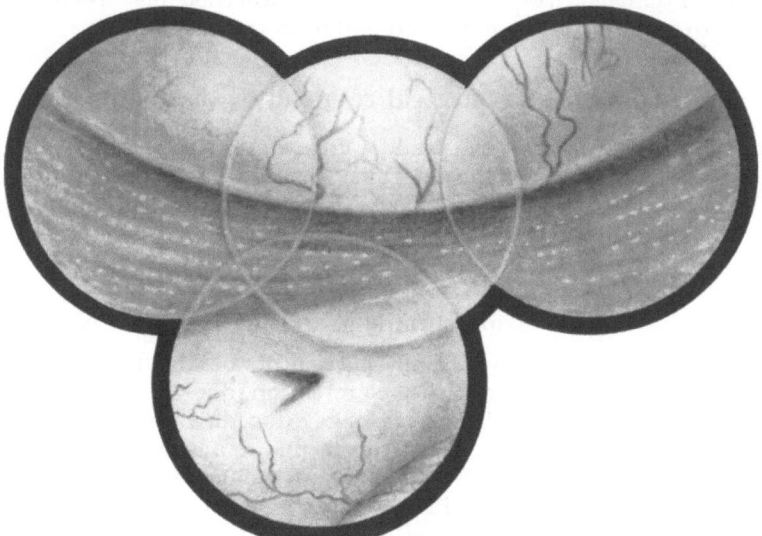

Fig. 56. Protrusion of fundus of bladder due to pressure from a large uterine fibroid. Distant view. Composite of four visual fields

or if it is located above but still adjacent to the bladder. When a protrusion of the fundus or dome is seen through the cystoscope, the examiner must be certain that the bladder is adequately distended before making a diagnosis of abnormal protrusion. Folds of normal bladder wall may appear to be due to extravesical pressure unless the bladder cavity is well filled.

I. From the uterus

An anteverted or anteflexed uterus is a common cause of inward protrusion of the dome and upper portion of the fundus of the bladder. Following fixation of the uterus to the anterior abdominal wall the bladder capacity is reduced and there is permanent depression of the bladder dome and fundus. When the uterus is enlarged due to fibromyomata, pregnancy or other causes, it presses on the upper portion of the bladder (Fig. 56). A small fibroid on the anterior portion of the uterus may cause a localized protrusion into the dome of the bladder (Fig. 57).

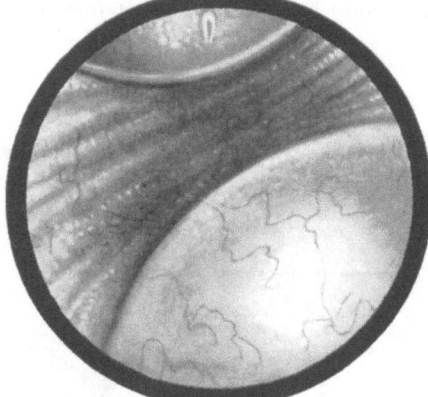

Fig. 57. Protrusion into dome due to small fibroid on anterior wall of uterus. Close view. Air bubble is seen in upper edge of field

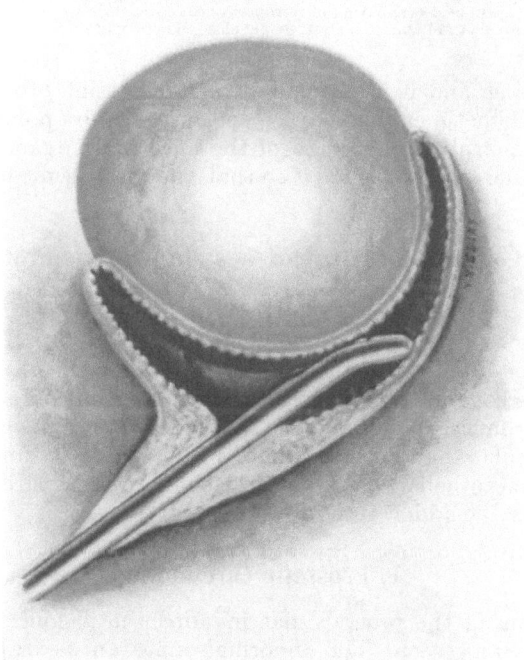

Fig. 58 A

II. From extravesical masses

Any mass, intraperitoneal or extraperitoneal, which is near the bladder may press on this viscus and depress its wall inward. Sometimes it is almost impossible to differentiate an extravesical mass of this nature from one which originated in the bladder wall and is covered by normal or nearly normal bladder mucosa.

Large tumors of the bowel, ovarian cysts, retroperitoneal tumors, cysts and abscesses must be considered when making a differential diagnosis. In some countries, large hydatid cysts occur in the retroperitoneal space dorsal to the bladder. Rarely the mass may be so large that it forces the bladder to be almost

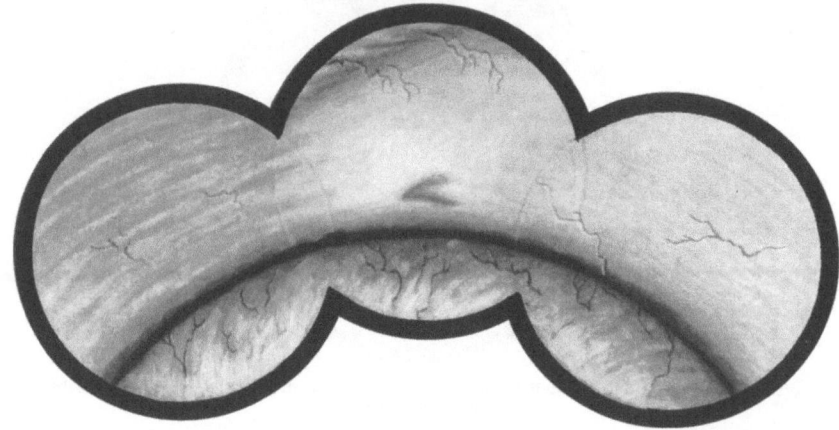

Fig. 58 B

Fig. 58 A and B. Large hydatid cyst originating in retroperitoneal space dorsal to bladder. A Schematic view of position of cystoscope in bladder. B Cystoscopic close view; composite of three visual fields

empty all of the time and it is difficult, if not impossible, to pass a cystoscope into the bladder. Sometimes this may be accomplished by pointing the inner end of the instrument sharply to one side of the mass and advancing it a short distance into the bladder, then rotating it so that the view is medialward (Fig. 58 A and B).

G. Irregular flat or sessile protrusions

I. Invasive malignant neoplasms

Irregular flat or sessile protrusions may be due to neoplastic disease, even though the mass is covered by nearly normal bladder mucosa. Neoplasms which involve the bladder muscle or submucous tissues are sometimes covered by mucosa which is more vascular than normal (Chap. VI). Malignant tumors originating in the bladder wall (mesothelial) are rare. Invasive prostatic carcinoma is the most common submucosal bladder tumor.

1. Prostatic carcinoma

When carcinoma of the prostate has invaded the trigone, irregular rounded protrusions (Fig. 59) are seen. The abnormal appearance extends from the prostatic urethra upward into the bladder. The nodules are more prominent about the bladder neck, and the prostatic urethra is rigid, holding the cystoscope as in a vise.

2. Sarcomata

Sarcoma and mixed sarcomatous tumors (fibrosarcoma, leiosarcoma, myosarcoma, rhabdomyosarcoma, myxosarcoma etc.) may produce no change in the bladder which is recognizable by cystoscopic examination while the neoplasm is small. As they enlarge there is first an inward bulge of the normal mucosa (Fig. 60), then a more prominent protrusion which may have a few dilated and tortuous blood vessels coursing through the overlying mucosa. The protrusion is smoother than that from invasive prostatic carcinoma.

3. Squamous cell carcinoma

This originates in the bladder mucosa but may invade the submucosa and muscle before there is much visible change in the mucosa. In its very early stages,

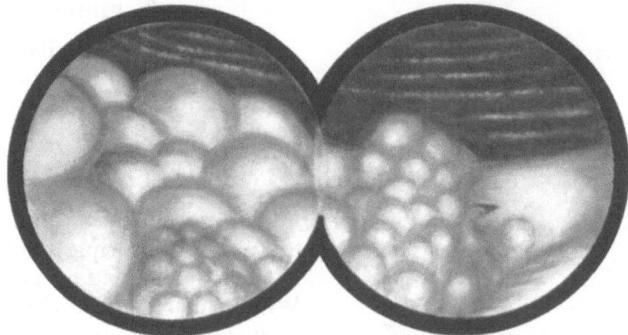

Fig. 59. Irregular rounded protrusions of trigone due to invasion from prostatic carcinoma. Distant view; composite of two visual fields

it may not be visible by cystoscopy, but a little later its appearance is similar to that of the trigonal portion of invasive prostatic carcinoma; there are small

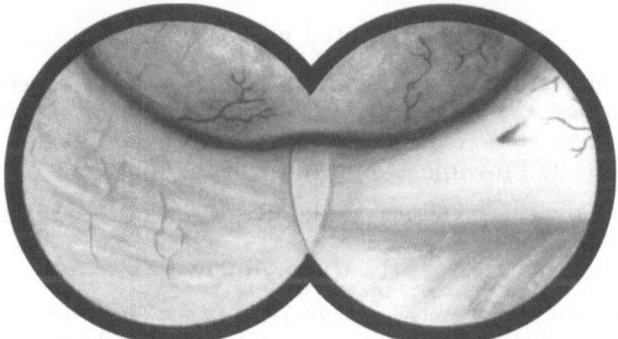

Fig. 60. Smooth protrusions of moderately advanced sarcoma of bladder. Distant view; composite of two visual fields

bulging areas covered by normal mucosa. As the tumor extends, it soon necroses through the mucosa and the appearance is similar to the more common bladder tumors which produce mucosal changes first.

II. Nonmalignant neoplasms

1. Myogenic and congenital

Nonmalignant tumors in the bladder wall are also rare. Those arising in the muscle *(myoma, leiomyoma[1], myofibroma, and rhabdomyoma)* and those arising from embryonic rests *(adenoma and cystadenoma)* are usually at first protrusions

---

[1] KRETSCHMER; KUTZMAN.

covered by normal mucosa. There are no prominent tortuous blood vessels as sometimes seen on malignant tumors. The masses may project prominently into the bladder cavity, but there is in early cases, no change in the mucosa covering them.

2. Neurofibromata

These are multiple, usually oblong elevations 3 mm. to 1 cm. in size, scattered over the floor of the bladder and about the vesical neck[1].

3. Fibromata

These may be similar in appearance but are more likely to be pedunculated (SMITH).

III. Papular cystitis

Chronic infection and inflammation often cause multiple small papular elevations of the bladder mucosa (Fig. 61). When this type of chronic change is seen, an upper urinary tract infection should be suspected. When the papules are grouped around one ureteral orifice, the primary infection is likely to be in the kidney on that side.

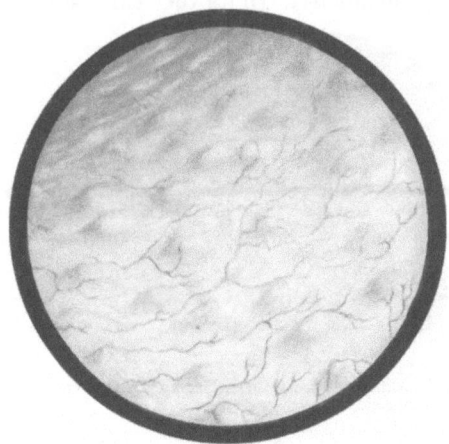

Fig. 61. Papular cystitis. Multiple small elevations due to chronic infection. Close view

H. Pedunculated protuberances

Pedunculated protuberances into the bladder which are covered by normal bladder mucosa are rare. They are usually benign tumors originating in the bladder wall.

I. Fibroma (Koll) and fibroadenoma[2]

These are usually single and may be located in any area of the bladder.

II. Mygoenic

Some benign tumors of muscle origin *(leiomyoma, rhabdamyoma, myxoma*[3]*)* are usually pedunculated protuberances; they may be sessile or only an elevation of the bladder mucosa. Pedunculated and sessile protuberant tumors covered by normal bladder mucosa may be similar in appearance to cysts. The latter, however, have semitransparent and avascular walls.

I. Ridges in the bladder wall

I. Hypertrophy of the interureteric ridge

Ridges in the bladder wall are common. The interureteric ridge, sometimes called MERCIER's bar, is a transverse ridge extending between the ureteral orifices (Fig. 62). It is often hypertrophied in the presence of chronic urinary obstruction at the bladder neck or in the urethra. Following removal of the obstruction the ridge itself may obstruct urinary outflow unless it is incised.

---

[1] MINTZ; FARBER.   [2] MAKAR; URQUEHART.   [3] O'CROWLEY; MARTLAND.

II. Trabeculation

Trabeculation consists of ridges of varying height, length and irregularity of distribution in the fundus and dome of the bladder, often extending down to the trigone (Fig. 63). The ridges forming the trabeculation vary in height from one millimeter to one centimeter or more. They crisscross or branch usually at nearly right angles, and some of the depressed spaces between the ridges may become wider, forming cellules or diverticula. The ridges of trabeculation are composed of the stronger muscle bundles which become hypertrophied from forceful contraction to expel urine from the bladder. They are covered by normal mucosa which dips down between the ridges. Trabeculation and hypertrophy of the interureteric ridge usually coexist.

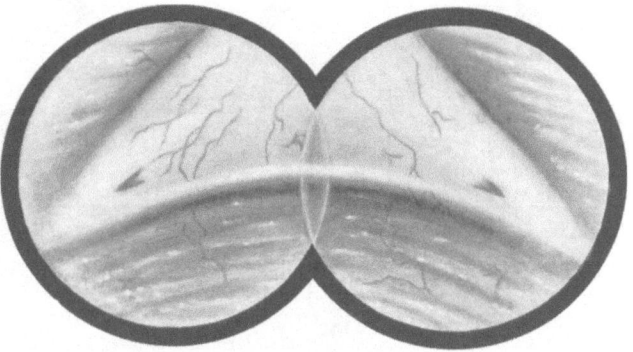

Fig. 62. Hypertrophy of interureteric ridge (MERCIER's bar). Distant view; composite of two visual fields. Retrograde lens

Causes

Trabeculation almost always accompanies *obstruction* to the outflow of the urine from the bladder. It may occur

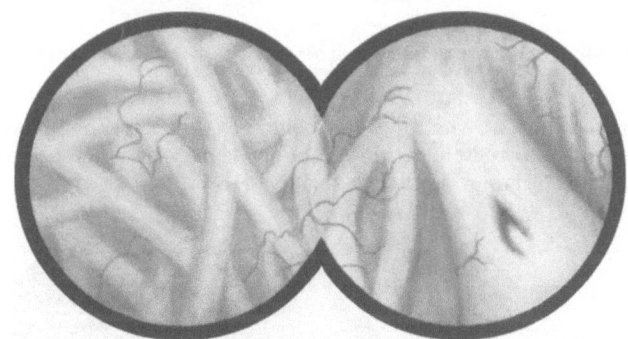

Fig. 63. Trabeculation in fundus of bladder in a patient with prostatic hypertrophy causing urinary obstruction. Close view; composite of two visual fields

in urinary retention due to a *central nerve lesion*. The ridges are fine and the depressions shallow, similar in appearance to those of very early urinary obstructive disease. *Bladder spasm* may produce temporary trabeculation which is similar in appearance to that seen in chronic urinary obstruction. The urine is ejected through the cystoscope with great force when the obturator or telescope is removed, or it may be forced out through the urethra around the cystoscope. Chronic bladder spasm from infection or other continued cause of irritation may result in trabeculation. In *cystitis vetularum*, a chronic bladder infection occasionally occurring in aged women, there are ridges and depressions covered by anemic mucosa; the bladder capacity is small.

III. Undermined or floating trigone (Ferrandiz)

Undermined or floating trigone is rare. The separation of the trigonal muscle (BELL's muscle) from the bladder wall may be only along its upper edge, an undermining of the hypertrophied interureteric ridge. The deformity is best

6*

visualized with the retrograde optical system which shows the depression beneath the elevated interureteric fold (Fig. 64). Very rarely the trigonal separation is more extensive; it may remain attached at only three points — at each ureteral

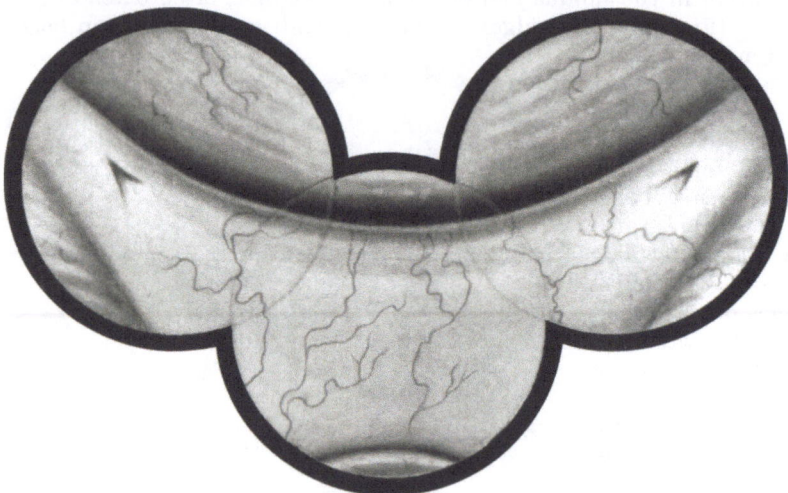

Fig. 64. Trigone undermined at its upper margin as seen through rights angle optical system. Elevated dorsal lip of bladder neck is seen in lower edge of field. Composite of three visual fields

orifice and at the bladder neck (YOUNG). When the cystoscopic examination reveals marked elevation of the interureteric ridge and of each ureteral ridge also, floating trigone should be suspected.

Fig. 65. Ridge in ventral wall of bladder extending from bladder neck to the dome. It resulted from reconstruction of the urethra with a tube made from a flap of bladder wall

IV. Postoperative

Ridges may be seen following surgical procedures on the bladder. When there is inaccurate approximation of the bladder edges or following some plastic operations on the bladder, ridges covered by normal bladder mucosa are seen (Fig. 65).

J. Septa in the bladder wall

I. Hourglass

Septa in the bladder wall are rare and are usually congenital in origin. A transverse septum extending around the entire circumference of the bladder divides it into an upper and a lower portion similar to an hourglass[1]. When the connecting passage is small it may be mistaken for a diverticular orifice; when it is large, the appearance is similar to a sacculation of the dome secondary to chronic urinary obstruction or to a central nerve lesion.

II. Septate

The septum sometimes extends longitudinally in the midline from the dome downward, and is usually wide at its upper portion. The lateral recesses of the bladder are deep, and the deformity is similar to a bicornate uterus. The septum may be longitudinal, dividing the bladder into a ventral and a dorsal portion and extending from the trigone upward[2].

Multilocular

Septa may divide the bladder into multiple compartments[3]. This congenital deformity usually accompanies other serious malformations which result in stillbirth or death soon after birth; a cystoscopic examination is usually not possible.

Chapter VIII

Color abnormalities of the bladder mucosa without change of contour

Many of the abnormalities discussed in this chapter show changes in bladder contour as well as in mucosal color. The color change is often an early manifestation of a lesion and the change in contour occurs later in the course of the disease.

A. Red and pink discoloration (predominating)

I. Generalized red discoloration

A generalized red discoloration involving all or most of the bladder mucosa is nearly always due to acute inflammation. Bacterial infection is the usual cause, but trauma from bladder overdistention or from instillation of irritating fluids may be the causitive agent. The red discoloration is often velvety in appearance in one area, and increasedly vascular in another.

Acute cystitis

An acute cystitis of this general inflammatory type is usually primary in the bladder but may be secondary to a renal infection. When it is due to an upper urinary tract infection, the inflammatory change is usually more severe about the ureteral orifice of the involved kidney. Many *infections and infestations* are

---

[1] KEARNS; ZELLERMAYER; CAULSON.   [2] HINMAN; WESSON.   [3] EDWARDS; KOHLER.

primary in the bladder and may cause a generalized red discoloration when they are extensive and severe. HODGKIN'S *disease* rarely involves the bladder; when it does there is extensive red discoloration and thickening of the mucosa (LEBO-WICH). When inflammation is severe there may be bleeding from the mucosa, and when it has been present for several days bullous edema may occur. When the involvement is less severe, the areas of discoloration are localized and often patchy.

II. Patchy areas of red and pink discoloration

1. Acute cystitis

This causes multiple patchy areas of red and pink discoloration of the bladder mucosa more frequently than it does a generalized inflammation. Often there are enlarged and more numerous capillaries visible through the cystoscope lens; these

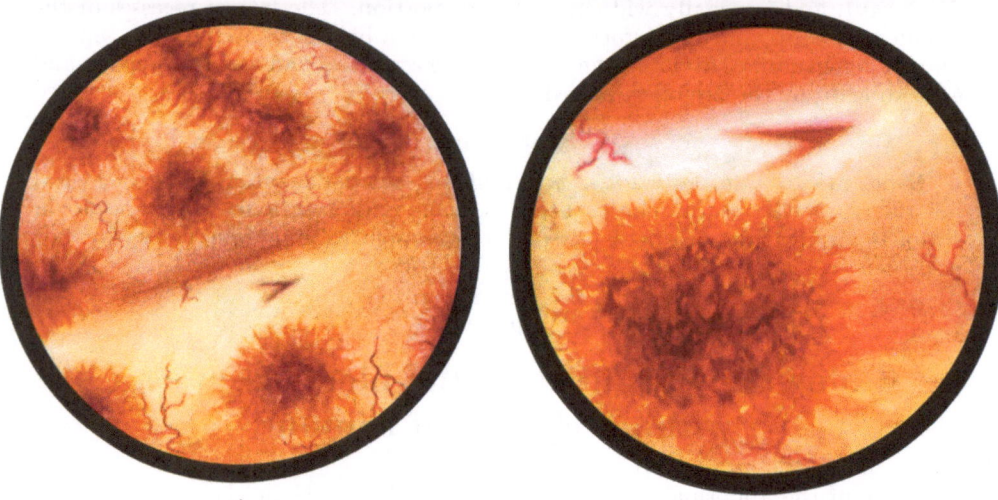

A   B

Fig. 66 A and B. Acute hemorrhagic cystitis. Multiple patchy areas of red discoloration. A Distant view. B Close view showing capillary network at edge of ecchymotic area

appear in some areas, and in others there is a more homogeneous velvety red and pink discoloration. In order to visualize the capillary pattern, the objective lens of the cystoscope must be close to the bladder mucosa.

2. Ecchymotic areas

Multiple discrete areas of red discoloration are seen in patients with *acute hemorrhagic cystitis* (Fig. 66). The bladder mucosa is normal between these areas. Around the edges of the ecchymosis numerous small capillaries are frequently seen by close vision. Sometimes slight oozing of blood is visible from the center of the area. A mild *reaction to irradiation* may be manifest by a single localized area of increased redness surrounded by a fringe of capillary prominence. *Contusion* of the bladder mucosa results in an irregular dark red hemorrhagic area; frequently a small irregular tag of light membrane, which is torn mucosa, is near the center of the red area (Fig. 67). No capillary network is seen at the periphery. Contusion of the bladder mucosa may occur on or above the trigone from insertion of the cystoscope or other instrument too forcibly or too far. *Purpura* of the bladder mucosa is similar in appearance to hemorrhagic cystitis except that the red discolored areas are much smaller than the ecchymotic areas of hemorrhagic cystitis (Fig. 68). The discoloration is due to spontaneous hemorrhages into the mucosa and submucosa. The condition is rare and is usually a local manifestation of generalized purpura. *Allergic reaction* in the bladder mucosa is characterized by ecchymotic areas, usually more numerous on or near the trigone. *Familial telangectasia* is a rare condition which sometimes involves the bladder. The lesions in the bladder mucosa are small areas of increased injection somewhat similar in appearance to purpura. When it occurs in the bladder, there are always lesions elsewhere, notably in the buccal mucosa. There are no bladder symptoms which accompany these lesions.

3. Cystitis granulomatosa

In some cases of chronic cystitis, areas of diffuse flat or slightly elevated granulation (Fig. 69) are scattered over the bladder mucosa. The patches of red discoloration vary to some degree in size; they

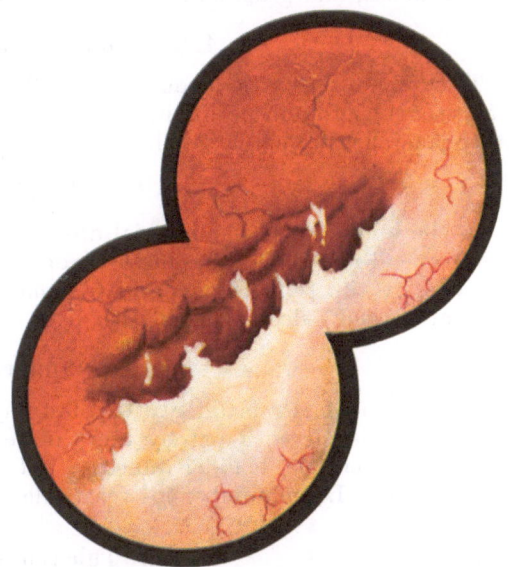

Fig. 67. Contusion of bladder mucosa. Light colored shaggy membranous tag is a bit of torn mucosa. Close view. Composite of two visual fields.

are smaller and involve a greater extent of the bladder mucosa than do the ecchymotic areas of acute hemorrhagic cystitis. Sometimes the granulation

Fig. 68. Purpura of the bladder. Distant view

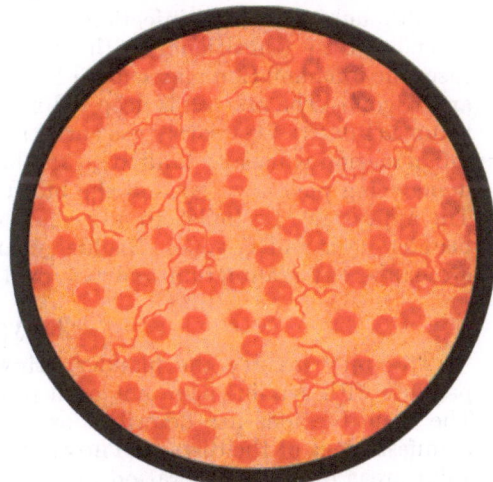

Fig. 69. Diffuse granulation occurring in chronic granular cystitis (cystitis granulomatosa, granularis, etc.). Distant view

tissue is so profuse and discrete that the appearance is identical with that of neoplasm; biopsy is necessary to make a differential diagnosis. When the areas of granulation are thick, a close view reveals tiny bits of red tissue interspersed

with equally small dots of more normal bladder mucosa, giving the appearance
of fine sand scattered evenly over an inflamed bladder mucosa. A number of
terms have been applied to this condition — cystitis granularis, granulomatosa
and granulosa, and granulomas or granulations.

4. Trichomonas vaginalis

This often invades the bladder, especially in women, and causes inflammatory
changes similar to nonspecific infections (NITSCHKE). There is increased redness,
either generalized or patchy. When severe, granulations and edema may be found.

5. Bilharziasis

In the initial stage of bilharzial disease of the urinary bladder there is only
one or several small areas of redness surrounded by increased vascularity. Usually,
however, by the time these patients come to cystoscopy, there is ulceration,
granulation and calcareous deposit.

6. Blastomycosis

Infestation of the bladder with blastomycosis is rare. It causes an acute
generalized cystitis with patchy areas of red and pink discoloration scattered
irregularly over the bladder mucosa, not unlike acute hemorrhagic cystitis
(MARQUARDT). Large numbers of yeast cells are found in the urine (RHAMY).

7. Tuberculosis

Early tuberculosis of the bladder may cause one or two small areas of increased
redness due to inflammation. These usually occur near the ureteral orifice on the
side of the involved kidney.

8. Gonococcus infection of the bladder

Although gonococcal infection of the urethra is common, the infection rarely
extends to the transitional epithelium of the bladder mucosa. In severe acute
cases the inflammation may extend from the posterior urethra to the bladder neck
and trigone. The mucosa is deep red in color and bleeds easily from the slightest
trauma (PUGH). Very rarely the entire bladder mucosa is involved, but the intense
inflammation is usually limited to the bladder neck and trigone, the fundus and
dome being pink in color and the blood vessels more prominent and numerous.
The appearance is similar to any acute cystitis which is more severe about the
bladder neck and trigone. Bladder changes from chronic gonococcic infection
are exuberant granulation and polypoid formation (Chap. IX).

9. Syphilis

This may involve the bladder. However, lesions found in the bladders of
patients with syphilis have not been proven to be a direct result of the disease.
The vesical lesions are described as being somewhat similar to the cutaneous
manifestations of the disease (THOMPSON). In the secondary stage there are irregular areas of red discoloration and an increased vascular pattern (THOMPSON).
(See Chap. IX.)

10. Stellate areas of red discoloration

A faintly visible stellate area of inflammation is almost diagnostic of *Hunner
ulcer*. When a close view of the area is obtained, an increased vascular pattern
is seen to surround a small central light colored area (Fig. 43, p. 68) (HIGGINS).

There is slight oozing of blood from this center area when the bladder is over distended. The vessels may coalesce, forming a light red ring about the ulcer. The typical Hunner ulcer is usually single and located in the bladder dome (ORMOND). They occur more frequently in women than in men and have been seen in girls (McDONALD et al.). Multiple ulcers are not uncommon. There may be generalized interstitial cystitis which involves most or all of the bladder wall and its mucosa.

11. Irregular pink discoloration of the trigone

This is commonly seen and is due to mild subacute inflammation. The term *pseudomembranous trigonitis* or cystitis has been applied to this condition. The

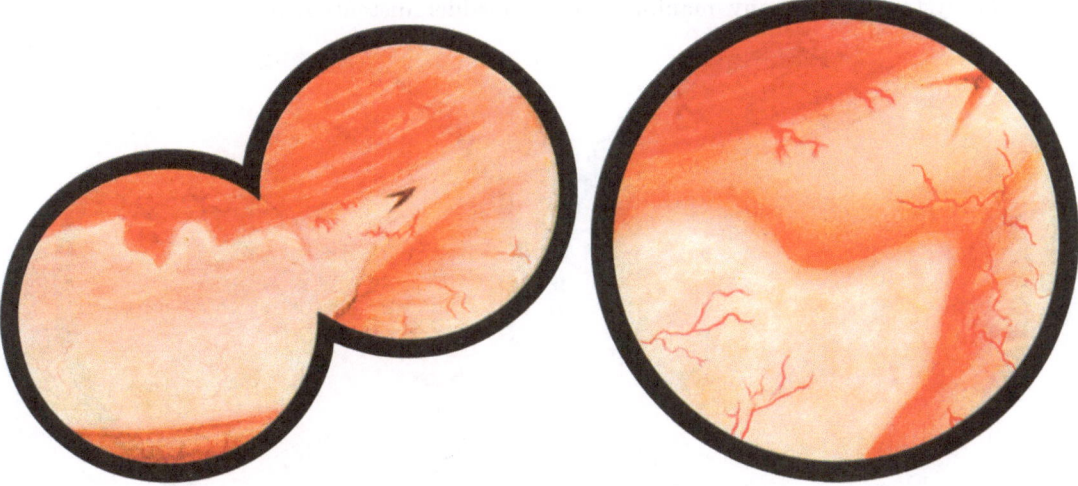

A   B
Fig. 70 A and B. Pseudomembranous trigonitis. A Distant view. Composite of two visual fields. B Close view

appearance is that of a thin white veil or membrane covering a moderately reddened mucosa. The edges of the pink discoloration are well defined, irregular and resemble the outline of a map (Fig. 70). The thin pseudomembrane may become detached from the trigone when inflammation is severe. This type of discoloration due to mild inflammation is more often seen in women than in men. There may be pus and bacteria in the urine, but often there is no evidence of infection. Mild frequency and urgency are usual symptoms; often there are none. Trichomonas vaginalis invasion of the bladder may be a cause of pseudomembranous trigonitis.

12. Red area in dome

This may be due to patent urachus (CHERRY). Pressure on the cyst may cause a ribbon of pus to appear (ATCHESON).

B. Red, white, light grey and light pink discoloration

White, light grey and pink discolorations due to mucopurulent exudate or other causes may be found overlying any inflamed red mucosa. There is often elevation of the mucosa due to bullous edema or other causes, but sometimes no change in bladder contour is evident.

I. White irregular areas

1. Alkaline incrusted cystitis[1]

This is a severe infection of the bladder characterized by flat greyish white deposits of calcareous material on a red inflamed mucosa (Fig. 71). Color changes are extensive. Usually the entire bladder mucosa is a deep red color except where the flat layer of greyish white calcareous material clings to it or is embedded in it. There may be only one or two patchy areas of deposit on or near the trigone; in severe cases most of the bladder mucosa is covered. Cystoscopy is almost impossible without general or spinal anesthesia, for the bladder is extremely tender and of small capacity. Mucus, pus, blood and debris becloud the view through the cystoscope and any manipulation or bladder distention increases bleeding

Fig. 71. Alkaline incrusted cystitis. Close view   Fig. 72. Severe irradiation reaction (burn) of bladder mucosa. Distant view

from the mucosa. Sometimes superficial ulceration and necrosis occur. Granulation tissue may build up or calcareous deposit become thick in severe cases of long standing, making differentiation from tumor difficult if not impossible without biopsy study. Symptoms are always severe; there is pain in the bladder, frequency and urgency, and blood in the urine; calcareous material is often passed through the urethra, increasing pain and bladder spasm.

2. Irradiation reaction

Severe reaction from x-ray or radium therapy[2] causes an irregular greyish white area surrounded by deep red discoloration (Fig. 72). It most often appears several weeks or months following application of the modality for treatment of neoplasm of the cervix. The bladder mucosal lesion is usually single, from 1 to 2 cm. in diameter, and located on or near the trigone. The greyish white center is necrotic tissue with more or less calcareous deposit on its surface. The surrounding red area may be slightly thickened and appears puckered toward the center. In less severe irradiation reaction there is an irregular, somewhat indefinite reddened area with increased vascularity about its periphery.

---

[1] HAGER; RANDALL; CAMPBELL.   [2] SMITH; DEAN; SLAUGHTER.

3. Leukoplakia

This is a term applied to grey and white discoloration of the bladder mucosa secondary to chronic infection of long standing. There are one or more plaques with irregular well defined edges located on or near the trigone (Fig. 73)[1]. A close view of the light area reveals a dull, somewhat wrinkled surface surrounded by a red area of inflammation. The entire bladder mucosa is also inflamed and in some areas is deep red; other areas may be lighter in color and almost avascular due to fibrosis. The greyish white plaques are cornified epithelium; they may become detached and pass out in the urine. True leukoplakia is not common, but the term is often used loosely, and is applied to any light pink or greyish white patchy areas of discoloration, such as pseudo-membranous trigonitis or alkaline incrusted cystitis. Sometimes true leukoplakia is a precancerous lesion[2].

Fig. 73. Leukoplakia of bladder. There is cornification of epithelium causing greyish-white areas of discoloration. Distant view

4. Thrush infection

Infection of the bladder with VINCENT's organism (thrush) is rare (MOULDER). When it occurs the mouth and other mucosal surfaces may be involved. The whitish grey discoloration of the bladder mucosa is irregular, and there are islands of reddened mucous membrane scattered through the involved area. The appearance of the lesion in the bladder is similar to that in the mouth (SAUER and METZGER).

II. Sloughing tissue

1. Severe infection

Sloughing or necrosis involving the bladder mucosa or deeper tissues appears as a grey or dirty yellow discoloration.

2. Gangrenous cystitis

Bladder infection may be so severe as to cause gangrenous cystitis in which there is extensive sloughing of bladder mucosa. Bullous edema and pieces of detached mucosa are seen in severe cases, but in milder ones there is no change in bladder contour.

3. Trauma

Sometimes sloughing occurs from injury to the bladder wall during pelvic surgery or as a result of other trauma. The involved area is an irregular greyish white discoloration surrounded by reddened mucosa (Fig. 67, p. 87).

III. Ulceration

Ulceration of the bladder mucosa often occurs without appreciable change in contour of the viscus. Some ulcers are dark red throughout, but the majority have more or less greyish white material scattered over the red base.

[1] THOMPSON; STEIN.   [2] YOUNG; SCHOLL.

1. Tuberculosis

Tuberculous ulceration (Fig. 74) is usually found just below the ureteral orifice on the side of the involved kidney. The ulcer sometimes has slightly undermined edges, and the deep red base may be partially covered by small patches of light pink exudate or yellowish white sloughing tissue[1]. Unless involvement of the bladder mucosa is extensive, there are usually only one or two small ulcers which vary in size from one millimeter to one centimeter in diameter. Small yellow circular dots, which are tubercles, are often seen near the ulcers. More severe and extensive tuberculous involvement of the bladder mucosa causes changes in the mucosal contour and is described in Chapter IX.

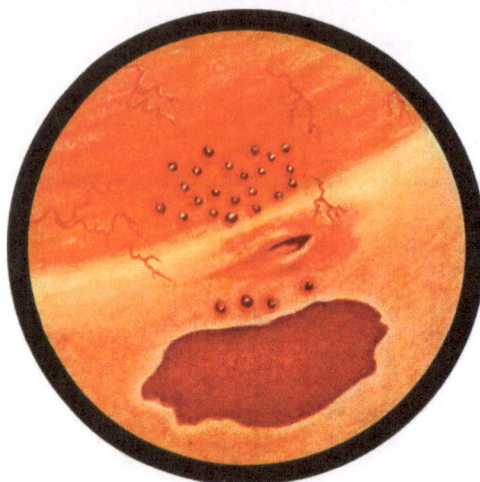

Fig. 74. Tuberculous ulcer of bladder. Distant view. Small yellow spots near ulcer are tubercles

2. Nontuberculous ulcerations

These may simulate the ulcers of tuberculosis. They are more frequently located in the fundus or dome of the bladder and are often secondary to a staphylococcic infection. Small pustules may be mistaken for tubercles. Nonspecific ulcerations are often a part of generalized cystitis, but may be solitary with no other change in the bladder mucosa. The appearance of these ulcers varies. There may be only very slight redness covered by a few small flakes of light pink exudate, or the ulcer may be deeper and the edges undermined and sharply defined. In severe cases, there is considerable exudate and yellowish white sloughing tissue partially covering the deep red ulcer. The so-called simple chronic ulcer or *solitary ulcer of* FENWICK is described as being located on or just above the trigone, and may be simple or more extensive and part of a generalized cystitis[2].

3. Actinomycosis

Infection of the bladder with actinomycosis may cause ulceration similar to that of tuberculosis, except for the absence of tubercles. Sometimes, however, no ulcerations are present, and a diffusely reddened discoloration of the entire bladder mucosa, more marked on the trigone, is found[3].

IV. Invasive malignant neoplasms

Carcinoma or *adenocarcinoma* of adjacent structures such as the cervix (BEACH) or prostate or *squamous cell carcinoma* originating in the bladder mucosa, may invade the bladder wall and ulcerate through the mucosa before causing any change in normal contour of the interior of the bladder. The ulceration is usually ragged in appearance, dark red in color and dirty white areas of sloughing tissue are often scattered about the edges or over its surface. It may be somewhat excavated, and is usually surrounded by an area of lighter red discoloration. Enlarged tortuous blood vessels are sometimes found about the periphery of the lesion.

[1] KAUFMAN; FERRIS.   [2] FENWICK; POLLACK.
[3] CECIL; HILL; HERGER; McCREA et al.; HATCH et al.

C. Blue discoloration

Varicosities

Large submucosal veins cause blue discoloration of the bladder mucosa. They are somewhat tortuous, and are most frequently located on or near the trigone but may occur in any area of the bladder. Unless they measure more than 4 or 5 mm. in diameter they are considered normal. The larger veins cause a tubelike elevation of the bladder mucosa (Fig. 48, p. 71).

Chapter IX

Abnormalities of both color and contour within the bladder

The endoscopic appearance of most bladder diseases varies in different individuals and at different times in the same individual. Therefore, in the classification used here, several different descriptions of the same lesion may be given depending upon its appearance at the time of endoscopy.

A. Smooth regular red protrusions

I. Benign bladder tumors

The intravesical protrusion of benign bladder tumors is covered by bladder mucosa which is either normal (Chap. VII) or a deeper red or lighter yellow than normal. The protrusion is most often a smooth rounded sessile mass, although it may be pedunculated. The vascular pattern is not changed from that of the normal bladder mucosa, and the appearance may be identical to that of an extravesical tumor which is pressing the bladder wall inward. Benign bladder tumors are rare, usually single, and are often impossible to differentiate without biopsy examination.

1. Arising from the bladder wall

These include *leiomyoma, myoma, myxoma, rhabdomyoma, fibroma, ganglioneuroma, neurofibroma and pheochromocytoma.*

2. Arising from embryonic rests

These include *adenoma, cystadenoma, osteoma* and *teratoma.* Some may be secondary to a parasitic infestation (KLEITSCH).

II. Granulomatous tissue

Exuberant granulation may appear as one or more smooth regular discrete red protuberances into the bladder. These granulomata occur in the dome of the bladder when a suprapubic catheter has been in place for several months (Fig. 75). Sometimes they form about a nonabsorbable suture which has penetrated into the bladder mucosa. Extensive granulations are more irregular and shaggy in appearance.

III. Malignant invasive tumors of the bladder

In their early stages these may be similar in appearance to benign tumors. The protrusions into the bladder may be covered by normal mucosa (Chap. VII).

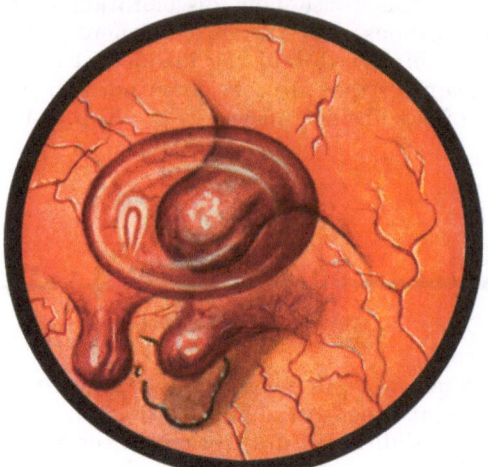

Fig. 75. Granulomata in dome of bladder resulting from indwelling suprapubic catheter and black silk suture (lower edge of visual field) through the bladder mucosa. Note bubble in dome. Distant view

Later there is a tendency for the mucosa covering the malignant neoplasms to contain tortuous vessels and to be deeper red in color.

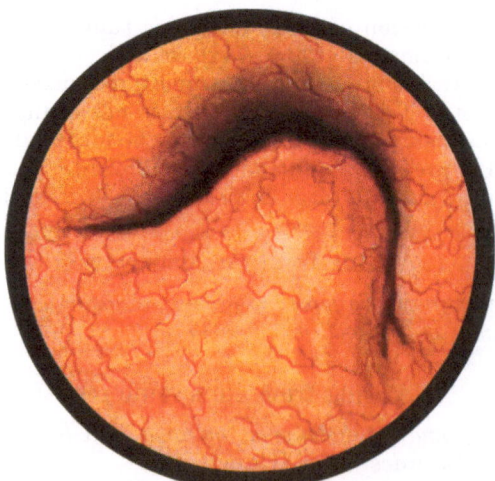

Fig. 76. Early squamous cell carcinoma which has invaded submucosal tissues but has not ulcerated through mucosa. *A* Distant view

1. Sarcoma and mixed sarcomatous tumors

These originate in the bladder wall and at first invade beneath the bladder mucosa. At this stage there are only slight protrusions of the mucosa. Soon, however, the mucosa overlying the sarcomatous masses becomes deeper red in color, and sometimes blood vessels in these areas are increased in number and are tortuous.

2. Squamous cell carcinoma

When this originates in the bladder mucosa it may present an appearance similar to that of sarcoma (Fig. 76). Later there is ulceration and necrosis.

3. Teratoma

Whether benign or malignant, this is very rare in the bladder, and may also create the mucosal changes described above. Sometimes it causes bullous edema, ulceration and sloughing of the bladder mucosa; however, usually it invades and metastasizes extravesically without involving the bladder mucosa to any great extent.

IV. Malignant extravesical tumors

Malignant tumors of adjacent organs such as the cervix (GEMMELL), uterus, rectum or prostate often invade the bladder wall, resulting in protrusions which

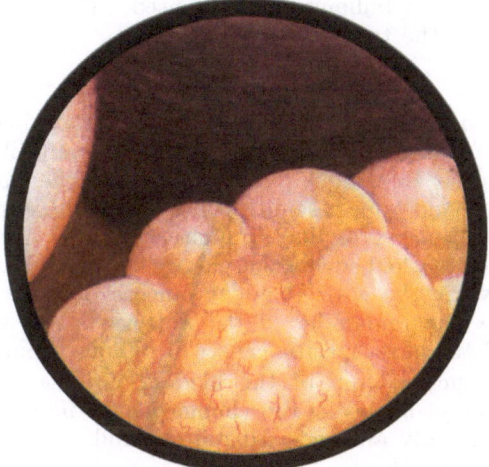

Fig. 77. Bullous edema from carcinoma of prostate invading trigone. Invasive neoplasm is seen in lower portion of field of vision

are at first covered by normal bladder mucosa which appears cystic or translucent. Sometimes there are tortuous vessels of the mucosa covering the involved area. Later in the course of the disease increased redness, ulceration and sloughing occurs. Invasion of the trigone by prostatic carcinoma usually causes multiple rounded elevations extending from the bladder neck upward (Fig. 77). When the invasive tumor emanates from the cervix, the usual first site of bladder involvements is the trigone; when it originates in the uterus or rectum, the fundus above the trigone is first involved.

B. Smooth red multilobulated protrusions

I. Edema

Inflammatory edema of the bladder mucosa is a red discolored multilobulated elevation of the mucosa. In its mild form the areas of protrusion are small, sessile and irregular in size and distribution. When the inflammation is minimal, the edematous area has a translucent appearance and is light pink.

1. Diffuse edema

Involvement of most of the bladder mucosa is usually due to severe *acute cystitis*. *Chemical trauma* from injection of an irritating or corrosive liquid into the bladder, causes edema of the entire mucosa. *Irradiation reaction* from massive x-ray therapy directed to the bladder or adjacent organs results in generalized edema of the bladder mucosa. When the application is concentrated to any one point, the edema may be limited to the portion of the bladder wall nearest the area of concentration. Infection is more likely to cause multiple irregular patchy areas of edema which is more severe on and around the trigone.

2. Localized areas of edema

Localized areas of edema of the bladder mucosa may be secondary to almost any bladder disease or to an extravesical lesion which involves the bladder wall. The edematous change in the mucosa may be very mild and the reddened protrusions of edema only slightly elevated. When the protrusions extend farther into the bladder, the term bullous edema is used. There is no definite line of differentiation between the two types of edema.

3. Bullous edema

When viewed from a distance through the endoscope this has a red, irregular somewhat mottled appearance. On close view, the multilobulated protrusions are seen to vary in color from light pink to dark red. The tips of the individual protrusions are lighter and appear semitranslucent; the base of the area of edema is darker and may be finely granular. Bullous edema may be due to any of the lesions causing the less marked edematous change of the bladder mucosa. Seldom, however, does it involve the entire bladder mucosa, but may be found in delimited areas whereas the remainder of the mucosa is less severely edematous or normal. A few of the most specific causes of bullous edema will be described.

a) **Allergy.** Allergic reaction of the bladder mucosa may rarely produce bullous edema. Usually, however, the change in the mucosa due to allergy is much milder; there is only an increased redness and some increase in size and number of the blood vessels.

b) **Amebiasis.** Infection of the bladder with *Entamoeba histolytica (amebiasis)* is rare. There is bullous edema of the mucosa about the bladder neck and on the trigone. Sometimes cysts or blisters form in this region, and are more likely to be at the vesical orifice (van Duzen). The surrounding area is hemorrhagic, and there may be some increased redness throughout the bladder mucosa.

c) **Intestinovesical fistula.** Intestinovesical fistula (Barnes and Hill) and fistulae from adjacent abscesses (Lower and Farrell) cause a localized area of bullous edema located in the fundus or dome of the bladder above the trigone (Fig. 51, p. 74). The appearance may be similar to that of a neoplasm (Fagerstrom). It is usually not possible to identify the fistulous orifice because it is small and the edematous protrusions cover it. The mucosa in the remainder of the bladder is either normal or only slightly inflamed; there is seldom an area of increased redness immediately surrounding the edematous patch. The occurrence of such a well defined area of bullous edema above the trigone is almost pathognomonic of intestinovesical fistula.

d) **Indwelling urethral catheter.** However, the tip of an *indwelling catheter* which has been in place for several days causes irritation and resulting bullous edema of a localized area in the midline of the fundus of the bladder. It may be similar in appearance to that from an intestinovesical fistula.

e) Invading neoplasm. If the change in the bladder mucosa is due to any other lesion, such as an *invading neoplasm* of the rectum or uterus, there is nearly always protrusion of the bladder wall, red discoloration and sometimes tortuous vessels about the edematous area, and often sloughing tissue. The involved area usually includes the trigone as well as the bladder wall above it. When the invading neoplasm is primary in the cervix, the base of the bladder is involved before the fundus is affected.

II. Neoplasms

1. Benign

a) Chronic cystitis. Long continued *chronic cystitis* predisposes to neoplastic changes.

b) Cystitis glandularis. Cystitis glandularis is a benign neoplastic change. There are smooth red multilobulated protrusions similar in appearance to edema, but the bleblike elevations appear fleshy and lack transparency (SAUER and BLICK). Sometimes they project more acutely and resemble papillomata (CULP). The mucus which they secrete is seen before the bladder cavity is washed out with irrigating fluid.

2. Malignant

Malignant neoplasms, either those originating in the bladder, or those invading from a nearby organ, may produce smooth, red multilobulated protrusions into the bladder, but they are more frequently irregular and the surface rough. Rhabdomyosarcoma is most often seen as a grapelike tumor mass (YOUNGBLOOD et al.); the protrusions may be pearly white and semitranslucent (SMITH).

C. Irregular red intravesical protrusions

Red protrusions into the bladder which are irregular and rough are most likely either bladder carcinoma or granulation tissue. The changes in the bladder mucosa from each of these lesions is comparatively common.

I. Neoplasms

1. Carcinoma

This usually protrudes farther into the bladder cavity than granulation tissue, although sometimes it may be almost flat. The surface of most bladder carcinomas of medium and high grade malignancy is similar to that of a cauliflower (Fig. 78); it is finely irregular, displaying numerous small elevations which cover most of the surface of the protrusion. There are also larger irregularities of the surface of the neoplasm due to varying degrees of protrusion of the tumor into the bladder. Most carcinomas of the bladder of medium and high grade malignancy are sessile; some are attached to the bladder by a broad pedicle and a few by a narrow pedicle.

In general, the more malignant and invasive tumors are more sessile and flat, while the less malignant ones are pedunculated. Bleeding readily occurs from the surface of the tumors: it is often spontaneous; slight trauma from passing the cystoscope may start profuse hemorrhage. Malignant tumors vary in size from one to two millimeters in diameter up to those covering most of the bladder mucosa. The surface of the larger ones is more likely to be ulcerated, necrotic and sloughing. The very small ones are occasionally found incidentally during cystoscopy for diagnosis or urinary tract disease which is not associated with the tumor. Carcinomas of the bladder are most frequently single, but multiplicity commonly occurs.

2. Osteogenic sarcoma

Some other malignant tumors such as osteogenic sarcoma are similarly variable in their pedicle attachments[1].

3. Amyloidosis (SENGER et al.)

This is a benign neoplastic change which is similar in appearance to carcinoma (ROEN and WIEMER) and is probably secondary to long continued chronic infec-

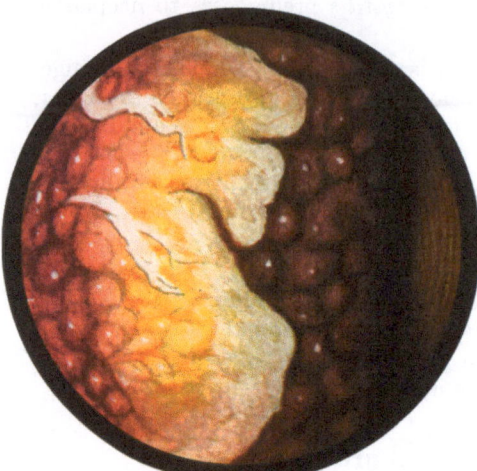

Fig. 78. Carcinoma of bladder. Typical cauliflowerlike appearance of surface of a tumor of medium or high grade malignancy. Close view. White discoloration is due to sloughing tissue

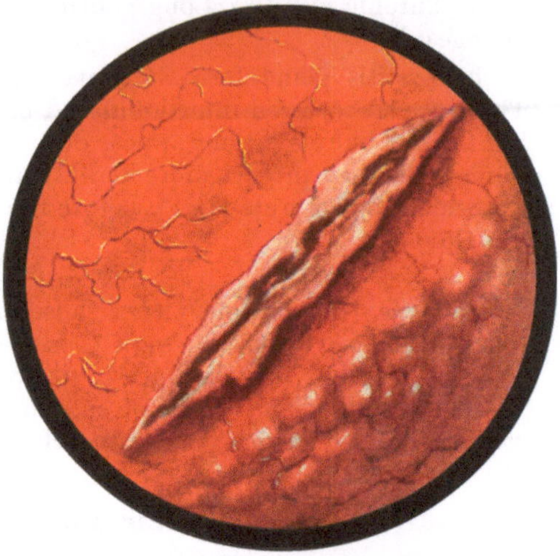

Fig. 79. Granulation tissue with accompanying edema in severe subacute cystitis secondary to neurogenic bladder dysfunction. Note area of necrotic ulceration Distant view

tion. When there is extensive involvement of the bladder, its capacity is reduced and large fungiform irregular red protrusions extend from the mucosa into the bladder cavity (CORBITT et al.).

II. Granulation tissue

This and accompanying edema may protrude into the bladder as much as one centimeter or more. Usually, however, the protrusion is only one to two millimeters, and sometimes there is no change from normal bladder contour. The surface is deep red in color, usually darker than that of carcinoma, and is more finely granular (Fig. 79). The broadest part of the protrusion is its base. Elevation above the normal bladder contour varies, causing irregularity of the surface. Edema nearly always accompanies granulation tissue and is interspersed with it. The lobulations of the edema are lighter in color, smoother and somewhat translucent. The extent of the bladder mucosa involved by granulation tissue may vary from a minute area to patches several centimeters in diameter. Involvement is usually more marked about the bladder neck and on the trigone, and is sometimes diffuse rather than confined to well defined areas. When the granulomatous protrusions are circumscribed they are often referred to as granulomata (EWELL).

[1] NOURSE; CRANE and TREMBLAY.

Areas of ulceration frequently occur on the protuberances of granulomata, or ulcers may be surrounded by granulation tissue.

1. Nonspecific infections

Cystitis granulomatosa occurs in some cases of chronic cystitis. Any abnormal condition which irritates the bladder mucosa over a long period of time causes granulations.

2. Subacute and chronic specific infections, stones, foreign bodies, neoplasms and trauma

These are the usual causitive agents of granulation tissue. Invasion of the bladder with *actinomycosis* (HERGER) with *trichomonas vaginalis* (HECKEL), with *amoeba*, with *Schistosoma haematobium* (bilharziasis), with *tubercle bacilli* or with any other organism which persists for several months or longer, results in the formation of granulation tissue (STERLINE and ASH). It is sometimes difficult, if not impossible, to differentiate granulations from early carcinoma when the latter is flat. As the malignancy extends, however, it becomes more characteristic in appearance.

D. Irregular red and white protrusions

I. Material causing white discoloration

1. Calcareous deposit; sloughing tissue

A distinct white discoloration on red protuberances is usually due either to calcareous deposit or to sloughing tissue. It may be difficult to identify the substance which causes the white discoloration by its appearance through the endoscope (Fig. 80). A very close view, however, may magnify the material sufficiently to make identification possible. Calcareous deposit is a slightly irregularly white or greyish white or tan color when viewed from a distance. The close view shows the tiny crystals or sandlike appearance of the mineral deposit. Similarly, sloughing tissue is white or greyish white as seen from a distance, but the close view reveals the shaggy edges and the fine undulations of the surface of the slough.

2. Mucopurulent and epithelial exudate

Mucopurulent exudate often causes a less distinct white or pink discoloration; it is more likely to be spread thinly and irregularly over the protuberances. Collections of mucus and desquamated epithelium may cling to granulation or tumor tissue and cause white or light pink discoloration; it is sometimes limited by a sharp line of demarcation. Mucus exudate mixed with pus and epithelial cells appears pinkish white from a distance (Fig. 81); the close view shows the irregular thin covering of the deep red slightly protruding granulation of chronic or subacute cystitis.

3. Miscellaneous

a) **Combination of substances.** Often the material causing the white or light colored change which appears on red protrusions is a *combination of several substances.* Calcareous material is sometimes deposited on sloughing tissue and an exudate may partially cover the entire area. Rarely the light colored changes are from other material.

b) Gauze sponge. A *gauze sponge* left in the pelvis near the bladder during pelvic surgery may be sloughing through the bladder wall (Fig. 98, p. 117).

c) Fragment of bone. A *fragment of bone* from a fractured pelvis may project through the bladder wall. Foreign bodies such as these are surrounded by bullous edema; they gradually emerge from the wall into the bladder cavity.

II. Lesions causing red and white protrusions

1. Neoplasm

This is the most common lesion causing red and white protrusions into the bladder (Figs. 78 and 80). Any *malignant neoplasm* originating in the bladder

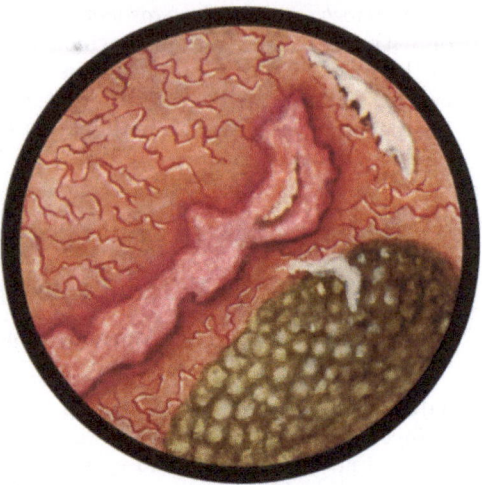

Fig. 80. Calcareous deposit and slough on a bladder tumor of medium to high grade malignancy. Medium close view. The small irregular light tan areas are calcareous deposit and light area with ragged edges is slough

Fig. 81. Mucous exudate mixed with pus and epithelial cells on granulation tissue resulting from a vesical calculus. Distant view

or in an adjacent organ may necrose, resulting in white areas of slough and/or calcareous deposit (WEINER et al.). This change occurs much more frequently in tumors of high grade malignancy than in more benign lesions. The surface of a rapidly growing or ischemic carcinoma often becomes necrotic, resulting in slough, then calcareous material is deposited on the dead tissue. The white discoloration is interspersed in one or more irregular patches over the surface of the typical cauliflowerlike tumor. When the red and white protrusion is in the dome of the bladder, it may be *adenocarcinoma of urachal origin* (PASQUEIR and LOPEZ). An *embryonic rest* may occur at any site in the bladder and give the same appearance (COLBY).

2. Severe chronic infections

These may result in red and white protrusions into the bladder. Ulceration and necrosis occur; the dead tissue turns white or dirty white and calcareous material may be deposited. Any severe infection may cause this change in appearance of the bladder mucosa. Calcareous deposit is most common when the infecting organsim is one which changes the reaction of the urine from acid to alkaline. In *alkaline incrusted cystitis* (Fig. 71, p. 90), a large portion of the bladder mucosa is covered by calcareous deposit. Some of the incrusted areas are flat, and others are elevated due to underlying granulation tissue. Infestation of the bladder with *Schistosoma haematobium (bilharziasis)* is frequently the cause of profuse granulations and calcareous deposits in countries where the disease is endemic (MAKAR).

E. Red and pink papillary projections

I. Bullous edema

Abnormal tissue often extends sharply into the bladder and is more of a projection than it is a protrusion. Bullous edema may take the form of papillary

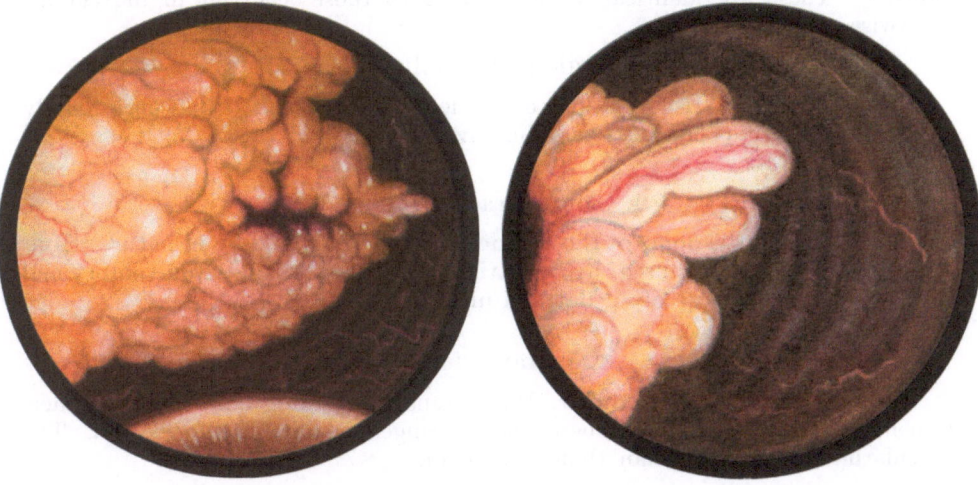

A   B

Fig. 82 A and B. Papillary carcinoma of bladder. A Distant view showing cluster of papillary growths.
B Close view showing individual projections

projections, but usually it is a lobulated protrusion. When it does project more sharply into the bladder in papillary form, it is difficult if not impossible to differentiate from papillomata, unless a specimen is removed and biopsy performed.

II. Papillary tumors

1. Papillomata

These are the most common cause of tissue projections into the bladder. They vary in size from those which are barely visible through the endoscope to huge ones which fill the entire bladder. They are frequently multiple and are most often located on the floor of the bladder. The individual fingerlike projections are from one-half to $1^{1}/_{2}$ mm. in diameter, and vary in length from a small elevation to several millimeters; the end is tapering. They occur in clusters of varying size, and the cluster is usually attached to the bladder by a pedicle, although the individual projections may arise separately from the bladder mucosa. When viewed through the endoscope from a distance, the cluster only is seen (Fig. 82 A); but a close view shows the individual projections (Fig. 82 B). One or more blood vessel can usually be seen coursing irregularly along each papillary projection. When the irrigating fluid is run in through the endoscope the papillomata wave in the current. Pieces often break off and hemorrhage occurs from the central blood vessel. There is no sharp distinction between papillary and nonpapillary bladder tumors; they vary from narrow projections into the bladder to broad irregular protrusions; the same tumor may appear papillary in one area and cauliflowerlike in another. Papillary tumors seldom necrose or ulcerate, and calcareous material is almost never deposited upon them. As a general rule, the more papillary and pedunculated a tumor is, the less malignant is its cell structure and the less likely it is to invade the bladder wall.

2. Aniline tumors

Bladder papillomata occur frequently in workers in aniline dye products industry. These are identical in appearance to those occurring in individuals otherwise occupied.

3. Leukoplakia; bilharzia

Papillomata may arise from long continued irritation such as that occasionally seen in leukoplakia (CONNERY) or in bilharzial infestation.

4. Colloid urachal tumors

These occur in the dome of the bladder and may be papillary in nature (HORWITZ et al.). Tumor caused by irritation and those arising from embryonal rests are, however, more often flat and ulcerative.

5. Hamartoma

This is a mixed benign tumor. DAVIS reported finding one in the bladder. It consisted of a delicate fibrous stroma arranged in papillary formation. The vascularity was greater than that of papilloma.

F. Discolored cystic, vesicular and polypoid elevations

Cysts, vesicles and polyps are characteristic of a few inflammatory and neoplastic lesions of the bladder mucosa. They may vary in color from almost transparent to dark red or blue.

I. Entamoeba histolytica

Amebiasis of the bladder is rare. The bladder mucosa, especially the trigone, in amebiasis of the bladder, is inflamed and edematous. There are usually numerous cystic or vesicular elevations on the trigone which vary in color from pink to deep red.

II. Cystitis cystica

This is an unusual lesion which is occasionally found in chronic infection of the bladder which has existed over a long period of time (NESBIT). There are numerous cystic elevations of the bladder mucosa which vary in size from one millimeter to more than one centimeter in diameter (Fig. 83). The trigone is most often involved, but the cysts may occur singularly or in groups at any location in the bladder (CRAIG). Often there is no surrounding red discoloration, and a cluster of cysts may be raised above the normal bladder mucosa (HOYT). Sometimes the cysts have a slight yellowish discoloration (KRETSCHMER).

III. Cystitis emphysematosa[1]

This also occurs in chronically infected bladders. The cystlike elevations of the mucosa are clear, thin walled and contain gas. The appearance is that of numerous tiny air bubbles under a very thin layer of mucosa. The size of the individual bubble varies from one which is barely visible to those from one to two millimeters in diameter. Between and around the cysts, there are patches of deep red discoloration due to inflammation and mucosal hemorrhage. Bleeding is easily provoked by the mild trauma of cystoscopy. A minute quantity of gas escapes from the blebs when they are punctured.

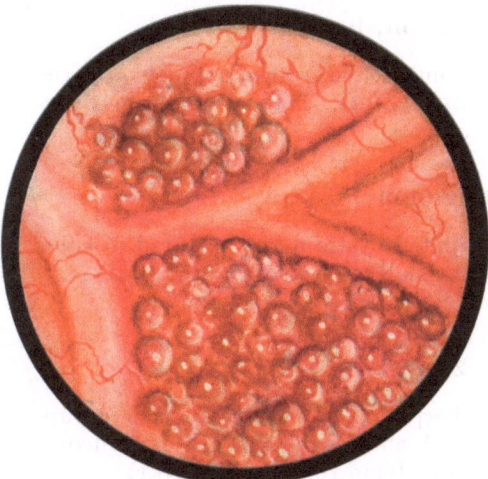

Fig. 83. Cystitis cystica. Cystic elevations vary in size and are often surrounded by normal bladder mucosa. Distant view of a cluster of cysts

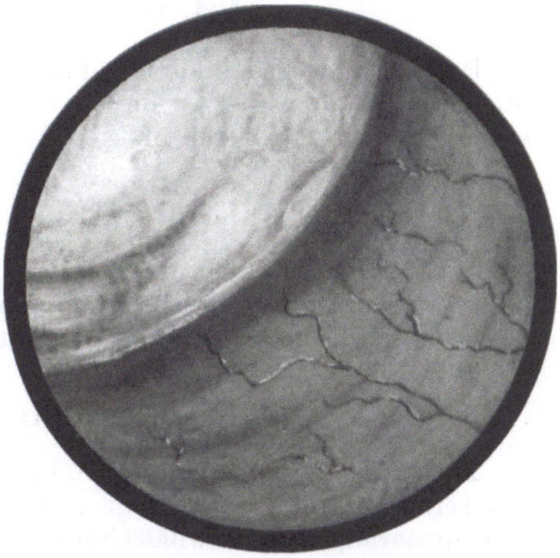

Fig. 84. Echinococcus (hydatid) cyst which had become detached from the bladder wall and was found floating in the bladder. Close view of edge of cyst

IV. Dermoid cysts[2] and teratoma

These very rarely involve the bladder wall. They may produce a cystic protrusion into the bladder cavity but more often the mass is irregular, deep red in color and may be identical in appearance to a well encapsulated carcinoma. When they arise in the bladder wall, the protrusion is at first covered by normal bladder mucosa.

V. Echinococcus disease

This is also very rare in the bladder, and produces light grey, thin walled cysts projecting into the bladder from the mucosa (Fig. 84). These hydatid cysts may occur in any area of the bladder and are either single or multiple. When one is located near the vesical orifice, it may cause urinary obstruction (DEMING). They may become detached from the mucosa and float in the distending medium during cystoscopy. A close vision of the partially collapsed cyst shows spiderweb-like strands of thin membrane similar to that seen in the perivesical space when the cystoscope has perforated the bladder wall.

---

[1] BURNS; BURRELL; LAKE; LUND; MARQUARDT; ORTMAYER; REDEWILL; WELLS.
[2] CAUFFIELD; SHIH.

VI. Endometriosis

The characteristic appearance of endometriosis of the bladder is a tumor mass which, on close view, is seen to consist of a group of cysts of varying size (MOORE). Most of the cysts are dusky blue in color (KRETSCHMER) and the group is surrounded by edema and reddened bladder mucosa. The entire mass is larger and more hemorrhagic (FITZGERALD et al.) during the menstrual periods than it is between them; the contour is similar to that of a blackberry, but the color is blue. There are only a few bluish cysts seen during the intermenstruum, and at that time the entire mass is yellowish red rather than deep red.

VII. Gonococcal infection, healed

Healed gonococcal infection of the bladder may leave cysts of polypoid growths about the bladder neck or on the trigone (BIRNBAUM). There is sometimes persistent granulation tissue of the adjacent mucosa.

VIII. Hemangioma

A very rare bladder tumor, this is a loculated cystic mass, dark red in color (RATHBUN). The mass is usually covered with small multicolored cysts (BALLLENGER et al.). There may be hemangiomata associated with carcinoma.

IX. Herpes zoster

This may rarely involve the bladder (herpes vesicalis); there is a group, or a series of groups, of small greyish vesicles which occur on the bladder mucosa. The appearance is somewhat similar to that involving the skin (DARGET).

G. Yellow or greyish yellow elevations of the bladder mucosa

I. Cystitis follicularis

Lymph follicles lie directly beneath the epithelium of the bladder mucosa (Fig. 85). They are seen through the endoscope as small greyish yellow solid elevations which are discrete, closely packed together, and usually involve the floor of the bladder (HURMAN). They may vary somewhat in size but do not become larger than 2 to 3 mm. in diameter (STIRLING). The surrounding bladder mucosa is inflamed and sometimes slightly edematous. There may be engorged capillaries coursing between or around the follicles.

II. Lipomata

Deposits of fat beneath or in the bladder mucosa may be seen through the cystoscope as yellow elevations of varying size, scattered over normal mucosa. The condition is rare, produces no symptoms and is found incidentally during routine cystoscopy or at autopsy.

III. Leiomyomata

Small bundles of smooth muscle are not rare (GRAUER). They are small yellow elevated areas which usually occur on the trigone (Fig. 86). The tissue surrounding the small elevations is normal. Larger tumors may be located at the vesical outlet and cause urinary obstruction, or may be covered by severe inflammatory changes (KATZEN).

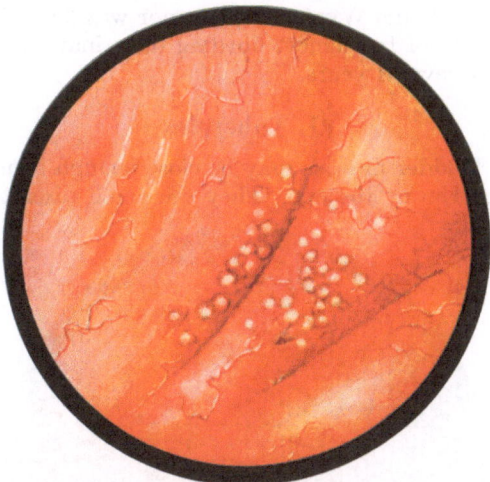

Fig. 85. Cystitis follicularis. Multiple small yellow discrete elevations on or near trigone. Distant view

A   B

Fig. 86 A and B. Small yellow elevations of bladder mucosa characteristic of leiomyomata. Surrounding mucosa is normal. A Distant view. B Close view

IV. Malakoplakia

The greyish yellow or canary yellow (ROLNICK and RAGINS) elevations due to malakoplakia of the bladder are larger and more sparsely scattered than those of cystitis follicularis. The rounded or oval elevations are spherical cell masses resulting from chronic infection involving the submucosa (CHRISHOLM and TUDHOPE). The surrounding area is usually deep red due to inflammation of the mucosa.

V. Osteoma

This is exceedingly rare in the bladder, and appears as a yellow elevation of the bladder mucosa surrounded by a narrow margin of deep red edema. In the case reported by COLLINS and WELEBIR the tumor was 3 cm. in diameter and the patient had complained of frequency, burning on urination, dribbling and some blood in the urine for several years.

VI. Tubercles

Tubercles in the bladder mucosa may be close enough beneath the epithelium to be visible through the cystoscope. They are small discrete yellow elevations

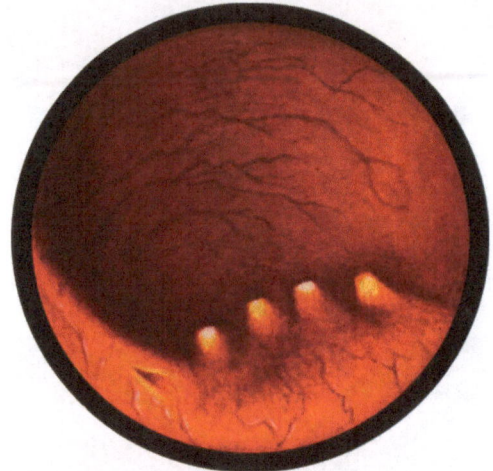

Fig. 87. Tubercles, when close under mucosal epithelium, are seen through cystoscope as discrete yellow elevations. The tubercles shown here are larger than are usually seen, and there is slight inflammation of the adjacent mucosa but no ulceration. Often there is slight edema and superficial ulceration

often surrounded by slightly reddened mucosa and usually located near the ureteral orifice on the side of the involved kidney (Fig. 87). There is often adjacent superficial ulceration, but when there is widespread ulceration and fibrosis, the individual yellow tubercles are obliterated by these more extensive pathological changes.

H. Blue elevations

I. Varicose veins

Large varicose veins are seen as tortuous cylindrical blue elevations of the bladder mucosa. They are easily identified (Fig. 28, p. 55).

II. Endometriosis

This may produce an elevated dusky blue tumor with intact mucosa which resembles a varix but is not as elongated (KRETSCHMER).

III. Metastatic melano-epithelioma

This is a bluish protrusion of the bladder mucosa when there is vesical involvement from a primary skin lesion. There are dark granules on the surface (MORROW et al.); the appearance is not unlike that seen on the skin.

I. Reddish brown elevations

Lichen planus

The nodules of lichen planus are rarely seen in the bladder. When they do occur they accompany the more common skin lesions. As seen through the cystoscope, the nodules are reddish brown in color and discrete, are sparsely scattered over the mucosa — more often in the trigone — and there is some inflammation of the mucosa which causes frequency and urgency of urination.

J. Discolored depressions

Discolored depressions in the bladder wall are usually due to trauma, to electrosurgical procedures or to ulceration.

I. Lacerations and rupture (WISHARD)

A laceration of the bladder mucosa without perforation may cause an irregular deep red depressed area with ragged edges (Fig. 61, p. 81). When the injury extends through the bladder wall, the black hole of the perforation can be seen (Chap. VII). There is often a light edge to the tags and the entire area may be surrounded by irregular dark red spots due to hemorrhage into the mucosa. If the rupture is large, it may be impossible to distend the bladder sufficiently to obtain visualization of its interior, or the rent may be plugged by a loop of bowel (NEMSER and WEINBERGER). Occasionally, hemorrhage is severe enough to obscure the site of rupture (CAMPBELL). When a perforation of the bladder wall occurs during endoscopic surgery the black hole is surrounded by white strands of resected bladder muscle and there are no surrounding ecchymotic dark red spots (Chap. XIX).

II. Following electrosurgical procedures

Whether the approach be endoscopic or open, there is a depression covered by white sloughing tissue corresponding to the resected or coagulated area. The edges of the white discoloration are sharp and there is a surrounding red area due to inflammation (Fig. 88).

III. Ulceration

This often causes a depressed area in the bladder mucosa.

1. Chronic infections

a) Tuberculosis. Tuberculous ulcers usually have well defined edges which may be undermined (Fig. 14, p. 92). The base is depressed when infection is severe, and is usually covered by mucopurulent exudate. Ulceration is sometimes single and is most often found near the ureteral orifice on the side of the involved kidney; it may be multiple and located in the bladder area. There are often irregular deep red elevations of the bladder mucosa due to granulations and edema near the ulcerations. In severe cases of long standing, irregular lighter areas due to fibrosis are also found and the bladder capacity is small. Cystoscopy is painful whenever tuberculous cystitis is severe enough to cause ulcerations. Depressed ulcers may occur in any severe or subacute bladder infection.

b) Nontuberculous. Nontuberculous ulcerations are more likely to be less well defined and are accompanied by more generalized edema and inflammation of the bladder mucosa. When solitary they are more likely to be in the fundus.

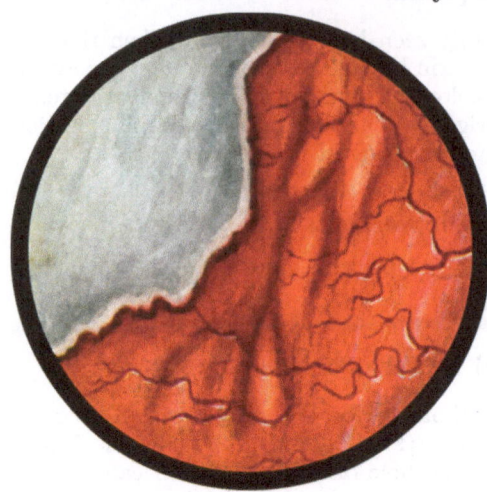

Fig. 88. White depressed area following transurethral resection of bladder tumor. Edge of area and surrounding red discoloration due to inflammation and edema are shown. Close view

c) Actinomycosis. Actinomycosis infection of the bladder is a rare cause of depressed ulceration. The appearance is similar to that of tuberculous cystitis.

2. Neoplastic

Depressed ulcers sometimes occur on malignant tumors which have a broad base. The surface of the protruding tumor is depressed by the ulceration, and the characteristic irregular deep red protrusion of the tumor is seen surrounding the edge of the depression. There are nearly always irregular white or greyish white sloughing areas on the ulcer, and the base is covered irregularly with thin mucopurulent deposit.

K. Lesions showing all types of color and contour abnormalities

A few bladder lesions show many types of simultaneous changes in the bladder mucosa. These may range from a small area of red inflammation to extensive discoloration and contour variations.

I. Vesical bilharziasis

Infestation of the bladder with Schistosoma haematobium (bilharziasis) produces a great variety of changes in the bladder interior (Fig. 89). HAYWARD has reported similar changes due to Schistosoma japonicum.

1. Hemorrhagic areas

Patchy areas of red discoloration located near the ureteral orifices are the earliest changes. They are similar to the hyperemia of acute hemorrhagic cystitis (Fig. 42, p. 68).

2. Edema

Soon the area becomes slightly edematous and elevated, and is similar in appearance to edema from any inflammatory process (Fig. 79, p. 98).

3. Discolored white areas

These are similar in appearance to leukoplakia from other chronic infections (Fig. 73, p. 91).

4. Ulcerations

The ulcers due to bilharzial infestation are usually superficial and single. The base is a light pink, and is lighter than the surrounding inflamed mucosa. The margins are not distinct or undermined.

5. Tubercles

Bilharzial tubercles are small yellow elevations of mucosa surrounded by moderate red discoloration and prominent, tiny blood vessels. The appearance is similar to tuberculosis tubercles (Fig. 87, p. 106). The surface of the tubercles is at first shiny but later becomes dull, and the elevations flatten out and have the appearance of "sand under water".

6. Protrusions

a) Nodules. Rounded *nodules* occurring in groups and which are light pink in contrast to the dark red of the surrounding mucosa are similar to the "cystitis

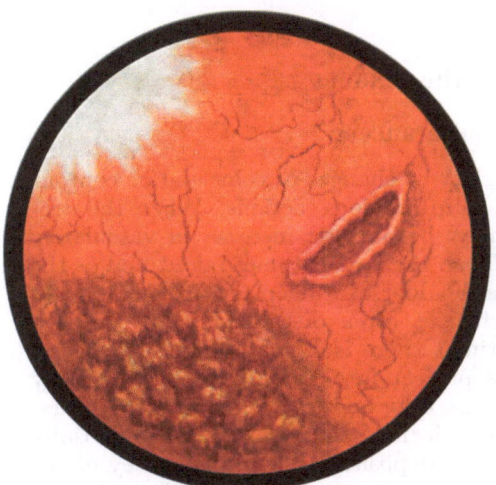 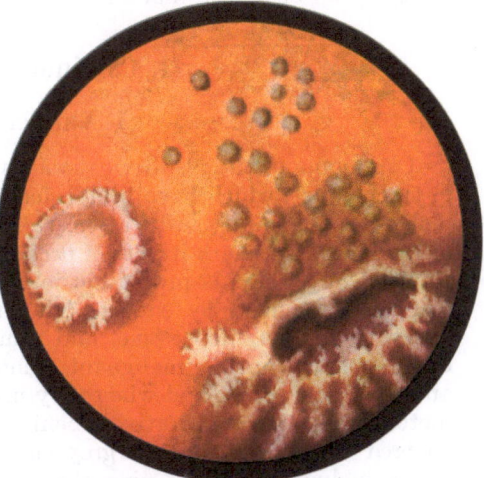

Fig. 89. Bilharzial infection of bladder. Distant view. White discoloration is leukoplakia. An uncomplicated ulcer at right; below, slight elevations of cystitis glandularis

Fig. 90. Bilharzial infection of bladder. Distant view. Grey nodules are above. Calcified bilharzial papilloma at left; at right, an ulcerating villous bilharzial carcinoma

glandularis" of other chronic infections. Higher and more solid protrusions into the bladder are the characteristic spherical nodules which are dark grey in color (Fig. 90). They are usually discrete and occur in groups, but may coalesce to form a larger protrusion, the "bilharzial node". These elevations often persist after the active infection has subsided; then the surrounding mucosa is light colored, due to anemia and fibrosis.

b) Cystic. Sometimes *cystic* elevations which are identical to the cysts of "cystitis cystica" are seen.

c) Papillomata. Most bilharzial *papillomata* are sessile and dark red in color simulating granulomata. Some are more papillary and pedunculated.

7. Complicating lesions

Bilharziasis may be the cause of other bladder lesions which complicate the disease. Secondary infection with mixed organisms is usually present and the mucosal changes are similar to those found in chronic cystitis (OCKULY). Calcareous material is often deposited on the lesions, and breaks off resulting in vesical calculi[1]. Carcinoma of the bladder is not infrequently a complication of bilharziasis and is probably secondary to the infestation (FAIRLEY). The effects

---

[1] CULVER and HOEPPNER.

of the disease may persist for many years after the organisms have been eliminated[1]. In fact the fibrosis, the small bladder capacity and the upper urinary tract damage often persist throughout life. Occasionally cases of bilharziasis of the bladder are reported outside the areas in which it is endemic (CHRISTOPHERSON and WARD).

8. Diagnosis

The diagnosis of active bilharzial urinary tract infection is made by finding the eggs of the Schistosoma haematobium in the urine. In most cases, they are numerous and easily found and identified in a fresh specimen of urine. When distilled water is added to the wet mount of the urine under the microscope, the eggs can be seen to hatch.

II. Carcinoma of the bladder

1. Variable appearance

Carcinoma of the bladder causes a variety of changes of color and of contour in the bladder, which have been referred to in previous sections. When the neoplasm is malignantly infiltrative, there may be only an increase in vascularity of the mucosa to mark its site of origin. Then the area becomes raised and as the tumor grows, there is protrusion into the bladder of a deep red irregular mass, usually surrounded by edema. The surface is often discolored white or light grey due to sloughing tissue, and sometimes there is ulceration and/or calcareous deposit. The less malignant tumors project into the bladder cavity early and sometimes are found when they measure only one or two millimeters in diameter. Even the comparatively benign growths may slough and become partially covered with white or light grey calcareous deposit. A thorough survey of the tumor includes a distant view through the cystoscopic telescope. Then the objective lens is moved closer to the lesion as visualization is continued. A close view of the edges of the discolored areas and of the contour abnormalities helps to identify the lesion. The bladder may be so full of tumor that it is almost impossible to obtain a distant view of the growth through the cystoscope. In such cases, a search is made to locate an area of normal bladder mucosa, and from there the edge of the mass can often be identified. Sometimes hemorrhage from the tumor is so profuse that vision is obscured. The addition of 15 to 20 drops of adrenalin solution to the irrigating fluid may help to control the bleeding sufficiently to permit visualization. The view is also enhanced by placing the objective lens very close to the lesion while the distending fluid is running in through the cystoscope; the space between the objective lens and the lesion is thus kept comparatively clear and visualization is improved.

2. Differential diagnosis

It may be impossible to differentiate carcinoma from other lesions such as tuberculosis and severe nonspecific cystitis, without biopsy examination. A satisfactory specimen can best be obtained with the resectoscope loop; the small biopsy forceps used through the cystoscope are seldom adequate to obtain sufficient tissue to make an accurate diagnosis (Chap. XV). A tumor arising in a diverticulum may not be visible through the endoscope or it may fill the diverticular orifice hiding it from view[2].

---

[1] MAKAR; HAYWARD.   [2] ABESHOUSE and GOLDSTEIN; BOYLAN et al.

III. Gangrenous cystitis (STIRLING and HOPKINS)

This is a severe extensive infection involving the mucosa of the entire bladder interior. Numerous terms have been applied to it, such as cystitis exfoliativa, cystitis gangrenosa desiccous, croupous, diphtheric, membranous or plastic, and necrosis of the bladder. There is red discoloration ranging from light to dark red or greyish black. Extensive slough produces large greyish white patches, the edges of which become detached and float in the distending medium. Sometimes large pieces of sloughing tissue are loose in the bladder and may occlude the vesical orifice (CRISTOL and GREENE). The contour of the interior bladder wall is irregular, due to protrusions of severe edema and to granulation. Ulceration and sloughing tissue may cause depressions in the irregular surface. Calcareous material may deposit on the surface of sloughing tissue in cases which survive the acute stage. Bleeding may be profuse and pain severe if cystoscopy is attempted without anesthesia.

IV. Syphilis

This has been described as the cause of various types of bladder lesions. The disease rarely involves the bladder, however, but should be considered when an unexplained inflammatory or neoplastic change in the bladder is seen by cystoscopy. During the second or exanthematous stage of the disease, the bladder lesion consists of irregular red patches due to increased prominence of small blood vessels. The center of the patch is deep red and the small vessels radiate outward from it (THOMPSON). The appearance is not unlike that of acute hemorrhagic cystitis (Fig. 42, p. 68). Elevated ulcers are sometimes seen. They are surrounded by the edema and granulation which are the cause of the elevation (FINESTONE). These ulcerating papillary syphiloderms have been reported as being the syphilitic lesion which is most commonly found in the bladder. The base of the ulcer is depressed and hemorrhagic and the edges discrete. Rounded elevations (papules) of various sizes may be scattered irregularly over the bladder mucosa (ZIMMERMAN and LEVY). Gumma of the bladder (ORMOND and HEMMING) may be an irregular red protrusion or may have areas of white discoloration due to sloughing tissue. Most gummas are ulcerated, causing a depression on the surface. The base is a greyish yellow color and the edges elevated and discrete. Gummas are usually multiple, are located near the ureteral orifices, and vary in size from one millimeter to two centimeters. Sometimes calcareous material is deposited on the sloughing tissue. The appearance of gumma may be identical to that of carcinoma; biopsy study is necessary to differentiate between them.

V. Pemphigus vulgaris

This involves the bladder very rarely when the skin lesion is severe and extensive. It is described as causing irregular dark red elevations of the bladder mucosa due to edema, congestion and granulation. Areas of yellow sloughing tissue or partially necrotic plaques varying in size up to 15 mm. are scattered over the vesical mucosal surface (HYAMS and BOTOINICK). Involvement is most severe in the fundus and dome of the bladder; the trigone and bladder neck are engorged but not sloughing. Hemorrhage may be so severe that clear visualization is difficult.

VI. Vesical tuberculosis

This has been described in several previous sections. Its appearance is so varied that almost all changes of color and contour of the interior of the bladder may be seen. The lesions are nearly always located on or near the ureteral meatus on the side of the involved kidney. The earliest change in the bladder mucosa is only a slight increased redness similar to that found in mild acute hemorrhagic cystitis (Fig. 42, p. 68). Small yellow elevations of the tubercles (Fig. 87, p. 106) or superficial ulceration (Fig. 74, p. 92) indicate slightly increased involvement. Smooth elevations of red edema and irregular protrusions of granulation tissue are first minimal, but later in the course of the disease may be large and extensive. A fungating ulcerating greyish green mass which bleeds from the surface, a tuberculoma, may be the only remaining evidence of a nearly quiescent lesion (BOWEN and BENNET). In other cases, the entire bladder interior may be involved by edema, granulations, ulceration, sloughing tissue and sometimes calcareous deposit. Fibrosis helps to shrink the bladder cavity and a contracted, rigid, white, bloodless vesical mucosa may remain after the infection has burned itself out (WEAR). Sometimes it is impossible to differentiate tuberculous cystitis from vesical neoplasm without microscopic examination of a biopsy specimen.

Chapter X

Abnormal bladder contents

It may be difficult at times to differentiate an object which is loose in the bladder from one attached to the bladder wall. Calcareous deposit upon a tumor, for example, may be mistaken for a stone in the bladder. Sloughing tissue may be loose or attached to a tumor. A foreign body such as a gauze sponge or a spicule of bone may project into the bladder, but be firmly fixed in the bladder wall. Differentiation may be accomplished by moving the object with the beak of the cystoscope, or with a ureteral catheter or other instrument passed through the cystoscope. An attempt may be made to manipulate the inner end of the cystoscope to the other side of the object. When the material in question is lightweight the distending fluid, as it runs in through the cystoscope, moves it.

I. Blood clots

1. Location

These are frequently found loose in the bladder or attached to some lesion of the mucosa.

2. Identification

When the clot is fresh it is medium red in color (Fig. 91 A and B); when old, it is dark red or almost black. The edges which are thin may be light greyish red due to fibrin. Light colored streaks of membranous like fibrin are usually on the surface of, or mixed through, the clot. When the clot is old and organized, it may be almost entirely covered by the light dull red fibrinous envelope. Identification of clots through the endoscope is usually easily made, but sometimes when they are well organized — especially when they are firmly attached to a lesion of the bladder mucosa — it may be difficult to differentiate them from tumor or sloughing tissue.

3. Evacuation

Clots in the bladder often obscure vision through the cystoscope to such an extent that their evacuation is necessary before it is possible to identify anything else. They can usually be aspirated out through the cystoscope sheath by the use of a piston syringe such as the Toomey (Fig. 92), which fits into the outside opening of the sheath. By suction with the syringe and irrigation through the sheath, the larger clots can be broken up and evacuated. When the bladder is full of clots, the process of removing them may require 20 minutes or longer;

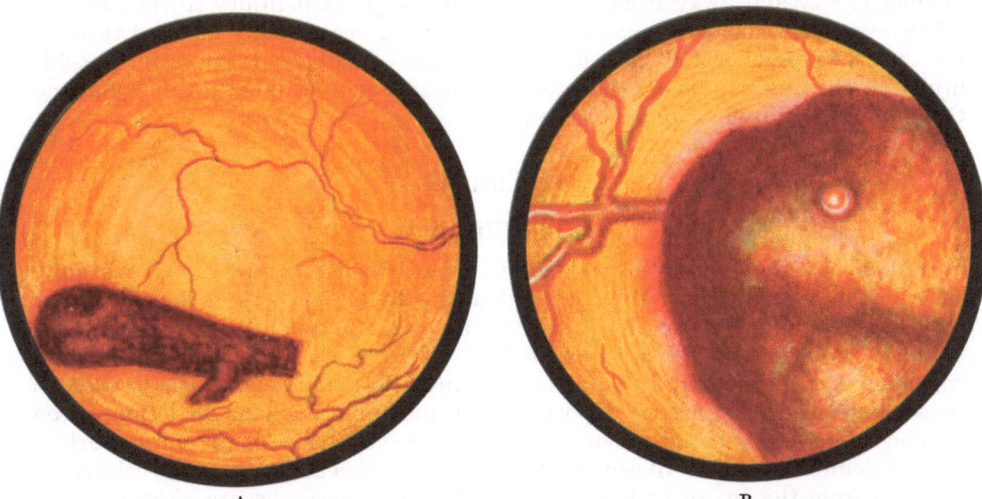

Fig. 91 A and B. Small fresh blood clot in bladder. A Distant view. B Close view. Thickest portion of clot is darker red than thin portion. Lighter streaks are fibrin

aspiration is, however, preferable to removing them through a suprapubic cystostomy. When the clots are old and firmly organized it may be difficult to break

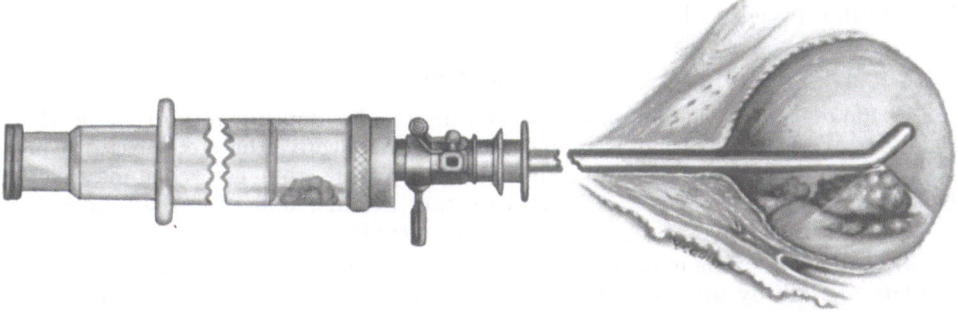

Fig. 92. Evacuation of clots from bladder through cystoscope sheath with a piston syringe (TOOMEY) which fits into end of sheath

them up into pieces small enough to aspirate. An effective maneuver is to push the plunger of the syringe in for one or two centimeters, then quickly and with considerable force pull it out. Repeat the maneuver rapidly until clots begin to come out through the sheath. Rapid aspiration of the clot into the fenestra tends to break off a piece. If this does not succeed, advance the sheath farther into the bladder, or pull it out a little; this may place the fenestra at the edge of the clot where it is more friable. Care must be exercised to avoid aspirating bladder mucosa or lesions of the mucosa into the fenestra of the sheath. When the patient is not under anesthesia, he will have pain when the mucosa is drawn into the fenestra;

when there is pain during aspiration, more fluid is injected into the bladder and the inner end of the sheath is manipulated away from the mucosa. There is more danger of damaging the bladder mucosa when the patient is under anesthesia. Adequate distention of the bladder with irrigating fluid during aspiration of clots, and care in keeping the fenestra of the sheath close to the center of the bladder cavity and away from its walls, helps to prevent injury to the mucosa. While evacuating clots, the examining telescope is occasionally inserted through the sheath to discover if all clots have been removed and also to check for possible injury to the mucosa during the process of evacuation (ROBERTS, COUNCILL and COUNCILL Jr., FOWELL and McLEAN).

II. Calculi

1. Identification

Calculi commonly occur in the bladder. Identification by cystoscopy is usually easy, but may be difficult. When the calculus is huge, the beak of the cystoscope may strike it and prevent the advancement of the instrument farther into the bladder. Pointing the beak to one side, then rotating the sheath medialward will allow the instrument to be advanced into the bladder and visualization to be accomplished. Calcareous deposit on a tumor or other bladder lesion, or on a foreign body may be mistaken for a stone. A close view of the entire surface of the calculus like mass will usually show an edge which demarcates the calcareous material from the tumor or foreign body. Bleeding may obscure vision; a close view while the irrigating fluid is flowing in through the sheath often clears the field. The bladder cavity may be reduced by fibrosis or by spasm to such an extent that distention sufficient to permit adequate visualization is impossible. Spinal anesthesia may be helpful in these cases. A calculus located behind the intravesical protrusion of an enlarged prostate, in a diverticulum or behind a tumor may not be visible through the cystoscope. The use of a retrograde lens may be helpful, especially when the stone lies behind an enlarged prostate.

2. Location

a) **Floor.** Most bladder stones gravitate to the dependent portion of the bladder. When the patient is in position for cystoscopy, the stone lies on the bladder floor above the trigone. When it is very large, the cystoscope may pass beneath it, and it becomes necessary to rotate the instrument so that the field of vision is directed ventrally in order to visualize it.

b) **Fundus.** A stone may be held higher in the fundus, either on the floor or on the lateral wall, by *trabeculation* or a *cellule*. It may be held in that position by a *diverticular orifice*; a part of the stone which cannot be seen is in the diverticulum.

c) **Dome.** When something which appears to be a stone is seen in the dome of the bladder, it is probably some light material such as paraffin or sloughed tissue covered by calcareous deposit; or it may be an encrusted urachal tumor.

3. Number and size

Vesical calculi are most often single, but multiplicity is not uncommon. Multiple stones usually originate and grow in a bladder which is chronically distended. They are usually phosphatic and are often multifaceted, the flat sides fitting together. Numerous small spherical stones are occasionally seen, and sometimes several large calculi are found in the bladder. Stones vary in size from minute ones to those eight to ten centimeters in diameter.

a) Estimation of size. The size of a stone can be estimated during cystoscopy. The position of the eyepiece is marked when the proximal edge of the stone is in the center of the visual field, then the cystoscope is advanced until the distal edge is in the center of the field. The position of the eyepiece is again marked. The distance between the two marks is roughly the diameter of the stone.

4. Shape and color

a) Composition in general. The shape and color of vesical calculi are sometimes indicative of their composition (WINSBURY-WHITE, JOLY). The majority of

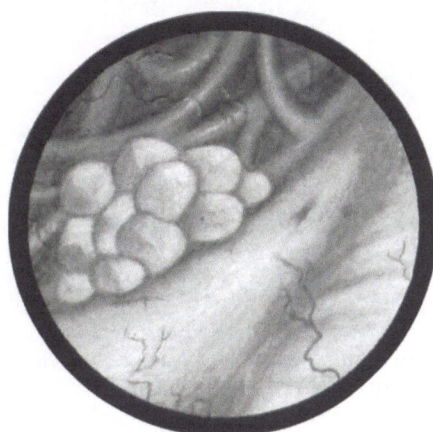

Fig. 93. Light tan slightly irregular bladder stone of mixed composition, predominantly phosphatic. Close view

Fig. 94. Multiple small faceted stones

stones are of mixed composition, and therefore do not follow the pattern of those whose appearance is due to a specific ingredient. As a general rule, hard stones have a shinier surface than soft stones. Irregular shaped calculi are more likely to form in a bladder which is chronically distended. When they are seen through the cystoscope, thorough search for a urinary obstructive lesion is made.

b) Light colored phosphatic. The most common bladder stone is greyish white or light tan, and smooth or only slightly rough (Fig. 93). Calcium phosphate or triple phosphates predominate in this group, although other ingredients — oxalates and sometimes urates — may be found.

c) Faceted phosphatic. When stones are faceted they are always multiple, usually light grey or tan, and composed mainly of phosphates (Fig. 94).

d) Brown spiculed ("mulberry") oxalate. Brown spiculed calculi are usually composed mostly of oxalates. They are sometimes referred to as "mulberry" stones. The distant view through the cystoscope may not show the spicules, but in the close view they are distinct (Fig. 95 A and B). The tips of the spicules are often light yellow.

e) Dark ("jackstone") oxalate. The so-called "jackstone" has long slender spicules which radiate outward from a central point similar to the jack with which children play. They too are mostly oxalate.

f) Pale yellow to deep brown mixed. Smooth or slightly rough stones of mixed composition may vary in color from pale yellow to deep brown; they may be round or oval.

g) Characteristic color. A few very rare bladder stones may have a characteristic color. *Cystine* stones are *yellow*, sometimes bright yellow. Stones composed

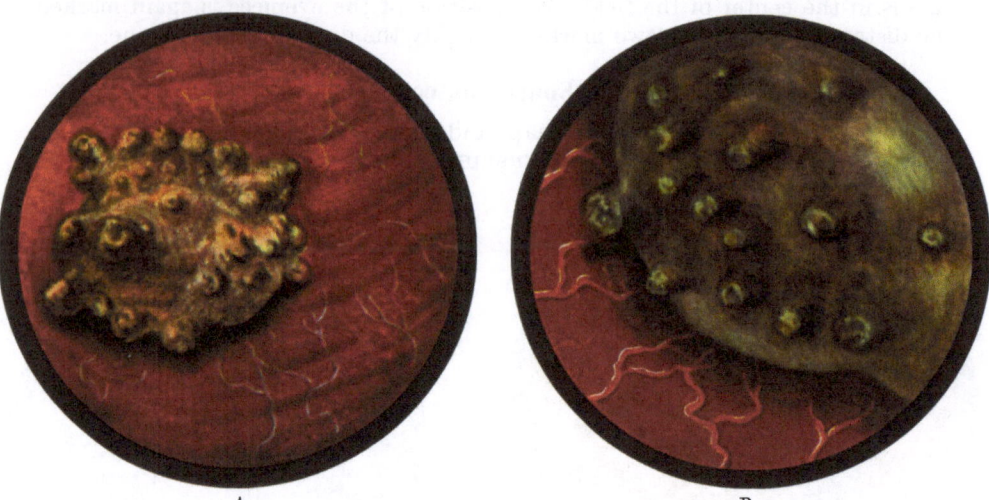

A   B

Fig. 95 A and B. Brown spiculed calculus composed mostly of oxalates (mulberry). A Distant view. B Close view

of *xanthine* are *red*. *Blue* stones composed of *indigo* have been reported (WINS-BURY-WHITE).

III. Foreign bodies

1. Inserted by patients

Most foreign bodies found in the bladder are inserted through the urethra by the individual himself. It is done for sexual gratification, dilatation of the urethra, taking the temperature or other purposes. The foreign bodies which have been removed from the bladder are numerous and varied. No attempt will be made to describe them all. Usually they are easily identified through the cystoscope lens, but occasionally their appearance or position is such as to make identification difficult.

2. Incrustation

Calcareous material is nearly always deposited on foreign bodies which have been in the bladder very long (Fig. 96). In some cases evidence of incrustation is seen after a week or two; in others, there may be none for as long as several months. The object may be completely covered by calcareous material, thus obscuring its identity.

3. Bone fragments

These may be thrust into the bladder from a fractured pelvis at the time of injury, or may gradually work their way in following splintering of a pelvic bone. Through the cystoscope lens these fragments are light yellow, and the rough edge of the broken bone can usually be seen. Incrustation has often taken place.

4. Floating objects

When the foreign body is lightweight, it floats and is found in the dome of the bladder, near the air bubble.

a) Debris and oil. Debris, especially when mixed with oil, is seen as an irregular light grey mass with numerous small air bubbles scattered over its surface. When

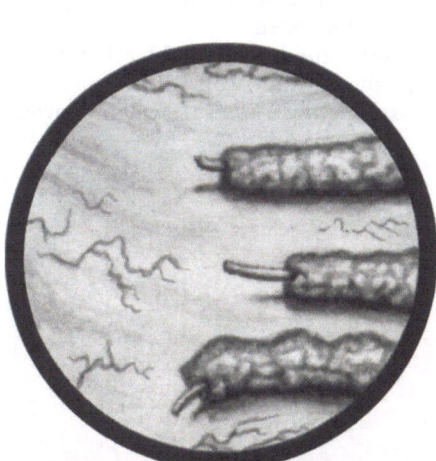

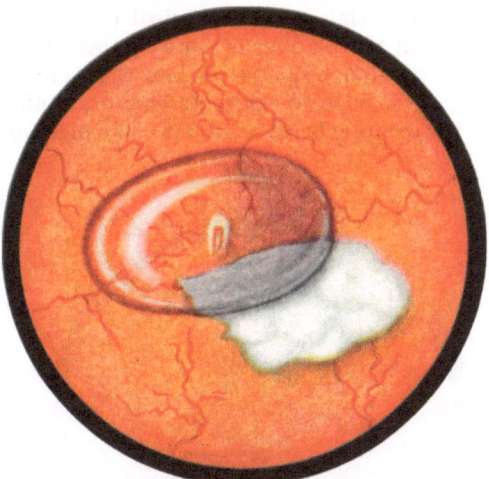

Fig. 96. Incrustation of broom straws which had been inserted by patient into bladder for purpose of dilating his urethral stricture.   Close view

Fig. 97. Paraffin floating in bladder. It was introduced by patient for sexual gratification. Note bubble in dome of bladder

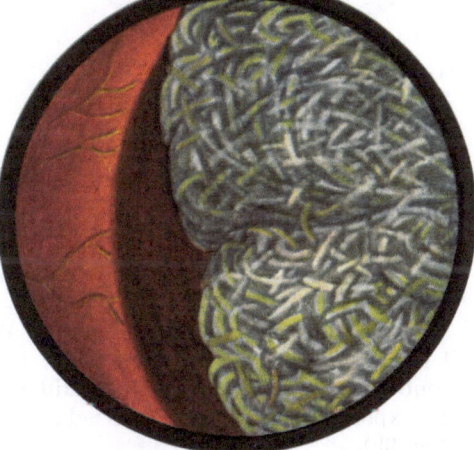

A   B

Fig. 98 A and B. Gauze sponge which migrated from bladder wall after it had been inadvertently left in postvesical space during removal of a huge hydatid cyst. A Distant view. B Close view

an oily solution is instilled into a bladder which contains persistent residual urine, it is not evacuated when the patient voids and may remain in the bladder for many months. The oil and debris become mixed and form a soft mass; calcareous material is deposited, and a soft floating light grey stone is formed.

b) Paraffin. This is a favorite material for sexual perverts to insert through the urethra. Cystoscopy shows it to have risen to the bladder dome in or adjacent to the air bubble (Fig. 97).

c) Wood. Lead pencils, slippery elm and other wooden foreign bodies float unless there is heavier material such as a rubber eraser attached, or an incrustation with calcareous deposit which causes them to sink to the bladder floor.

5. Following medical or surgical procedures

Material used in surgical or manipulative procedures may inadvertently be left in the bladder or may migrate through the bladder wall.

a) Gauze sponge. When this is accidentally left outside the bladder but near to it, it usually works its way gradually through the bladder wall into its cavity; the opening heals behind it. The mass may not be recognized as gauze when viewed

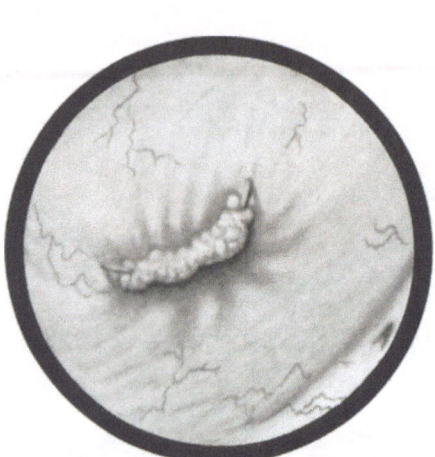

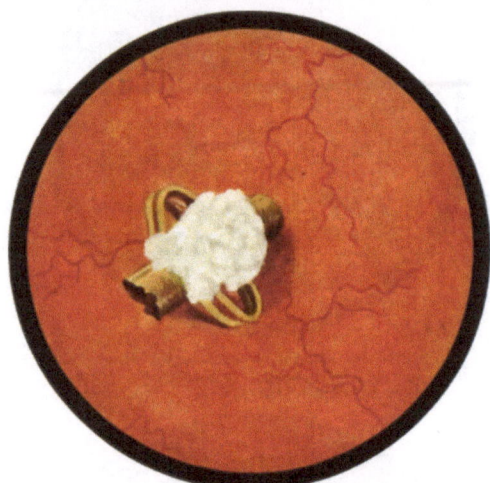

Fig. 99. Calcareous deposit on a linen suture exposed on floor of bladder. It had been used to close a vesicovaginal fistula. Close view

Fig. 100. Head of a mushroom type wing-tipped (Malécot) suprapubic catheter which has been almost entirely covered by calcareous deposit. Distant view

through the cystoscope lens. The appearance from a distance is an irregular, well defined, light grey mass which resembles sloughing tissue (Fig. 98A). Detailed observation, however, shows a faint trace of the crisscross lines of the gauze threads. The close view (Fig. 98 B) brings into vision the individual threads, but the criss-cross pattern is not distinct; instead, the ends of the threads fade away.

b) Nonabsorbable suture material. When this is exposed in the bladder cavity it is usually recognized. Black silk (Fig. 75, p. 94) cotton, or linen appear the same as when seen outside the bladder (PETREN). When incrustation has covered the exposed suture, it is visualized as a small, narrow, calcareous, cylindrical mass (Fig. 99).

c) Urological equipment. Pieces of catheters, bougies, filiforms and other urological instruments are sometimes broken off in the bladder or urethra. These are easily recognized unless they are covered by calcareous incrustation. The head of *a mushroom type catheter* which has been lying in the bladder through a suprapubic wound may become so covered with calcareous deposit that it will not collapse sufficiently to be removed. When an attempt is made to pull it out, the head breaks off. The distant view of this foreign body through the cystoscope may be identical to that of a stone. A close view of the entire object, however, will usually disclose the brick red color of the rubber where it is not covered by the light grey calcareous deposit (Fig. 100).

IV. Shreds of mucus, pus and epithelial cells

These are often so numerous in the bladder that they obscure vision. When distending fluid is running in through the cystoscope, this material, floating in the agitated fluid, gives the appearance of a snow storm. Numerous small white pieces move about in the visual field, and unless the inner lens of the telescope is placed very close to the object which is being examined, nothing can be seen except the cloud of floating pieces. The field of vision can usually be cleared easily by several irrigations with the distending fluid through the cystoscope sheath. When the clouding material comes from a diverticulum or from a pus filled ureter, it may be more difficult to clear the visual field.

V. Sloughing tissue

This may become detached from a tumor, from necrotic bladder mucosa, or from other intravesical lesions. It is seen through the cystoscope as a dirty white irregular mass which is moved by inflowing distending fluid, but which settles to the bladder floor when there is no agitation of fluid. By a close view of the material, small shreddy strands can usually be seen. The surface of the slough is smooth except at the edges where it is irregular. When covered by calcareous deposit, it may have the appearance of a stone. It is usually more irregular than most stones, and when touched by a catheter or other instrument passed through the cystoscope, it is found to be soft.

Chapter XI

Abnormalities of the bladder neck and posterior urethra in the male

I. Contracture of the vesical orifice

The vesical orifice may be contracted due to fibromuscular hyperplasia, to muscular hypertonia or to fibrosis. Each of these conditions may be either congenital or acquired, and may occur in either male or female children or adults. When the contracture is acquired it is usually a fibrous contracture, and is due to infection or is a sequela of operative procedures on the bladder neck.

1. Appearance at the margin of the vesical neck

The appearance of the bladder neck varies as viewed through the different optical systems. The posterior lip of the bladder neck appears higher when seen through the foroblique lens than it does through the right angle telescope. The direct forward view makes it appear to be still higher than the foroblique. Therefore, the optical system in use must be considered when estimating the amount of contracture of the bladder neck and/or the degree of elevation of its posterior lip above the apex of the trigone.

The "straight down" view of the rim of the vesical orifice is obtained through the right angle optical system. In order to obtain an adequate view of the dorsal lip of the bladder neck through this optical system, the cystoscope is advanced into the bladder, then slowly withdrawn while the view is directed dorsalward toward the six o'clock position. Normally the apex of the trigone fades off gradually into the rim of the bladder neck as the instrument is withdrawn. There is a gradual change from a clear distinct field of vision to the blurred field resulting from very close proximity of the tissue of the bladder neck to the objective lens of the cystoscope (Fig. 101 A). When contracture of the vesical orifice is due to fibromuscular hyperplasia, the transition from the distinct view of the apex of the trigone to the blurred field of the close range bladder neck is sudden instead of gradual. As the rim of the bladder neck comes into the lower portion of the visual field, the apex of the trigone can be seen at a little distance from the rim instead of close to it (Fig. 101 B). When bladder neck contracture is due to muscular hypertonia, the elevation of the dorsal lip is plainly seen but is not as marked as when it is due to fibromuscular hyperplasia. Fibrous contracture of the vesical

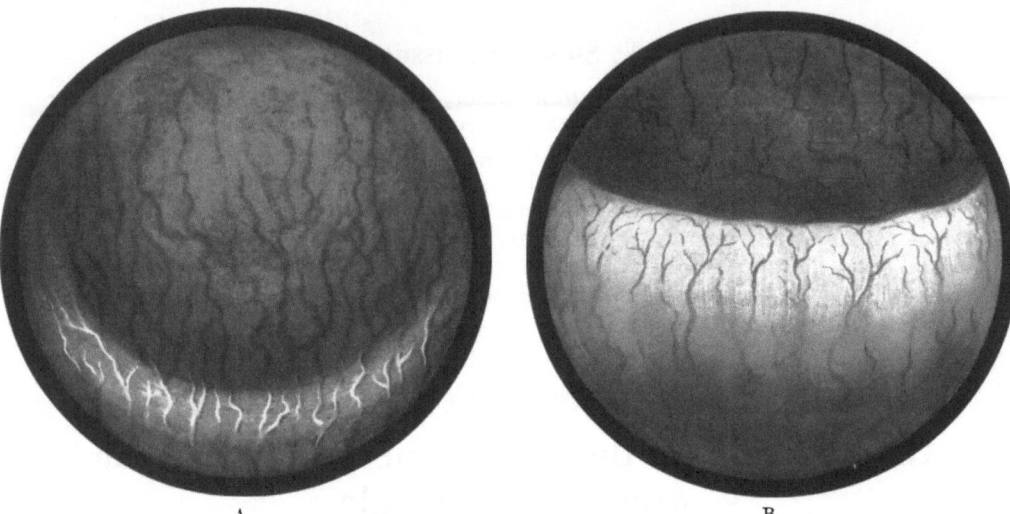

A   B

Fig. 101 A and B. "Straight down" view of rim of vesical neck through right angle lens at 6 o'clock position. A Normal. B Fibromuscular hyperplasia

orifice causes only slight elevation of its dorsal lip, but the apex of the trigone and the rim of the bladder neck are lighter in color and more avascular than normal. When an obliquely forward or foroblique optical system is used and the objective lens is placed at the rim of the bladder neck, the elevation is apparently more marked than when it is seen through the right angle system; the trigone is still farther away from the lens due to the oblique view. It is difficult if not impossible to diagnose bladder neck contracture through the right angle or obliquely forward lens when the view is directed lateralward or ventralward. At these positions the rim of the normal bladder neck as well as that of the contracted one comes into the field of vision suddenly while the instrument is being withdrawn.

2. Appearance from within the bladder

Bladder neck contracture can be more accurately diagnosed through the retrograde optical system than through the right angle or obliquely forward lens. This view is from within the bladder, and the entire circumference of the vesical orifice can be thoroughly examined both by distant and close vision. Fibromuscular hyperplasia is seen as an intravesical protrusion of the entire circumference of the rim of the bladder neck, and its appearance is that of a collar surrounding the vesical orifice; it is sometimes referred to as a collar contracture. As the cystoscope is slowly withdrawn the rim of the collar is seen as a cherry red protrusion into the bladder. When the objective lens is directed dorsalward, the rim is seen in the upper edge of the field of vision, and the shadow cast by the rim is seen on the trigone in the center portion of the visual field (Fig. 170 C, p. 225). As the cystoscope is rotated, the same intravesical protrusion can be seen entirely around the vesical orifice. The appearance of muscular hypertonia and of fibrous contracture of the bladder neck is similar to that of fibromuscular hyperplasia, but intravesical protrusion is not so great. In fibrous contracture there may be no intravesical protrusion, but the mucosa is lighter in color and less vascular than normal and the vesical orifice fits tightly around the cystoscope sheath.

3. Appearance from the prostatic urethra

The external or distal aspect of the vesical orifice can be seen best from the prostatic urethra through a directly forward vision instrument, either with or

Fig. 102. Bladder neck contracture following transurethral prostatic resection. View through foroblique lens from prostatic urethra. Elevation at lower edge of field is tip of verumontanum

without an optical system. The obliquely forward telescopes also give a good view from the prostatic urethra. From this viewpoint the contracted bladder neck is almost closed. Even though the distending fluid is flowing through the instrument into the bladder, only a small aperture is visible. The mucosa may be deeper red than normal, due to inflammation. When fibrous contracture follows adequate removal of the prostate, the prostatic urethra is concave and the contracted vesical orifice can be seen as a small aperture at the far end of the prostatic urethra (Fig. 102).

II. Intrusion into the bladder neck and prostatic urethra

1. Median bar

This is an elevation of the dorsal lip of the bladder neck with very little or no constriction of the lateral and ventral portions of the vesical outlet. It is most often due to fibromuscular hyperplasia. Sometimes muscular hypertonia or fibrosis may involve the dorsal lip more markedly than the rest of the bladder neck and exist as more of a bar across the dorsal portion than as a contracture surrounding the orifice. The cystoscopic appearance is similar to that of contracture except that the dorsal segment only is involved. An aid in making the diagnosis and in estimating the thickness of the tissue is to palpate through the anterior rectal wall while the cystoscope is in place. A transverse ridge is felt between the cystoscope and the examining finger in the rectum (Fig. 103). This same maneuver helps to diagnose contracture of the vesical orifice due to fibromuscular hyperplasia. Median bar is frequently associated with hypertrophy of the interureteric ridge (p. 82). The ridge may curve downward so far that it

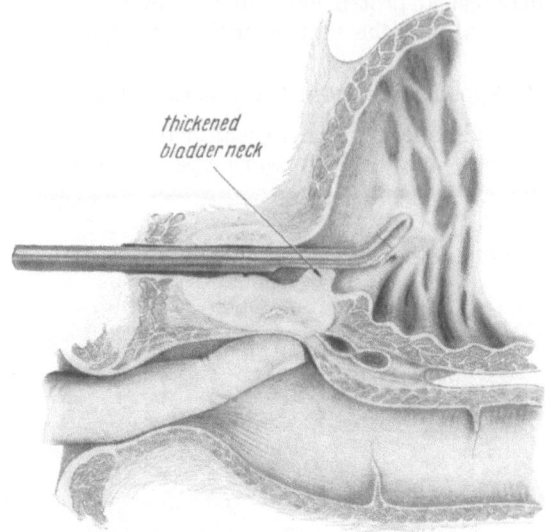

thickened
bladder neck

Fig. 103. Palpation of bladder neck between cystoscope and examining finger in rectum. "Median bar" or fibromuscular contracture can be felt as thickened tissue extending transversly across cystoscope sheath

joins with the median bar, causing an elevation of the entire trigone. Lich et al have reported this abnormality in patients who have posterolateral spinal cord disease; it is referred to as "straight line defect".

2. Median lobe prostatic hypertrophy

Enlargement of the middle lobe of the prostate is a common cause of dorsal intrusion into the vesical orifice and prostatic urethra. A marked elevation of the dorsal lip of the bladder neck is seen through the right angle and obliquely forward lens systems. The prostatic border is convex and covers part or all of the trigone. There is a groove or V-shaped depression or cleft between the lateral margin of the middle lobe and the medial margin of the lateral lobe on each side (Fig. 104). The intrusion of the middle lobe extends outward, elevating the floor of the prostatic urethra for a varying distance, sometimes as far as the verumontanum.

3. Lateral lobe prostatic hypertrophy

Hypertrophied lateral lobes of the prostate intrude into the vesical orifice and prostatic urethra. The smooth convex prostatic border of the intravesical intrusion is best seen through the obliquely forward or directly forward optical systems. The direct vision cystoscope also gives an excellent view of prostatic intrusion. The prostatic border of the intravesical intrusion may appear near the ureteral orifice as observed through the foroblique lens (Fig. 105). As the cystoscope is withdrawn into the urethra, the convex surfaces of the lateral lobes are seen to approximate each other in the midline when the prostate is larger than Grade I hypertrophy (Fig. 106). When the visual field is directed ventralward, the deep cleft or inverted "V" formed by the junction of the hypertrophied lateral lobes is seen at the 12 o'clock position on the bladder neck (Fig. 107 A). To complete

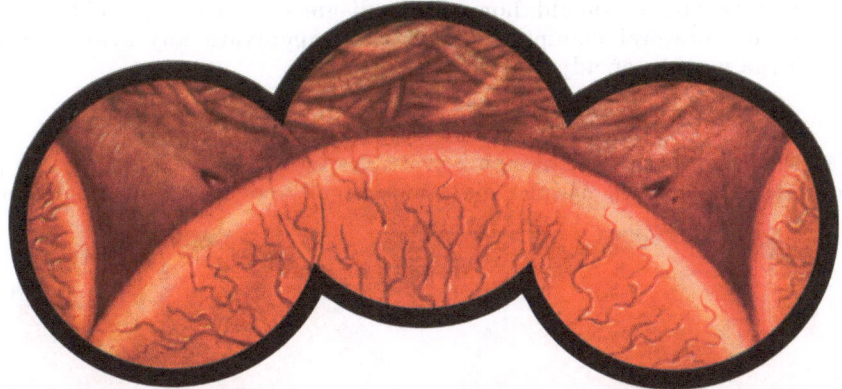

Fig. 104. Middle lobe prostatic hypertrophy. Intravesical protrusion covers most of trigone. Ureteral orifices can barely be seen above it. Note trabeculation above trigone and cleft separating middle lobe from lateral lobes on each side. View through right angle telescope; composite of three visual fields

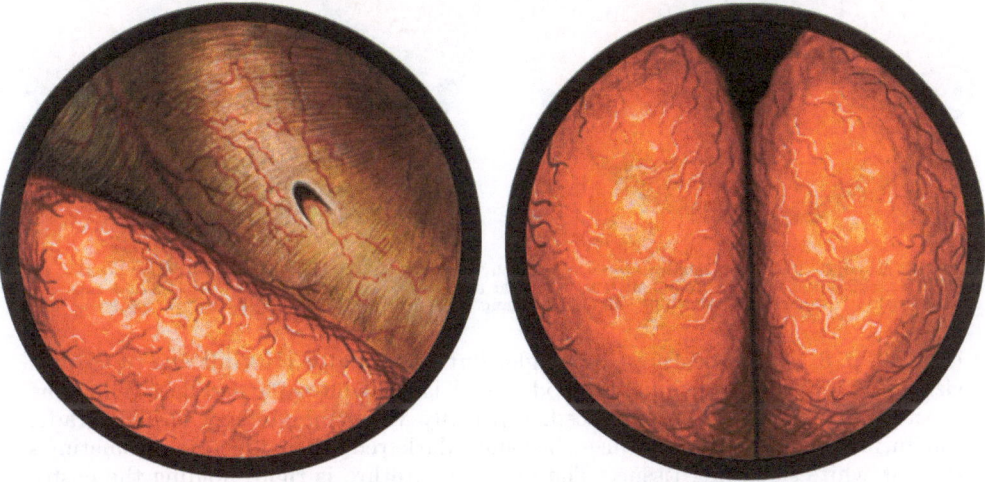

Fig. 105. Lateral lobe prostatic hypertrophy. Convex border of intravesical intrusion of right lateral lobe as seen through foroblique lens. Right ureteral meatus just above prostatic border. Close view

Fig. 106. Intraurethral lateral lobe prostatic hypertrophy (grade II in size) as seen through foroblique lens at a point midway between vesical orifice and verumontanum

the examination of the prostatic urethra the field of vision is directed dorsalward again and the cystoscope withdrawn until the verumontanum comes into view at the lower edge of the field. The intruding lateral lobes are seen to hang over the verumontanum (Fig. 107 B) and, as the cystoscope is withdrawn a little farther, to flatten out into the small tunnel of the membranous urethra.

4. Ventral lobe prostatic hypertrophy

This rarely occurs. When it is found, the appearance is similar to middle lobe enlargement except that the intrusion is on the ventral rather than on the dorsal portion of the prostatic urethra.

5. Prostatic abscess

This may cause a smooth regular intrusion into the prostatic urethra similar to lateral lobe prostatic hypertrophy. Abscess commonly causes the urethral wall to bulge on one side only, whereas hypertrophy usually involves both lateral lobes. Prostatic abscess should, however, be diagnosed without the aid of cystoscopy, because urethral manipulation tends to aggravate any acute infection involving the urethra or adjacent tissues.

6. Carcinoma of the prostate

Far advanced prostatic carcinoma protrudes into the vesical orifice and prostatic urethra as smooth nodular elevations usually covered by normal mucosa

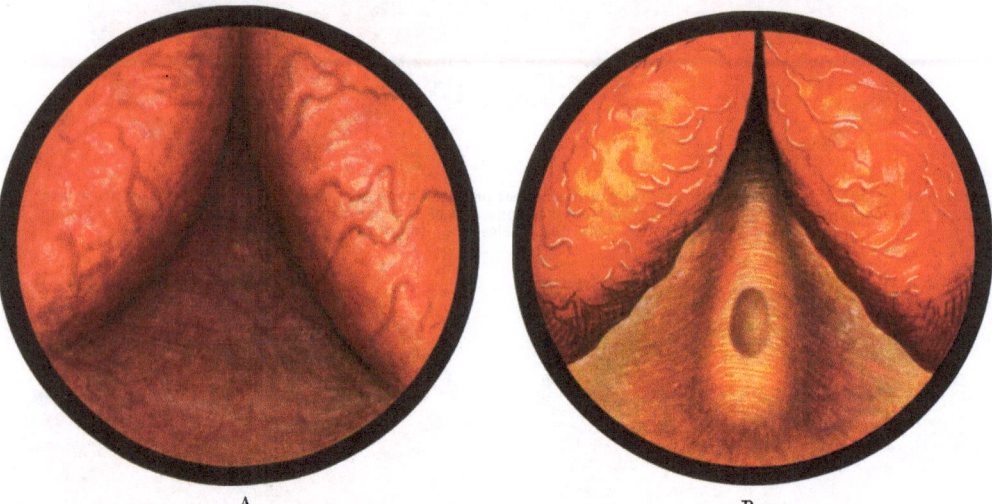

A   B

Fig. 107 A and B. A Lateral lobe prostatic hypertrophy. Junction of lateral lobes at 12 o'clock position on bladder neck. Vision through foroblique lens with field of vision directed ventrally B Lateral lobe prostatic hypertrophy. Intraurethral intruding lateral lobes hang over verumontanum. View through foroblique lens

(Fig. 59, p. 81). Nodules may be single, but are usually multiple and of various sizes. Sometimes a prominent blood vessel is seen on each nodule. When the prostatic neoplasm is far advanced, especially when malignancy is high grade, the surface of the protuberances becomes dark red, then ulcerates, sometimes there is white sloughing tissue. The prostatic urethra is rigid, holding the cystoscope firmly as in a vise (SMITH).

7. Carcinoma primary in the mucosa

The irregular protrusions of carcinoma arising from the mucosa of the prostatic urethra or bladder neck, or extending into them from a bladder neoplasm, have the same appearance as those seen in the bladder. It is not possible to obtain a distant cystoscopic view of the tumor when it is in the prostatic urethra, but that portion which protrudes into the bladder neck can be seen from a distance through the retrograde lens.

8. Polypi and cysts

Small polypi are sometimes seen in the prostatic urethra and bladder neck in men, but are much more frequently observed in women. Large polypi or cysts may present an obstruction to urinary outflow. They are visualized through the direct vision optical system as elongated projections or rounded protrusions covered by normal mucosa which has the appearance of being thinner than that on the bladder and prostatic urethral walls (Fig. 108) (DAVIES).

III. Rigidity of the prostatic urethra and bladder neck

This is usually due to prostatic carcinoma. An invasive neoplasm arising in the bladder or in the mucosa of the prostatic urethra may also make the area hard and rigid. Fibrosis causes rigidity but less marked than that due to invasive neoplasm.

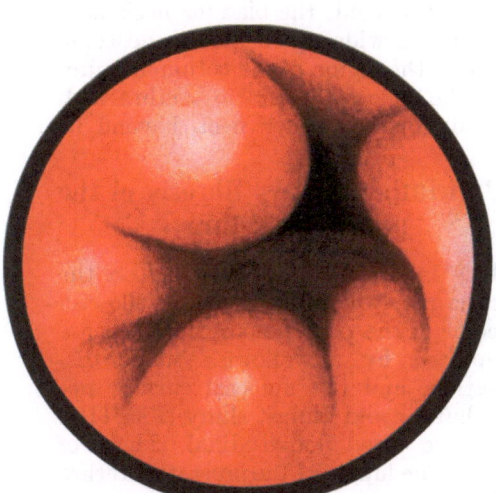 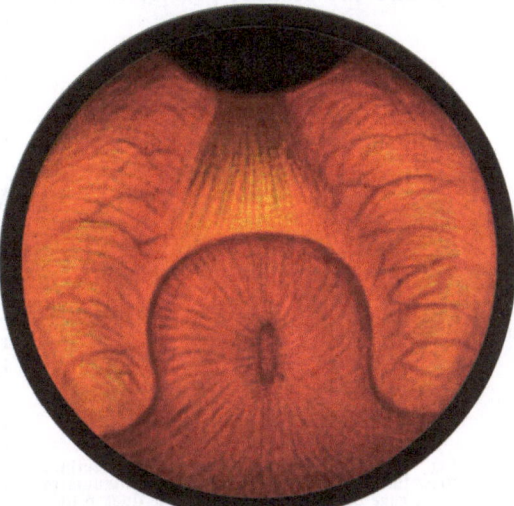

Fig. 108. Numerous large polypi in prostatic urethra of a man 19 years old. Urinary obstructive symptoms had been present for 9 years. View through direct vision optical system. Bladder cavity is seen beyond intruding polypi

Fig. 109. Abnormally dilated bladder neck due to spinal cord lesion. View through right angle telescope. Oberve verumontanum in foreground (lower edge of field), bladder cavity in distance (upper edge of field)

IV. Abnormal dilatation of the vesical orifice

This is best observed from the prostatic urethra through a directly forward vision telescope. The direct vision endoscope without a lens system and the obliquely forward vision telescope are also useful for this purpose. The normal bladder neck, as seen from the prostatic urethra through a direct vision instrument, remains slightly open when dilating fluid is being run in through the sheath. When the orifice is abnormally dilated its circumference is as great or greater than that of the prostatic urethra; the black cavity of the bladder can be seen beyond the dilated vesical orifice. These changes are sometimes called SCHROMM's phenomenon (NATANSON). Even through the right angle lens both the verumontanum and the edge of the bladder neck can often be seen in one visual field (Fig. 109), because the walls of the dilated posterior urethra are farther away from the objective lens, thus permitting a more distant view of the relaxed bladder neck.

1. Congenital defects and neurogenic lesions

The endoscopic appearance of bladder neck dilatation due to congenital defect and to central nerve lesion is the same. Hypertrophy of the vesical neck, which succeeds hypertonia at a certain stage, is particularly frequent in the course of neurologic disorders (TRUC, GUILLAUME and BAUMEL).

2. Urinary obstruction

Increased diameter of the vesical orifice secondary to urinary obstruction distal to it, such as congenital valves of the prostatic urethra or stricture of the membranous or anterior urethra, also has the same endoscopic appearance as dilatation due to a neurologic lesion.

3. Prostatic adenoma; postoperative

A large adenoma of the prostate which has extended through the bladder neck up into the bladder greatly dilates the vesical orifice. Even after the adenoma has been removed, the bladder neck usually remains wide open. The endoscopic view from the region of the verumontanum shows the concave prostatic urethra with the wide open bladder neck beyond it.

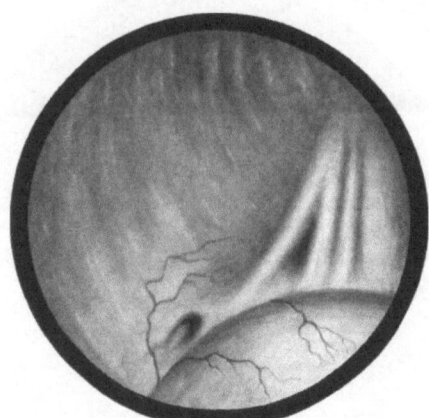

Fig. 110. Fibrosis of floor of prostatic urethra. View through foroblique lens. Verumontanum in lower edge of field, dilated prostatic duct orifice just above it

V. Post inflammatory fibrosis of the prostatic urethra

This is usually more marked on the floor than elsewhere. The normal yellowish red color of the posterior urethral mucosa is replaced by a yellowish white discoloration, and the mucosal surface appears harder and almost shiny (Fig. 110). There are often longitudinal ridges of fibrosis. The lumen of the prostatic urethra tends to remain rigidly open.

VI. Dilatation of prostatic duct orifices

Dilatation of these orifices, which are scattered irregularly along the floor of the urethra, is often seen to accompany fibrotic changes. There are usually several small orifices, which may remain open or which may open only when distending fluid is flowing through the urethra. Sometimes pus can be seen to exude from the dilated ducts; it retracts into the orifice when irrigating fluid is flowing into the bladder, then emerges again when the fluid is shut off. Occasionally a prostatic calculus can be visualized through a dilated duct orifice.

Orifice of prostatic diverticulum or abscess

Sometimes a larger opening is seen in the prostatic urethra, usually on the floor, which opens into a diverticulum or old abscess cavity (DE LA PENA).

VII. Abnormalities of the verumontanum

1. Congestion and infection

A slightly enlarged verumontanum which is deeper red than normal is evidence of congestion or of infection. The urethral mucosa is also a deeper red and bleeds easily. Congestion is usually seen in patients who have had excessive sexual activity.

a) **Granulation tissue.** When congestion or infection has persisted for several months, granulation tissue grows and the verumontanum becomes a deeper red and bleeds more easily.

2. Enlargement

A greatly enlarged verumontanum occurs as a congenital malformation and may be large enough to obstruct urinary outflow (EMMETT, BALDRIDGE). It may be difficult to visualize a greatly enlarged verumontanum through the endoscope because, being confined within a small urethra, it is necessarily so close to the objective lens that the visual field is filled with a red mass which cannot be identified. The enlarged verumontanum may be smooth and regular or may be irregular due to polypoid growths which cover it. When a transurethral prostatic resection has been adequately performed, the prostatic tissue is well removed from the area surrounding the verumontanum except at its distal edge. This allows the prostatic urethra to collapse, elevating the verumontanum closer to the resecto-scope lens so that it appears much larger than before resection — possibly larger than normal. Sometimes seminal fluid can be seen spurting from the ejaculatory duct orifices while tissue is being resected near the verumontanum.

VIII. Abnormal red discoloration of the mucosa of the prostatic urethra

This occurs in the presence of acute or subacute posterior urethritis from any cause. Mild edema similar to that found in the bladder is sometimes seen when the inflammation is severe.

IX. Calculi and foreign bodies in the prostatic urethra

These have the same appearance as in the bladder; the view, however, is always very close. It is not possible, within the narrow confines of the prostatic urethra, to remove the objective lens far enough from the object to obtain even a slightly distant view.

X. Posterior urethral valves

Congenital valves in the prostatic urethra may be the cause of urinary obstruction in infants or small boys (KEARNS). They are seen best through an obliquely forward close vision lens system. They appear as thin septa of normal mucosa on the floor of the urethra extending from the verumontanum to the side of the urethral wall. Several types have been described: those extending inward, those transversing lateralward and those going outward from the verumontanum to the side of the urethra. Sometimes only a single valve exists on one side, but those which cause symptoms and are usually discovered are bilateral, single or multiple. Dilation of the vesical orifices due to back pressure of urine from the obstructing valves is often found.

XI. Interpretation of findings in the prostatic urethra

Endoscopic examination of the prostatic urethra may lead to misinterpretation of the findings.

1. Close view

The view is always close, which makes all objects seem larger than they actually are — especially when seen through a lens system. A more accurate estimate of the size of the object can be obtained when it is viewed through the direct vision endoscope. When the cystoscopist becomes accustomed to the interpretation of findings as seen through a cystoscopic telescope and takes into account the optical system he is using at the time, he can accurately judge the actual size of the object and can interpret his findings correctly. For example, when a right angle lens is used, the object is very close to the objective lens even though distending fluid is being run in through the endoscope. Sometimes it is so close that no distinguishing features are visualized by which identification can be made. Therefore, an obliquely forward or straight forward optical system is preferable in most instances to the right angle lens for examining the prostatic urethra. A more distant view of the object can usually be obtained through the direct forward optical system than through the obliquely forward one.

2. Distortion due to passage of the endoscope

Another factor which may lead to misinterpretation, especially when estimating the size of the prostate, is the distortion which is caused by the presence of the endoscope; the encroaching lobes are flattened and the walls of the prostatic urethra are straightened (STEGEMAN). The smaller the calibre of the instrument, the less distortion there is. The endoscopist must, therefore, take into account the size of the instrument he is using when he interprets his endoscopic findings in the prostatic urethra.

XII. Cystoscopy for diagnosis of prostatism

There is a difference of opinion among urologists concerning the advisability of routine cystoscopic examination when symptoms of prostatism are present. Some believe that sufficient valuable information is obtained to make it worth while to perform cystoscopy in nearly every case (LAZARUS). Others hold that if a reasonably definite diagnosis can be made without cystoscopy, it is preferable to avoid it in these patients. Regardless of the gentleness and ease with which a cystoscopic examination is made, it always elicits some reaction: Obstructive symptoms are nearly always temporarily increased, and sometimes complete urinary retention and/or high fever may occur following instrumentation. Unless a cystoscopic examination is necessary to make a diagnosis of prostatism it is better omitted. Although it is true that a more exact estimate of the size of the gland can be obtained by cystoscopic examination, the experienced urologic surgeon who wishes to avoid resection in large glands will in any case reserve these for an open approach.

Chapter XII

Abnormalities of the bladder neck and urethra in the female

Careful endoscopic examination of the female bladder neck and urethra is very important when an attempt is being made to find the cause of bladder symptoms in women. This examination is often neglected; when the cystoscopist finds a normal bladder, he may conclude that the patient's symptoms are due to an extraurinary disorder or are functional. Examination of the bladder neck and urethra may disclose a cause for the symptoms.

I. Contracture of the vesical orifice

This condition in women is similar to that in men. The cystoscopic appearance at the bladder neck through the right angle or obliquely forward telescopes and the retrograde view from within the bladder are also similar to those in the male. The collar-like intravesical protrusion of the entire circumference of the vesical neck is best seen through the retrograde lens (Fig. 170C, p. 225). Elevation of the dorsal lip of the bladder neck is usually observed associated with cystocele; the floor of the bladder is depressed below the vesical neck (NELSON, BARNES, BERGMAN and HADLEY). The view inward toward the bladder through the objective lens positioned in the urethra, however, is not of value in making a diagnosis of bladder neck contracture, because the female urethra does not distend in the form of a fusiform cavity as does the male urethra, therefore, the contracted bladder neck is not always evident when viewed from the urethral aspect.

Contracture of the vesical orifice may or may not cause urinary obstruction in women. Obstruction is more likely to occur when there is protrusion of the vesical neck into the bladder than when there is only elevation of the dorsal lip. Trabeculation is usually found in the fundus of the bladder when there is obstructed urinary outflow at the bladder neck or in the urethra.

II. Dilatation of the vesical orifice

When this is definite enough to be seen through the endoscope, it is most likely due to a central nerve lesion. Trauma, particularly as a result of prolonged difficult labor, may produce a relaxed bladder neck. Dilation is more difficult to diagnose in the female than in the male because the straight forward view of the vesical orifice from the female urethra is obscured to some extent by the urethral walls which tend to collapse into the visual field. Sometimes, however, the vesical orifice can be seen through both the right angle and the forward vision lenses to be definitely dilated.

III. Increased curvature of the urethra

The curve of the female urethra is normally slightly concave ventrally. This curve is often increased — sometimes so much so that it is angulated. Elevation of the dorsal lip of the bladder neck increases the ventral concavity of the urethra. In the presence of cystocele the curve is still further increased. Therefore, when endoscopes or other instruments are passed through the urethra, it is necessary to point the tip first dorsalward then ventralward in order to follow the increased curvature. There may be some sacculation in the floor of the urethra at the bottom of the curve.

IV. Normal urethral mucosa

The normal female urethral mucosa as seen through a direct vision optical system is a deeper yellowish red color than the normal bladder mucosa, and the blood vessels are not as prominent. Slight mottling is usually seen and there are small areas of slightly deeper red. The instrument holds the urethral walls apart and the inward view shows them to converge and obliterate the lumen at a point about 1 cm. proximal to the inner end of the sheath (Fig. 111); the larger the sheath, the greater the amount of urethra which can be seen in one visual field.

V. Fibrosis and stricture of the urethra

Fibrosis of the female urethra changes the normal yellowish red color of the mucosa to light yellowish pink or white and reduces the calibre of its lumen. Fibrosis may be most marked at the bladder neck, causing fibrous contracture of the vesical oritice; usually, however, it involves almost the entire urethra. Sometimes the constriction is sufficient to cause retardation of urinary stream or even complete urinary obstruction. Fibrosis of the female urethra is called

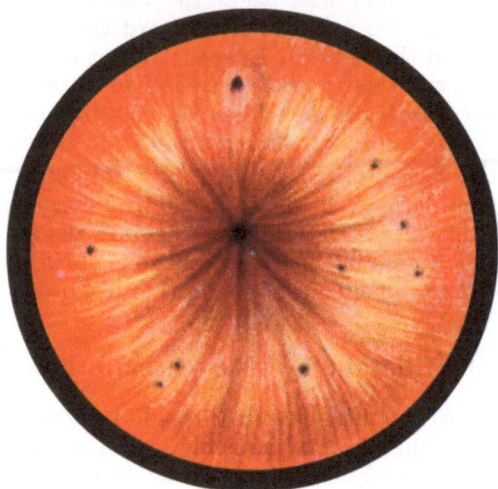

Fig. 111. Normal female urethra as seen through direct vision optical system using size 24 Fr. sheath. Distending fluid is flowing in. Several periurethral duct orifices are seen.

stricture by some urologists when there is only minimal narrowing of the urethral lumen; by others, the term is applied only when there is sufficient constriction to cause urinary obstructive symptoms.

VI. Increased redness of the urethral mucosa

This is usually due to infection. Patchy areas of red discoloration occur more often than a generalized increased redness, throughout the mucosa. In acute infection, such as gonococcal urethritis, the inflammatory red discoloration involves the entire mucosa, but this is seldom seen through the endoscope because instrumentation is contraindicated in the presence of acute infection. Patchy areas of red discoloration and submucosal hemorrhage may result from trauma. Granulation tissue is often a cause or patches of red discoloration and may not be profuse enough to cause elevation of the mucosa.

VII. Irregularities at the bladder neck

Irregularities in the bladder neck of women are common. In fact, a certain amount of irregularity is considered normal (Fig. 29, p. 56). When these irregularities cause a protrusion of more than one-half millimeter into the vesical orifice, they may be considered abnormal. It must be remembered that the objective lens is always very close to the bladder neck, except when the retrograde telescope is used, and that any irregularities may appear to be much larger than they actually are.

1. Edema

Multiple lobulated smooth irregularities are characteristic of edema. Their appearance is the same in the bladder neck as the close view of edema in the bladder.

2. Granulations

These are a deeper red and the surface is finely irregular or granular instead of smooth. Exuberant granulations may be mistaken for neoplasm or vice versa.

3. Neoplasm

This usually protrudes farther into the vesical orifice and is more irregular than granulation. The appearance of neoplasm in the bladder neck is the same as the close view of it in the bladder.

4. Polypoid growths

These are very commonly found on the bladder neck in women. They can be seen through the direct vision telescope or through any lens system, although the direct vision shows them best. Polypi are best visualized by first withdrawing

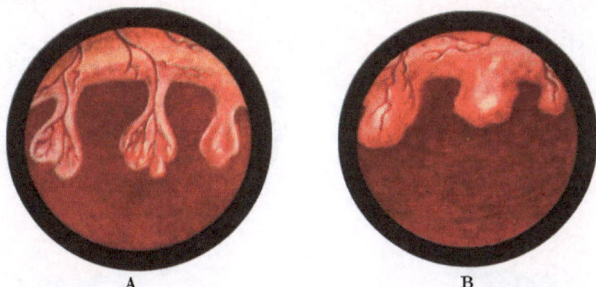

Fig. 112 A and B. Polypoid growths at ventral lip of vesical neck. A Pedunculated vascular and avascular type. B Sessile polypi and more solid polypoid growth or granuloma

the cystoscope until the objective lens is within the urethra just outside the vesical orifice, then advancing it again two or three millimeters. This maneuver displaces the tip ends of the growths inward so that they can be visualized against the contrasting more distant darker bladder wall. Polypoid growths are smooth, regular, oval, rounded or fingerlike protrusions or projections into the vesical orifice (Fig. 112). Their attachment to the mucosa may vary from a narrow pedicle to a broad base. Their most frequent location is at the ventral lip of the bladder neck but they may occur any place on its circumference. The mucosal covering is thin and sometimes appears almost transparent. Polypi are light yellowish pink — often considerably lighter than the surrounding mucosa. Some are almost avascular, while in others a prominent vessel can be seen extending from the base to the tip of the growth. Sometimes a polypoid growth appears more solid and is like a granuloma. Size varies from polypi which are almost too small to be detected to large ones which may entirely fill the outlet of the bladder. Polypoid growths are usually the result of mild inflammation continued over a long period of time. Sometimes, however, no cause can be found. Many produce no symptoms; others are a constant source of irritation.

VIII. Irregularities in the urethra

Irregularities in the urethra are similar to those found at the bladder neck, but occur much less frequently. Clear visualization is more difficult to obtain in the urethra than at the bladder neck. When the lesion intrudes into the vesical orifice, it stands out and is silhouetted against the darker cavity of the bladder. When it is located in the urethra, it may be flattened against the urethral wall or may be difficult to identify because the urethra tends to collapse even though distending fluid is flowing in through the endoscope. The larger sized sheaths help to keep the urethral lumen open.

1. Granulations

These are darker red than normal urethral mucosa and tend to occur in patches. Sometimes they are more extensive and cover most of the urethral walls. Occasion-

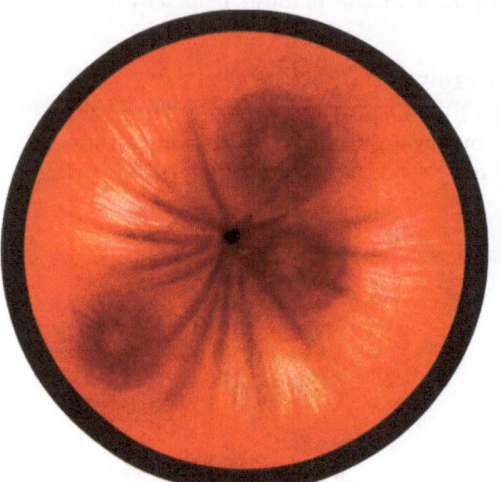

Fig. 113. Exuberant granulations in female urethra as seen through direct vision optical system

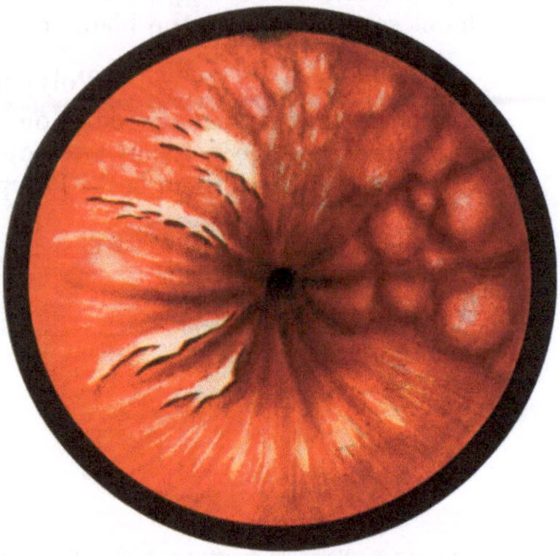

Fig. 114. Carcinoma of female urethra as seen through direct vision optical system. There are irregular areas of lighter colored tumor tissue and light streaks of slough

ally they may be exuberant enough to be slightly elevated above the normal contour (Fig. 113).

2. Neoplasm

Primary neoplasm of the female urethra is rare. By the time it begins to produce symptoms, it has usually constricted the urethral lumen so much that the passing of an endoscope through it is difficult if not impossible. The instrument may be passed as far as the obstruction, then the area of stenosis examined with the dilating fluid flowing; a fairly good view can be obtained unless the trauma of passing the instrument has started excessive hemorrhage. The surface of a urethral carcinoma is deeper red than normal, granular in appearance and protrudes irregularly into the urethral lumen (FAGAN and HERTIG) (Fig. 114). Sometimes lighter irregular areas of tumor or greyish white streaks of sloughing tissue are seen.

3. Polypoid growths

Urethral polypi are similar in appearance to those at the vesical neck, although they do not stand out as clearly and occur much less frequently. When they are located in the urethra, they are more likely to produce symptoms than when they are located on the rim of the bladder neck. Occasionally a urethral polyp may be long enough to project out through the meatus.

4. Sacculation

A sacculation or dilated area of the urethra distends easily when fluid flows in through the endoscope. The walls of the urethral lumen, instead of being in contact distal to the end of the instrument, are seen to separate, the lumen to expand, and the normal urethral folds to flatten out. The mucosa is usually a little lighter in color and has the appearance of being thinner than normal. RINGLEB has described a twin cystoscope for stereoscopic study.

IX. Diverticular orifice

The orifice of a urethral diverticulum is nearly always located on the floor of the urethra (COOK). When the opening is small it is difficult to identify because it is usually covered by a fold of mucosa. When it is more than two millimeters in diameter, it is visualized as a slightly irregular opening leading into a dark cavity (Fig. 115). There is often a surrounding area of fibrosis which is lighter in color and somewhat rigid in appearance.

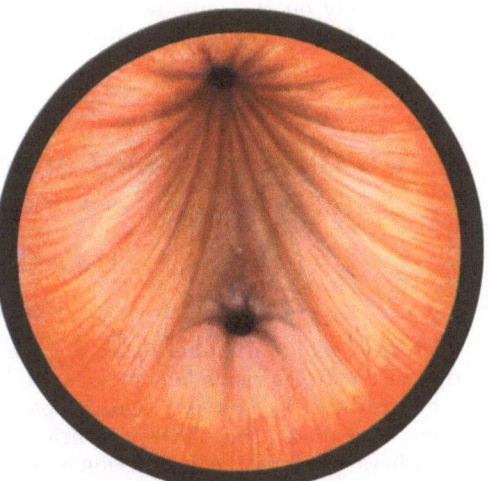

1. Calculus in diverticulum

When a calculus resides in the diverticular cavity, it may be dimly seen through the orifice.

2. Neoplasm in diverticulum

Rarely a tumor may occupy the diverticulum and a portion of it protrude out through the orifice.

Fig. 115. Orifice of a urethral diverticulum as seen through direct vision optical system. View is directed somewhat toward floor of urethra. Urethral lumen is seen in upper portion of field

X. Periurethral ducts

Periurethral ducts open into the urethra and can occasionally be identified. They are seen as small openings sometimes with a slightly elevated rim. When the periurethral gland is infected, there is a small area of reddened mucosa surrounding the orifice. If fibrosis of the mucosa is evident, the duct orifice remains open and the surrounding mucosa is smoother, flatter and lighter in color than normal. Most of them, however, are so small and so well covered by folds of mucosa that identification through the endoscope is not possible.

XI. Urethral meatus

The external centimeter of the urethra is best examined through the skenoscope which is a small, short, funnel shaped instrument with a fenestra on the side located about one-half centimeter from the tip. Some skenoscopes have a light carrier which places the bulb near the fenestra; others depend upon an external source of illumination. By the use of this instrument which is introduced into the urethral meatus for a distance of 1 to 2 cm., the orifices of SKENE's glands — one located on each side of the meatus and just within the meatal ring — can usually be seen. They are similar in appearance to duct orifices located deeper in the urethra. When the urethral meatus is fibrotic, the mucosa becomes slightly everted and the orifices of SKENE's glands are more likely to be located on the rim of the meatus or just outside it. Other irregularities of the urethra, especially exuberant granulation or solid polypoid growths, are of more frequent occurrence in this region than deeper in the urethra. A urethral caruncle, which appears at the meatus, may extend inward for a few millimeters and is seen through the skenoscope as a deep red granular area which protrudes slightly into the urethral lumen at this point. The growth of a urethral caruncle does not usually invade or destroy the identity of the urethral meatus ring. A prolapsed urethral mucosa, on the other hand, obliterates the ring because the mucosa is continuous with the mucosa of the introitus. Abnormalities of color and of contour of prolapsed mucosa occur entirely on the outside of the urethra, whereas most of the abnormalities of a caruncle are located just within the meatus.

Chapter XIII

Urethroscopy and miscellaneous endoscopic procedures

I. Urethroscopy

1. Urethroscopes

Urethroscopic instruments are described in Chapter I. The irrigating urethroscope with a direct vision lens such as the Lowsley instrument provides a good view of the urethra. Instruments can be passed through it for treatment of urethral lesions. Air distention of the urethra through the Swift Joly aero urethroscope and others of similar construction exposes a wide field of vision and provides good visualization (HARKNESS). Air embolism, however, is a hazard of this method of urethroscopy. Open tube urethroscopes are selected for the good visualization they provide of areas to be treated by endoscopic application of medicaments.

2. Technique of urethroscopy

The urethroscope with the obturator in place is passed into the bladder when a view of the posterior or membranous urethra is desired. When only the anterior urethra is to be examined, the instrument is passed to, but not through, the membranous urethra (external sphincter). The obturator is then removed and the urethroscope gradually withdrawn as the urethral walls are being examined. Water flows in during the examination when the irrigating instrument is in use, and air is pumped into the urethra through the aero urethroscope. Utilization of the open tube instrument requires a dry field; the water is aspirated with a long nozzled bulb syringe, then long cotton applicators are used to dry the field. Fulgurating electrodes, catheters and forceps are passed through the sheath for treatment. Silver nitrate stick, 10 per cent silver nitrate on an applicator and other medication can be used through the open tube instrument.

3. Normal urethra

a) **Prostatic.** The normal prostatic urethra is described in Chapter XI.

b) **Membranous.** In the male this is short and closes quickly as the urethroscope is being withdrawn through it. The mucosa can be seen when the urethral lumen is kept open by the endoscope; it is smooth and somewhat darker red than the remainder of the canal. With the objective end of the urethroscope in the membranous urethra, it is possible to obtain a distant view of the membranous urethra. When the bladder is distended and making an effort to empty itself through the urethra, the membranous portion remains open. The external sphincter is normally closed except during the act of micturition. While it is closed the urethral lumen is puckered, the folds of mucosa converging at a central point (Fig. 116).

c) **Bulbous.** The lumen of the bulbus urethra is considerably wider than any other portion of the channel, and when distending fluid is being run in through the urethroscope it appears to be a sacculation (Fig. 117). Sometimes a few small periurethral duct orifices can be seen on its floor. When the bulbous urethra is not being distended

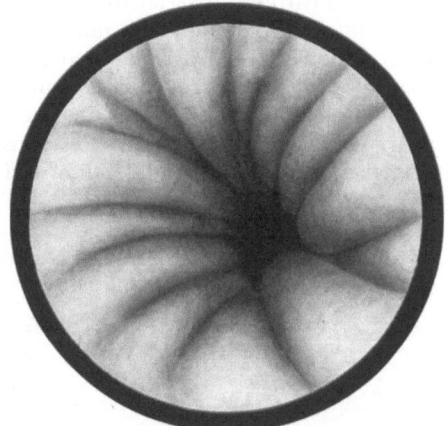

Fig. 116. External sphincter (closed) as seen through a direct vision lens from the bulbous urethra

with fluid, the mucosa collapses into folds which converge to a crescent shaped or transverse closed orifice.

d) **Penile.** The urethral mucosa in this section is smooth when distended, and the lumen narrows to a slit as the urethroscope is slowly withdrawn. Periurethral duct orifices are scattered irregularly along its walls, chiefly on the floor (ventral surface). The lumen of the navicular portion, which is the distal centimeter of the penile urethra, is a little larger and the mucosa has a firmer appearance.

4. Abnormal urethral contour

a) **Constriction.** Stenosis of the urethral lumen is common, due to inflammatory stricture. The mucosa of the strictured area is light pinkish yellow, avascular, slightly irregular and firm. The fibrosis may extend for varying distances inward or outward from the stenosed portion of the lumen. Congenital stricture of the urethra is rare except at its two extremities —

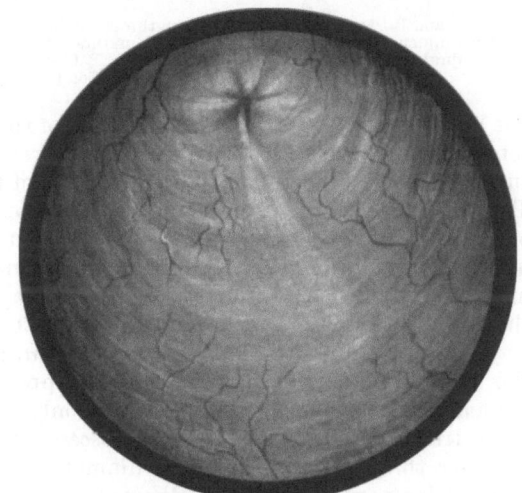

Fig. 117. Normal bulbous urethra as seen through a direct vision lens irrigating urethroscope while distending fluid is flowing in. Puckered mucosa at upper edge of visual field is a distant view of closed external sphincter

the meatus and the vesical orifice. The stenosed area is slightly lighter in color than normal, and is smooth and symmetrical. Constriction due to carcinoma is granular in appearance, is usually a darker red color than the normal mucosa and bleeds easily when an attempt is made to pass an instrument. The irregularities characteristic of neoplasm of the bladder can sometimes be seen.

b) Depression, sacculation and dilatation. In the anterior urethra these may be sequellae of trauma followed by infection, or may result from straining to force urine through a stricture. The dilated area is easily recognized through the irrigating urethroscope; the mucosal folds are flattened out by the distending fluid. When the open tube endoscope is used the pliable urethral walls are collapsed, and the large folds of mucosa are seen.

The lumen of the membranous urethra may be wider than normal, may remain open in some cases of neurogenic vesical dysfunction associated with urinary incontinence and no residual urine. When the mucosa of the membranous urethra

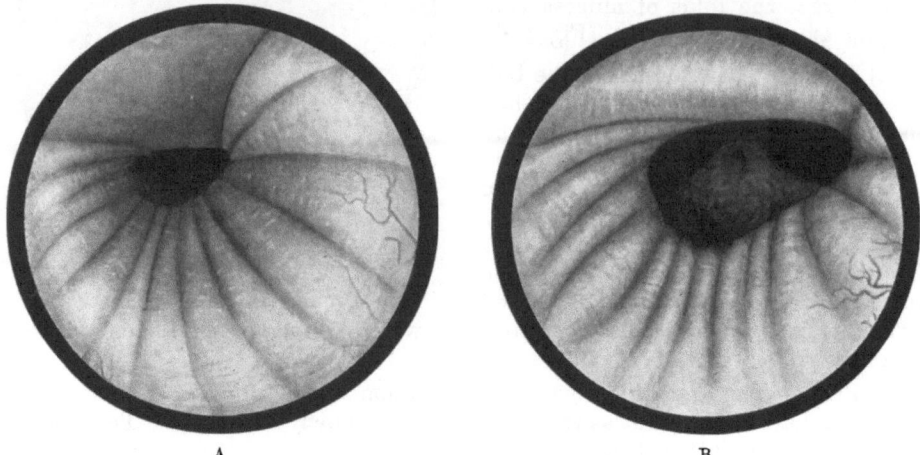

A   B

Fig. 118 A and B. External sphincter as seen through a direct vision lens from bulbous urethra, showing fibrosis following injury to the mucosa of membranous urethra. A loopful of tissue was inadvertently removed during transurethral prostatic resection. A Distant view. B Closer view; verumontanum is dimly seen through the orifice. Compare fig. 116

has been stripped out during enucleation of the prostate or has been inadvertently removed with the resectoscope loop, the resulting fibrosis causes a flat fibrous surface involving part of the circumference of the membranous urethra (Fig. 118). The stiff area of cicatrix prevents snug approximation of the mucosa, and leakage of urine is prone to occur. When more extensive fibrosis of the membranous urethra exists, and when prostatic carcinoma invades it, the resulting rigidity may hold it open continuously. Fibrosis may cause rigidity at any point along the urethral mucosa and distort the normal circular contour of the lumen.

c) Intrusions. Intrusions into the urethral lumen may be recognized through the urethroscope. *Edema* has the same appearance as when seen in the bladder. *Granulations* are deep red and flat or only slightly intrusive; they bleed easily when traumatized. A *periurethral abscess* presses on the urethra; if the abscess is larger than one centimeter in diameter, it causes a regular curved intrusion covered by normal mucosa. *Polypoid growths* are similar to those found at the bladder neck (p. 131). *Papillary tumors* of the urethra appear the same as the close vision view of the same tumor in the bladder (p. 101). *Carcinoma* is rare; when it occurs it usually intrudes into the urethra. It is often difficult to identify because it constricts the urethra to such an extent that only the external or distal aspect is visible. The tissue is irregular; it may have the same appearance as bladder tumor tissue, but its color is more often light pink or almost white, which makes it difficult to differentiate from stricture. Bleeding often beclouds the visual field. When the urethral constriction is tight, the distending fluid will not flow through it to keep the field of vision clear.

5. Abnormal color of the urethral mucosa

a) **Increased redness.** *Granulation* of the mucosa is deep red, and is usually scattered irregularly throughout the entire urethra or a portion of it. Sometimes elevated granulomas are found but usually the granulations are flat. *Acute inflammation* is deep red in color and is usually accompanied by some edema. This is rarely seen through a urethroscope, however, for acute inflammation of the urethral mucosa contraindicates all instrumentation including urethroscopy. Lighter red patches of *subacute or chronic inflammation* may be found scattered irregularly over the urethral mucosa. This occurs more commonly around the periurethral duct orifices. *Ulceration* of the urethral mucosa is rare and is similar in appearance to the close view of bladder ulcers. The base is deep red and sometimes the edges are undermined. Chancre and chancroid have a granular base, are often covered by mucopurulent exudate and are difficult to differentiate from other forms of ulceration.

b) **White or light colored areas.** Mucosa which has undergone *fibrosis* is lighter in color than normal and may be almost white. HARKNESS reports that patches of *Leukoplakia* not associated with stricture may occur in the penile urethra. Sometimes *infiltrating neoplasm* appears light pink or white. *Mucopurulent exudate* is light pink in color; it is seldom seen because it occurs in acute urethritis in the presence of which intrumentation is infrequently performed. Moreover, the exudate may be removed by the instrument as it is passed. *Tuberculosis* of the urethra is rare. It may cause color and contour abnormalities similar to those found in the bladder[1].

6. Orifices

a) **Urethral diverticulae.** In the male urethra these orifices are usually easily seen through the urethroscope. They are small round openings varying in size from one-half to two millimeters in diameter, and are most frequently located on the floor. In the female urethra, however, they are not so easily visualized (p. 133).

b) **Periurethral gland duct.** These orifices are often visualized and may be enlarged up to one millimeter in diameter. In the presence of subacute infection, they are surrounded by an area of red discoloration due to inflammation. When chronically infected they are rigidly open, and pus can often be seen to exude through them. The surrounding mucosa is often light in color due to fibrosis.

c) **Ectopic ureteral orifice.** This may be located on the floor of the prostatic urethra in the male (p. 60), or at any point along the floor in the female urethra (p. 60). It does not occur in the male anterior urethra.

II. Miscellaneous diagnostic endoscopy

1. Endoscopy of the intestinal bladder

Endoscopic examination of the rectum or other portion of the intestine following uretero-intestinal anastomosis may be indicated. KLEIMAN has designed a special sigmo-uretero stomoscope for this purpose. The routine cystoscope can, however, be used in most cases. The usual purpose of such an examination is to identify the ureteral orifices and sometimes to pass a catheter through it. Following mucosal anastomosis, the ureteral orifices are seen as round openings somewhat similar to "golf hole" ureter. Folds of intestinal mucosa often cover

[1] LAZARUS and ROSENTHAL; SEELIG.

the orifices, obscuring them from view. Distention of the rectum to smooth out the folds may be difficult to achieve because the distending fluid runs up into the colon. An isolated rectal or other intestinal bladder distends more easily. In cases in which a portion of the trigone, including the ureteral meatus, has been sutured to the intestinal mucosa, the orifice is usually located on a rounded protrusion (Fig. 119). Ejection of indigocarmine from the orifice helps to identify it. A catheter may be passed up the ureter unless there is excessive angulation of the lower ureteral end.

2. Endoscopy of the kidney

This can be performed through a nephrostomy opening. A straight or slightly convex sheath is used, and direct vision, foroblique and right angle telescopes should be available. Recurrent renal calculi may thus be detected and sometimes removed endoscopically. Diagnosis of tumor in a kidney pelvis which has held a nephrostomy indwelling tube for a long period of time may be made by removing a piece of tumor tissue through an endoscope for biopsy study (LEADBETTER).

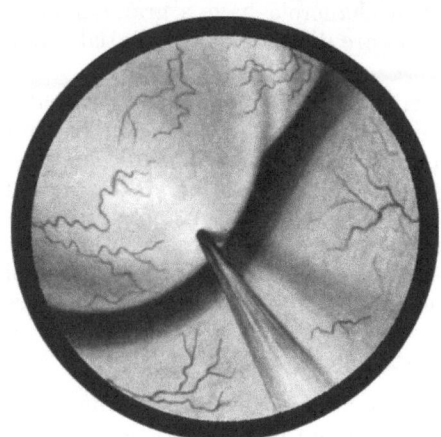

Fig. 119. Ureteral orifice ejecting indigocarmine. Each half of trigone, including ureteral meatus, had been transplanted into an isolated rectal bladder for treatment of extrophia vesicae

3. Endoscopy of the vagina

Vaginoscopy of infants and girls is easily performed by utilizing one of a number of endoscopes used in the urethra. The Butterfield instrument (p. 3) is made for use both as a urethroscope and as an infant vaginoscope. The open tube urethroscope and the Kelly cystoscope provide good vision of the cervix and vaginal walls. Medicinal applications can be made through these instruments in the same manner as they are made to the urethra: foreign bodies in the vagina can be identified and removed. The irrigating urethroscope or the panendoscope is preferable for endoscopy of the small vagina when a search is being made for an ectopic ureteral orifice or other small abnormalities or lesions which might be covered with mucosal folds; the vaginal cavity can be distended to smooth out the folds. By a close view of the vaginal walls through the endoscopic telescope a small ectopic ureteral oritice may be identified and sometimes catheterized. Ejection of indigocarmine through the orifice may help to locate it. Often, however, the segment of kidney above the orifice does not function sufficiently to excrete the dye.

Vaginoscopy may be done to discover the cause of persistent vaginal discharge. The mucosa in the presence of acute vaginitis is deep red in color and sometimes edematous; mucopurulent exudate may cover it in places. When the discharge originates from an extravaginal source through a sinus, it may be possible to force some out through the opening by pressure over the suprapubic region or with the finger in the rectum.

4. Intraperitoneal and gastric endoscopy

Before the development of special endoscopes for examination of the peritoneal cavity through the anterior abdominal wall, for survey of the pelvis through the vaginal cul-de-sac, and for visualization of the interior of other cavities, the cystoscope was sometimes used for these examinations. When these special instruments are not available, the cystoscope may still be used, and a diagnosis made or a piece of tissue removed for biopsy study. An unusual use for a cystoscope is to examine the interior of the stomach through a gastrostomy opening (HARA).

Part II. Endoscopic surgery

Chapter XIV

Miscellaneous endoscopic surgical procedures and treatments

I. Endoscopic ureteral treatment

1. Ureteral dilation

Ureteral dilation for the treatment of ureteritis and of stenosis of the ureteral lumen is practiced extensively by many urologists and is condemned by others. A middle road between the two extremes is the preferable course (HUNNER; MILLIN). When definite dilatation of the ureter and/or renal pelvis exists above a ureteral stenosis, progressive dilation of the ureter is indicated. Many patients present symptoms characteristic of obstruction to the flow of urine through the ureters, although there is no evidence of ureteral obstruction or of renal pelvic back pressure. A few will be helped by ureteral dilation, but it is difficult to predetermine which individuals these are. Reproduction of the pain by a catheter passed up the ureter or by distention of the renal pelvis with fluid injected through the catheter is believed by some to be an indication for treatment by ureteral dilation. Such treatments are best spaced at intervals of ten days to two weeks. More frequent application is likely to cause edema and excessive trauma. Large cystoscopes are available for use by urologists who practice dilation of the ureters to a large size (Chap. 1, p. 21). BRAASCH bulb catheters cause less trauma than dilating instruments which are a uniform diameter all along the shaft. The calibre is increased one or two sizes at each treatment; when the ureter does not dilate easily, it is necessary to repeat the same size for several treatments. Many urologists lavage the renal pelvis with 1:500 silver nitrate or other antiseptic solution injected through the catheter which is used to dilate the ureter.

2. Renal pelvic drainage by ureteral catheter

Indwelling ureteral catheters for continuous drainage of the renal pelvis may be indicated in cases of ureteral calculi and/or stenosis, especially when the kidney is infected. The largest size catheter which can be passed is inserted through the endoscope in order to obtain free drainage through it. Plastic polyethylene or other special catheters may be useful (VLIETSTRA).

II. Endoscopic manipulations for removal of ureteral calculi

Endoscopic manipulative procedures for removal of ureteral calculi are many and varied. It is beyond the scope of this volume to discuss the indications for the different types of treatment. Only the more commonly used endoscopic methods will be briefly mentioned.

The importance of avoiding trauma, especially when the stone is located in the upper two thirds of the ureter, cannot be overstressed (RUSCHE). Trauma to the ureter, including perforation, is a hazard which may result in serious complications (IWANO et al.; BENEVENTI).

1. Ureteral dilatation

a) Catheters. Simple dilatation of the ureter is preferable in most cases when the calculus is high. Repeated dilations at ten-day intervals with increasing sizes of catheters is applicable when the stone does not obstruct urinary outflow. Passing one or more catheters by the stone and leaving them in place for a few days dilates the ureter effectively and with little trauma (NATION). Multiple catheters, after having been passed by the stone, may be twisted and slowly

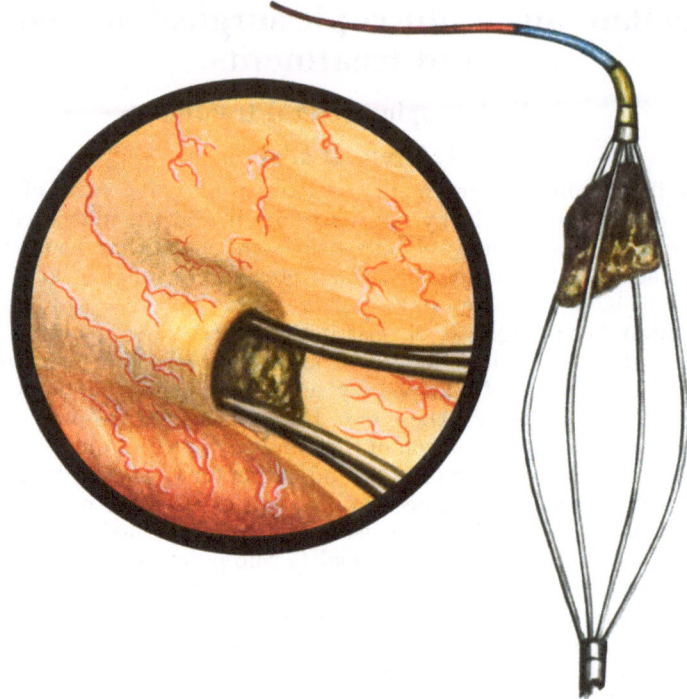

Fig. 120. Ureteral calculus being removed through meatus with Johnson wire basket. Cystoscopic view. Basket with calculus after removal

withdrawn; the calculus may be caught in the mesh of catheters and removed with them (ALYEA).

b) Bag distention. The ureter may be dilated with a distensible bag on the end of a catheter which is passed up the ureter (DOURMASHKIN).

2. Injection into ureter

Paraffin oil or other lubricants may be injected above the stone through a catheter following dilatation (WHITE). Other medicaments such as a smooth muscle stimulant to help propel the stone downward, or a relaxing agent to dilate the ureter below it have been injected into the ureter through the catheter.

3. Instruments for extraction of calculi

Calculi in the lower third of the ureter are more amenable to extraction by endoscopic manipulation than are those higher up.

a) Filiform and dental floss. A filiform catheter or a bougie with several strands of dental floss or other type of threads tied to the tip and extending along the shaft to the outside end, causes little trauma and may be successful in removing ureteral calculi[1].

b) Looped catheter. A loop of ureteral catheter (p. 25) may be manipulated to surround the stone, which is pulled down the ureter by gently applied traction[2]. ELLIK has used this method to extract stones from the ureter in children.

c) Corkscrew catheter. This may be used to extract a stone impacted in the intramural ureter (IMBERTS).

d) Wire basket (p. 25). This may be useful for extracting stones from the lower third of the ureter (JOHNSON; BROWNE). Its four springlike wires keep the spaces between them open, and as the basket passes by the stone, the stone falls between the wires and becomes wedged into the upper end of the basket as the instrument is withdrawn (Fig. 120). Preliminary dilatation of the ureter may facilitate passage of the basket up the ureter (ALCORN; GARVEY). A distensible bag located on the shaft of the dislodger below the basket is sometimes helpful (COUNCILL).

4. Reactions and care

Reactions of fever, pain in the kidney, hemorrhage and nausea and vomiting are common following endoscopic manipulation for removal of ureteral calculi. The same methods used for prevention and treatment of reactions following ureteral catheterization (p. 53) are applicable. The passing of a catheter up the ureter immediately after extraction of the calculus may prevent obstruction due to edema and thus eliminate one source of postmanipulation reaction. The catheter is left in place for 24 to 48 hours for continuous drainage of the renal pelvis (ALCORN). It is, however, often difficult to pass a catheter up the ureter following manipulation.

III. Ureteral meatotomy

1. For calculus

Endoscopic ureteral meatotomy may be required for removal of a calculus larger than 0.5 cm. in diameter which is impacted at the ureteral meatus. A knife electrode or a loop electrode used through the resectoscope and high frequency current can be used to incise the meatus down onto the stone[3]. When the location of the calculus is more than 0.5 cm. above the ureteral meatus, it is preferable to insert the tip of a bougie or catheter through the orifice and make the incision down onto it as outlined above. Sometimes a stone is engaged in a ureteral calculus basket and pulled down to the meatus but is too large to be extracted; a meatotomy can then be performed either with scissors passed through the same cystoscope, or with a narrow electrode and high frequency current. When this is done, great care must be exercised to avoid severing the wires of the basket with the electric current.

2. For stricture

Endoscopic ureteral meatotomy is often indicated to treat a meatal stricture (AINSWORTH-DAVIS). Congenital stenosis of the ureteral orifice is usually accompanied by dilation of the intramural portion of the ureter *(ureterocele)*. Meatotomy is indicated for most of these, but some can be progressively dilated.

---

[1] DAVIS; McKAY; MOORE.
[2] BALKUS; DARGET et al.; ELLIK.
[3] KIEFER; WINSBURY-WHITE; DAVIS et al.

Cystoscopic scissors which are sharp can sometimes be inserted into the pinpoint meatus; an incision about 1 cm. long is made. Scissors sharp enough and tight enough to cut satisfactorily are, however, not often found. A small bougie or catheter may be inserted into the meatus, and then a narrow wire or knife electrode pressed firmly onto the mucosa over the bougie. The current is applied for only a fraction of a second; longer application burns a wider area and results in excessive postoperative fibrosis. If the electrode does not cut entirely through the mucosa of the ureteral meatus, cystoscopic scissors can be used to complete the incision; the tissue cuts more easily and there is less bleeding when the line of incision has been lightly electrocoagulated. Ureterotomes of various kinds have been devised, some of which are useful. The roof of a ureterocele may be removed by a snare activated by high frequency current (FOLEY) or by the resectoscope loop (ADAMS). Inasmuch as the intramural ureter is dilated, stricture is not likely to follow incision or excision of a ureterocele. Endoscopic meatotomy for relief of stenosis of the meatus in the absence of dilation of the intramural ureter is usually followed by recurrence of the stenosis. These cases are better treated by transvesical ureterovesicoplasty.

IV. Endoscopic renal treatment

1. Through nephrostomy opening

a) **Renal calculi.** Recurrent renal calculi visualized in this manner may sometimes be removed with forceps passed through the endoscope. Sometimes stones up to 2 cm. in diameter can be extracted by placing the end of the instrument close against the stone under vision; then the telescope is removed and an alligator jawed forceps passed through the sheath and the stone grasped. The endoscope with the forceps and stone in its jaws are all removed together, the stone being held with the forceps tight against the end of the sheath.

b) **Foreign body.** The broken end of a nephrostomy tube may also be removed in this manner.

V. Ejaculatory duct catheterization

The ejaculatory ducts are catheterized to obtain a seminal vesiculogram, to inject medications into the seminal vesicals and to dilate the duct orifices (HERBST and MERRICKS). Special endoscopes are available for ejaculatory duct catheterization (p. 17). When catheters made specifically for the ejaculatory ducts are not available No. 3 or 4 Fr. ureteral catheters can be used. The duct orifices usually are not visible through the endoscope. In order to pass a catheter into them it is necessary to probe gently with the tip of the catheter the area where the duct orifices should be. Their location is one on each side of the verumontanum, usually one millimeter or less distal to its crest (Fig. 32, p. 59). They are small slits which remain closed except when seminal fluid is being ejected. The ejaculatory duct catheter is directed sharply dorsalward at almost right angles to the axis of the urethra, and passed into the duct until it meets obstruction; this may be from one-half to 5 cm. above the point of entry. Sometimes it is impossible to identify the duct orifices or to pass the catheter any distance into them. Radiopaque fluids or medication is injected into the seminal vesicles after the catheters have been successfully passed; a vesiculogram may thus be obtained. Sometimes the fluid returns around the catheter instead of going into the seminal vesicle.

VI. Application and injection of medicaments

Cauterizing agents and other medicaments are often applied to the prostatic urethra, bladder neck and trigone through endoscopes (ORMOND). The open tube urethroscope (p. 12) is the most suitable instrument to use for this purpose. Solid silver nitrate on the end of an applicator stick or 10 per cent silver nitrate solution on a cotton applicator is applied either under vision or blindly through the endoscope. Other medicinal agents such as phenol and glycerine (COLLINGS), medicated powders and solid tablets may be endoscopically applied. The Kelly cystoscope is especially adaptable for use in administering these treatments to women (BEATTY). Injection of solutions in the prostate with a long needle passed through an endoscope has been recommended for treatment of prostatic infections (McCARTHY). Podophyllin, 1 per cent in liquid paraffin may be applied to small bladder papillomata through the cystoscope (SEMPLE).

VII. Application of radium or its elements to bladder tumors

1. Radiation element

The measured amount may be placed within the beak of a cystoscope and held in place with a cystoscope holder (p. 26) for the desired length of time. Special instruments have also been devised for placing the material in the right position through the cystoscope (BARRINGER; MOORE).

2. Radon emanation seeds

These are implanted into the bladder wall through the cystoscope by means of a needle implanter (BARRINGER 1952). Most adult size operating cystoscopes are suitable. When a sheath with the fenestra in the side is used, the needle must be flexible; a rigid needle is used when it is inserted through an open end sheath such as the McCARTHY panendoscope. The rigid needle can be inserted into the tissue with more force than the flexible one; the shaft of the latter tends to bend or buckle as it is pushed in. The radon seed is inserted into the lumen of the needle and positioned about one centimeter inside the point. The needle is passed through the cystoscope and the point is thrust into the bladder wall for a distance of approximately one centimeter. A mandrin acting as a plunger then forces the radon seed out of the needle into the tissue of the bladder wall; the needle is then removed. The number of seeds used and their spacing in the bladder wall depends upon the dose of irradiation in each seed and on the type and extent of the tumor which is being treated. The most commonly used dose is 2.0 millicurie seeds placed 1 cm. deep and 1 cm. apart throughout the tumor bearing area of the bladder wall.

VIII. Biopsy of bladder lesions

1. Indications

Tissue for biopsy study is often removed from the bladder through an endoscope. When open surgery for removal of a neoplasm is contemplated, preliminary biopsy study is mandatory if there is any doubt whatsoever about the nature of the lesion. If the intravesical growth is comparatively small and endoscopic removal is contemplated, preliminary biopsy is usually not necessary.

2. Armamentarium and technique

Small biopsy forceps used through the cystoscope (p. 24) may remove tissue which is satisfactory for microscopic examination (CHAPMAN et al.), but often it is impossible for the pathologist to reach a definite conclusion from a small bit of tissue removed in this way. The tissue is necessarily taken from the surface of the lesion, which may be normal mucosa or may be sloughing or ulcerative. The specimen may be crushed to such an extent that distortion is marked, or it may be so small that the structural outline cannot be recognized. It is preferable to use a larger biopsy forceps (p. 24); the jaws must be sharp to prevent crushing the tissue. Another method of biopsy, which is usually the most satisfactory, is the use of a resectoscope. A large deep piece of tissue can be obtained either with the cold punch resectoscope such as the Braasch or Thompson instrument (p. 2) (BEVY et al.) or with the loop activated by the cutting current. When the latter is used, the loop is drawn through the tissue rapidly, in order to prevent electrocoagulation (DEAN and ASH); if the specimen is removed properly only a thin layer a few cells deep is distorted by the heat of the electric current (MIL-NER). Bleeding can be stopped easily by the coagulating current.

3. Biopsy of intraureteral tumors

Successful biopsy has been reported by ROOME and FLETT. A flexible biopsy forceps is passed up the ureter to the obstruction, opened and a piece of tissue excised.

IX. Electrocoagulation

1. Indications

In the case of small lesions of the lower urinary tract endoscopic electro-coagulation is often indicated. Polypoid growths at the bladder neck and in the urethra are destroyed by electrocoagulation when they are small. Granulomatous tissue which protrudes into the lower urinary passages and does not respond to application of the silver nitrate stick can be eradicated by electrocoagulation. An enlarged and inflamed verumontanum which does not respond to other treatment usually shrinks after light coagulation. Light electrocoagulation of the tract of a small vesicovaginal fistula may result in sufficient fibrosis to close it (PETERSON). The modality may be applied from both the bladder and the vaginal orifices (O'CONOR and SOKOL). Small bladder tumors, especially those which recur after removal, can be destroyed by electrocoagulation. Occasionally a blood vessel in the normal bladder ruptures spontaneously; or sometimes bleeding occurs due to rupture of a dilated vessel in the mucosa covering a prostatic adenoma or from the central vessel of a papilloma which has broken off. In these cases the bleeding can be controlled, at least temporarily, by electrocoagulation through an endoscope (Fig. 121). Secondary hemorrhage following endoscopic surgery can sometimes be controlled in the same way, but it is usually preferable to use a resectoscope. Blood clots can be dislodged with the loop, the bleeding point identified and then coagulated with the loop. Larger lesions are best treated by endoscopic resection (Chap. XVII).

2. Armamentarium and technique

The flexible electrode (p. 24) is used for electrocoagulation through endo-scopes which carry the fenestra in the side of the sheath, and the rigid electrode through those with the fenestra in the end. Small electrodes down to No. 4 Fr. are available for use through infant endoscopes. Large ball electrodes are used for electrocoagulating larger growths. The electrode is placed against the lesion and the current applied until the growth is destroyed. Deep coagulation of tissue in the urethra and bladder neck is avoided unless the lesion being treated is a malignant neoplasm; otherwise, extensive slough may occur and involve the external sphincter, resulting in urinary incontinence. Healing of the sloughed area may be followed by bladder neck or posterior urethral constriction and urinary obstruction. A urethral catheter should be left in place for continuous drainage of

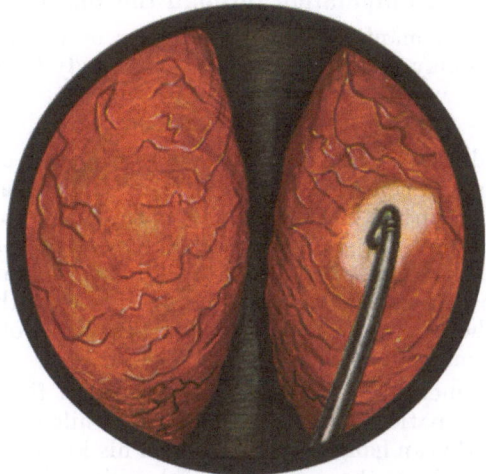

Fig. 121. Electrocoagulation of a bleeding vessel on the mucosa of an hypertrophical lobe of the prostate

the bladder for 24 hours following electrocoagulation in the prostatic urethra or at the bladder neck unless the coagulated area is very small; urinary retention is likely to occur if an indwelling catheter is not used. When coagulation is applied to a small lesion in the bladder, indwelling catheterization is usually not necessary.

X. Removal of foreign bodies from the bladder

Many ingenious methods have been devised for the endoscopic removal of foreign bodies from the bladder.

1. Forceps or cystoscopic rongeurs through the cystoscope

Small, firm objects such as pins, straws, and pieces of catheters or filiforms are grasped with cystoscopic rongeurs or forceps (p. 24) passed through the operating channel of the cystoscope (BODNER). If the foreign body, together with the head of the forceps which grasps it, is too large to be extracted through the fenestra and lumen of the cystoscope, it is pulled up against the fenestra. Then the entire sheath is removed with the object, still in the grasp of the forceps, trailing after it. An open end sheath such as the McCarthy panendoscope or resectoscope (SCHLOSS et al.) is best suited to this maneuver. Sharp corners or ends of the foreign body must be pulled into the fenestra to prevent injury to the urethral mucosa. When the foreign body is larger, such as a pencil, thermometer or stick of wood, a larger cystoscopic ronguer or foreign body forceps is more effective (KRUZMAN et al.). These instruments are inserted directly through the urethra rather than through a cystoscopic sheath, and have much wider and stronger jaws. The telescope, with the light at its tip, is inserted through a tunnel in the shaft of the instrument. Soft friable foreign bodies such as paraffin or chewing gum can be removed piecemeal by the larger rongeur. If the object such

as a glass rod is smooth, the forceps tends to slip off; smearing a sticky substance
like soft black tar on the jaws to give better traction may help. A snare, such as
the looped catheter used for extracting ureteral calculi, may be placed over
a slippery object for extracting it.

2. Manipulation through the vagina

In women patients, manipulation of the floor of the bladder through the
anterior vaginal wall may help to place the foreign body in the open jaws of the
rongeur (HOBERG).

3. Floating foreign bodies

These are more difficult to remove than those which lie on the bladder floor.
The foreign body forceps cannot reach them easily and there is no solid backing
to press the object against while engaging it with the jaws of the instrument.
Heavy pressure over the suprapubic region depresses the dome, making it more
accessible to instruments passed through the urethra. Women patients can be
placed on the table face down with the thighs flexed over the end of the table.
This position puts the more solid flat trigone on the ceiling of the operating area,
where the object is more easily grasped.

Paraffin. It is sometimes possible to remove small floating bodies such as
paraffin by having the patient empty a full bladder while he is in the *upside down
position* (SCHULTE). He can hang head down with his knees flexed over a trapeze
bar; the floating material is then at the vesical orifice when he voids. When the
bladder is distended with air instead of water, the floating foreign bodies remain
on the floor of the bladder and can be more easily grasped with forceps (JOELSON).

XI. Litholapaxy

1. Advantages, indications and contraindications

Skillfully performed litholapaxy or lithotripsy has certain advantages over
cystolithotomy (NESBIT). There is less shock, less after pain and much more
rapid convalescence. In fact, most patients who do not have urinary obstructive
disease can leave the hospital on the second postoperative day. When vesical
calculus and obstruction at the vesical orifice coexist, both can be removed during
the same operation unless the stone and/or the prostate are excessively large.
Inadequate endoscopic surgical experience on the part of the urologic surgeon
also contraindicates the combined procedure. Litholapaxy cannot be performed
on small children nor used to crush stones larger than 3 to 4 cm in diameter.
When combined litholapaxy and endoscopic prostatectomy is performed, it is
usually best to crush the stone before resecting the prostate, unless the adenoma
is so large that the stone falls behind it and is out of reach of the lithotrite. When
litholapaxy is performed first, there is less danger of trauma and bleeding in the
prostatic urethra and the fragments of stone can be evacuated with the resected
pieces of prostate.

2. Visual versus blind lithotrites

Litholapaxy is performed either under vision or by using the sense of touch
for crushing the stone; endoscopic examination is made before and after the blind
operation.

a) Litholapaxy performed under vision. This is done with a lithotriptoscope
(p. 23). The stone is visualized through the telescope of the instrument, is grasped
between its jaws and crushed. The bladder must be well distended with irrigating

fluid during the procedure. The larger fragments are recrushed until all are small enough to be evacuated through an evacuator tube or an endoscope sheath. The greatest advantage of lithotripsy under vision is that the calculus and its fragments can be accurately grasped and the bladder mucosa preserved from trauma; it is not necessary to manipulate the instrument blindly over the floor of the bladder to find the calculi. While using the lithotriptoscope, the jaws of the instrument can be pointed safely toward the floor of the bladder to pick up a stone located behind a projecting prostatic lobe. A calculus in this position might be out of reach of the blind lithotrite. A great disadvantage of the visual instrument, however, is the difficulty encountered and the length of time required to keep the visual field clear. It is usually necessary to empty the bladder and refill it, and sometimes to irrigate several times after crushing the stone in order to clear the field enough to visualize the stone again. Another disadvantage is the weakness of the instrument; it must be of lighter construction to allow room for the telescope. Stones which are very hard cannot be crushed with it, although some instruments can exert a force up to 300 lbs. (TWINEM).

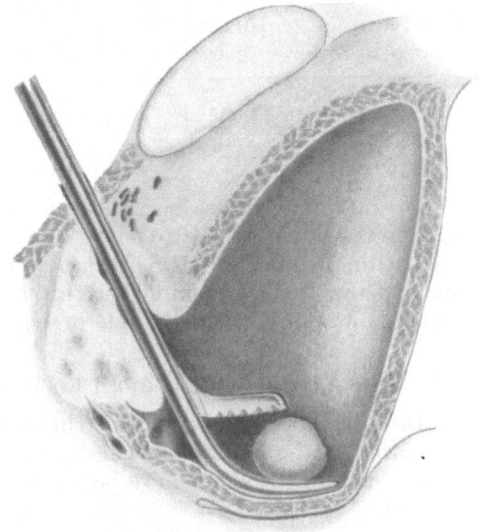

Fig. 122. Litholapaxy with blind lithotrite. The lower jaw is pressed against the bladder floor forming a depression into which the calculus rolls

b) **Blind litholapaxy.** This is performed with a BIGELOW, COTIN, WEISS, or other lithotrite. Cystoscopy is carried out just before passing the lithotrite. Note is made of the location of the stone, of the size of the prostate and of trabeculation or other irregularities of the bladder wall. The bladder is left distended, the cystoscope removed and the lithotrite passed. The jaws are opened about 3 cm. and the lower jaw pressed firmly against the floor of the bladder and slightly vibrated. The pressure of the jaw against the bladder floor makes a depression into which the calculus rolls (Fig. 122). The upper jaw is slid down against the stone and locked into the threads, fixing the stone between the two jaws. The instrument with the stone between its jaws is raised from the floor of the bladder and slightly rotated to make certain there is no mucosa caught with the stone. The wheel of the instrument is then rotated clockwise, screwing the upper jaw down onto the lower jaw and crushing the stone. If the stone does not crack easily, the wheel is turned until considerable pressure is exerted on the stone, then it is held for a few minutes; the stone may crack under the sustained pressure. The fragments are picked up and crushed in the same way.

3. Evacuation of fragments

When it is judged that all pieces are small enough to be evacuated through an evacuator, the lithotrite is removed and the evacuating tube introduced. A resectoscope sheath is convenient for this purpose, because a telescope can be inserted through it to ascertain that all fragments have been evacuated. Manipulation of the sheath in the same manner used for evacuating pieces of tissue during endoscopic surgery will remove most of the fragments from the bladder (p. 205). The use of an evacuator such as the McCARTHY or the ELLIK, or of a piston syringe which fits into the end of the sheath, may help in evacuating fragments. When it is probable that the majority has been removed, the telescope is inserted into the sheath through which the fragments have been evacuated. Observation of the interior of the bladder is made to determine if fragments still remain and whether or not they are too large to pass through the sheath. The lithotrite must be reintroduced to crush any remaining large fragments. An evacuator set, including a metal tube and a detachable bulb and glass container such as the BIGELOW, is often used for removing stone fragments after litholapaxy. The disadvantage of this instrument is that it has to be removed and a cystoscope passed in order to ascertain that all the fragments have been evacuated.

XII. Extracystoscopic endoscopic procedures

1. Through the urethra

Operative procedures within the bladder may be conducted through the urethra outside the cystoscope (COPPRIDGE et al.). The forceps, electrode, radon implanting needle or other instrument is passed along the side of or with the cystoscope and visualization is accomplished through the examining telescope. This technique is more adaptable to the female urethra than to the male.

2. Through a suprapubic cystostomy

Instruments may be passed through a suprapubic cystostomy opening and visualized in the bladder by means of an endoscope passed through the urethra. Stones or foreign bodies may be grasped, tumors fulgurated or biopsy specimens excised.

XIII. Endoscopic treatment of urethral strictures

1. Dilatation

A filiform bougie can sometimes be passed through a urethral stricture by manipulation under vision through an endoscope. The orifice is identified and the tip of the bougie passed into it.

2. Incision and resection

Linear incision of the stenosed portion of the urethra may be done under vision with a COLLING's knife, or the fibrosed area may be removed with a loop electrode through a small resectoscope (ANGLE and PFEIFER).

XIV. Intraperitoneal and gastric endoscopic treatment

In the presence of severe injury to the esophagus, a gastrostomy may be required, through which the patient is fed. After the acute stage has subsided it may not be possible to dilate the resulting esophageal stricture from above, but retrograde dilatations through the stomach may be successful. This may sometimes be accomplished by passing a straight cystoscope through the gastroscopy opening, up into the esophagus as far as possible. The strictured area is visualized and a bougie passed through the stenosed lumen. When there is complete closure of the esophagus, a fulgurating tip passed through the cystoscope can be lined up under a fluoroscope with a bronchoscopic forceps inserted from above (HARA and ROSENVOLD). A high frequency current is applied through the fulgurating tip as it is thrust through the fibrosis. When it emerges into the esophagus above the stricture, it is grasped with the bronchoscopic forceps. Cords can then be pulled through it, and after a few days active dilation of the stricture started.

Chapter XV

Endoscopic surgery—a specialty within a specialty

The introduction of a new surgical procedure or technique is always hailed by a volley of conflicting opinions regarding its application, and considerable time must elapse before its acceptance becomes general and its indications standardized. Factional disparity has been particularly rife in the history of endoscopic prostatic surgery, ranging from the operator who finds almost no indications for the transurethral approach to the one who treats practically one hundred per cent of his cases by this means. The younger generation of urologists is more enthusiastic about endoscopic prostatic surgery than are its elders, and men who have been schooled in centers where this operation is in current usage are more ardent supporters than those who have served their apprenticeship where the older types of prostatectomy prevail. Thus, a urologist's training is the prime influence which molds his judgment concerning the relative merit of the different operative methods.

The technique of endoscopic surgery has continued to improve and the number of urologists who are proficient in this type of surgery has rapidly increased during the past two decades. The problems with which the endoscopic surgeon has to deal and the techniques which he must master are very different from those encountered in open surgery. Experience in the performance of open surgery is of little help to him in solving these problems and in learning the technique. Endoscopic surgery is really a highly technical specialty within the specialty of urology, and requires different abilities and more intensive and prolonged training to master than open urological surgery.

I. Advantages and disadvantages of endoscopic surgery

The advantages of endoscopic surgery which is intelligently, skillfully and adequately performed far outweigh the disadvantages. On the other hand, if endoscopic surgery is carried out inaccurately and inadequately the open approach is more advantageous. Surgery performed by the endoscopic route is much more frequently inexpertly executed than are similar operations performed by an open approach. This is one reason why so many surgeons condemn transurethral surgical procedures. The following advantages apply only to endoscopic surgery which is performed expertly.

1. Advantages

a) **Better tolerated.** Endoscopic surgery is better tolerated than open surgery. Patients who are poor surgical risks and who could not stand open surgery can be operated upon through the transurethral approach without producing shock. When a poor risk patient has a large prostate or bladder tumor it may, in certain cases, be advantageous to remove it in two or more stages. Several comparatively minor operations may be tolerated better than one major one. In most cases, however, the patient tolerates one open operation better than several endoscopic procedures performed at close intervals.

b) **Less postoperative pain.** There is considerably less pain following endoscopic surgery. Many patients require no analgesics.

c) **Shorter hospitalization.** The period of hospitalization for an endoscopic operation is about one half that for the same surgery done through an open approach.

d) **No external wound.** There is no external wound to become infected and/or require a long period of time to heal.

e) **More accurate and more adequate removal of tissue.** This can be accomplished by endoscopic surgery in most cases. The tissue is magnified when a close view is obtained through the endoscope lens; it can be identified more easily than when viewed with the naked eye from a distance through an open wound. As a result of accurate identification of tissue, the lesion can be removed in its entirety. For example, it is possible by endoscopic surgery to remove the false or surgical capsule of the prostate. When the adenomatous hypertrophy is enucleated either suprapubically or perineally, this portion of the gland is always allowed to remain, because the line of cleavage for enucleation is between the adenoma and the surgical capsule.

f) **Shorter operative time for removal of small lesions.** These can be removed in less time through the endoscopic approach than by open methods.

2. Disadvantages

a) **Long apprenticeship and technical difficulty.** The greatest disadvantage of endoscopic surgery is the time required to learn it and the technical difficulty of performing it accurately and adequately. There is far greater danger of serious consequences resulting from inexpertly performed endoscopic surgery than there is from a poorly performed open operation.

b) **Requires large calibre urethra.** Endoscopic surgery cannot be performed through a urethra which is narrow, due either to the small size of the patient or to atresia of the urethra from any cause. Dilatation or perineal urethrotomy (NESBIT) may succeed in enlarging the urethra sufficiently to allow the electrotome to pass into the bladder.

c) **Longer operative time for removal of large lesions.** Removal of these by the endoscopic route is usually more time consuming than by an open approach. This is especially true when the surgeon has had only limited experience in performing endoscopic operations.

d) **Multiple stage operation.** Excessively large lesions cannot be removed endoscopally unless a two or more stage operation is done. The disadvantages of multiple stage endoscopic procedures in most cases outweigh the advantages of the approach. The length of hospitalization which is required when several such operations are performed for the removal of bladder lesions is usually greater than when an open approach is utilized. Repeated anesthetics tend to weaken the patient, and the psychological trauma of undergoing surgery several times may be more harmful than the greater shock sustained when one open operation is done to remove the entire lesion.

II. Training the endoscopic surgeon

1. Difficulties and importance

Endoscopic surgery, especially endoscopic prostatectomy, is a highly technical procedure, the successful performance of which requires considerably more preparation and practical application than do other forms of genitourinary surgery. It is, therefore, essential that the would-be resectoscopist submit to an intensive term of study, observation and supervised practice before attempting to work independently. No other surgical procedure is so difficult to teach. This is a broad statement, but surgeons who have mastered the techniques of many other surgical procedures and then have taken training in endoscopic surgery will attest to its veracity. They have learned to do other operations after performing them under supervision several times. The technique of the most difficult open surgical procedure can be mastered by most surgeons by the tenth or twelfth operation, but it is necessary to carry out one hundred or more endoscopic prostatectomies before the surgeon feels competent to do the operation correctly; even then he is not capable of doing it as well as he can an open operation.

2. Preliminary endoscopic training

Before starting to learn to perform endoscopic surgery, the trainee must have had preliminary training and experience in other endoscopic procedures. He must be very familiar with the use of cystoscopes, particularly the type of visual system he is to use for endoscopic surgery. Even after he has become an expert cystoscopist, the surgeon loses his bearings in a maze of blood and unfamiliar appearing tissue after he has removed a few pieces of prostate during his first endoscopic prostatectomy. The landmarks which were so familiar and so clear before he started the operation have all disappeared.

3. Instruction

The inability of the trainee to visualize the field of operation while the teacher is performing endoscopic surgery, and the inability of the teacher to see through the same lens as the trainee, makes the training process difficult and slow. Frequent observations through the resectoscope by both trainee and teacher while the other is operating is the best that can be accomplished. The use of a teaching attachment (p. 26) is cumbersome and impractical.

It is of benefit for the trainee to read available literature pertaining to endoscopic surgery before he begins to learn the technique. The study of a comprehensive treatise on the subject[1] will give him valuable information which he would otherwise need to learn by trial and error — a method which might, at times, prove disastrous.

The urologist whose duty it is to teach endoscopic surgery must give constant supervision to the trainee for at least the first fifty cases. He must be scrubbed, gowned and gloved and must examine the field of operation through the resectoscope at frequent intervals. Only by so doing, can he give adequate aid to the man who is learning. It is dangerous for the instructor to go off to some other part of the hospital while the operation is in progress; serious difficulty may arise suddenly and may rapidly develop into a catastrophy. It is much more important for the instructor to remain with and to assist the trainee during endoscopic surgery than during open surgery.

4. Who should be trained

a) **All trainees in preparation for the specialty of urology.** These men should be trained to perform endoscopic surgery. Younger men are usually more adept at learning the technique than are older persons.

b) **Trainees possessing abundant manual dexterity.** The trainee should also have considerably more than the average amount of manual dexterity. It is

---

[1] See references following Chapter XX.

impossible for some surgeons to learn to perform endoscopic surgery expertly; they may not have the native ability, they may be too old, or their patience may terminate before they have completed the arduous training period.

c) **Some urologists.** Those who have not had previous training in endoscopic surgery may learn to do it later in life, but usually it is better for them to continue doing the open operations with which they are familiar.

d) **Not the occasional endoscopic operator.** Inasmuch as numerous endoscopic procedures are necessary to learn the technique, no attempt should be made to teach it unless sufficient cases are available to allow the trainee to perform at least two endoscopic operations a week. After learning to do endoscopic surgery, the surgeon should perform at least two such operations a week in order to remain proficient; the urologic surgeon who carries out only an occasional operation on the bladder should not attempt to do it by the endoscopic route.

e) **Not general practitioners or most general surgeons.** These men do not have the time or the inclination to go through the long and arduous training period necessary to learn to perform endoscopic surgery properly. Even were they to submit to such a routine, few if any would encounter applicable urological disease in quantity sufficient to maintain adeptness in performing endoscopic surgery.

5. Preliminary practice

Practice in manipulation of the resectoscope can be maintained by cutting meat or a clay model with the instrument (BAUMRUCHER).

a) **Beef heart.** A practical method is to tie a beef heart, which is obtainable at most butcher shops, onto the indifferent electrode of the electrosurgical unit and place them on the cystoscope table. The resectoscope is inserted into the heart cavity through one of the large vessel openings and the heart muscle is resected. The inflow of water is manipulated and the loop alternately advanced and retracted. The inner end of the instrument is placed close to the tissue for clear visualization and for excision of a deep piece of tissue. Then it is swung clear to allow the piece to be washed out of the field of vision and the loop to be advanced without striking unresected tissue. Application of the cutting and the coagulating currents by depressing the corresponding foot switch is coordinated with manipulation of the loop. Practice in handling the instrument is thus obtained which, if continued long enough, will result in the development of an operating rhythm. Then, when the trainee starts resecting prostates, smooth, rhythmic manipulation of the instrument and coordinated movements will take place with almost no conscious effort, and he will be able to focus his concentration on identification of tissue, control of bleeding, and sustained orientation.

b) **Clay model.** Resecting a clay model fashioned like a prostate helps somewhat to develop orientation. Skill in the performance of endoscopic surgery can be acquired, however, only by operating upon many patients under the close supervision of a trained resectoscopist.

III. Armamentarium and supplies

Endoscopic armamentarium, including instruments for endoscopic surgery, is described in Chapter I. There are, however, a few special features, advantages and uses which are closely associated with the performance of this type of surgery.

1. Resectoscopes or electrotomes

a) **Stern-McCarthy electrotome.** This is the instrument for endoscopic prostatic surgery which enjoys the widest favor. It provides vision by means of

⅄ lens system, which makes it particularly adaptable for use by the urologist whose cystoscopic training has acquainted him with lens instruments. The working element is designed to afford accurate and complete control of the cutting loop, and its construction is sturdy enough to give continued service without frequent repair if it is carefully handled. This instrument has been made possible through the genius and technical skill of REINHOLD H. WAPPLER and his son, FREDERICK G. WAPPLER, in collaboration with JOSEPH F. McCARTHY and other urologic surgeons.

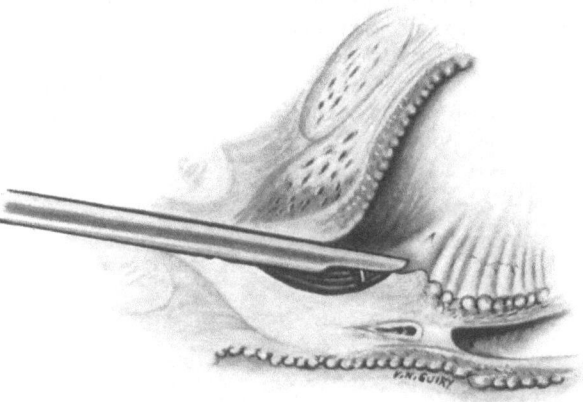

Fig. 123. Long bevel sheath impinges on edge of bladder neck and prevents proper excavation of prostatic tissue. It does however protect loop in its most extended position

The *sheath* of the standard electrotome is No. 28 F., and is made of bakelite which supplies insulation between the working element and the urethra. Thus, the cutting loop can be drawn into the sheath and will neither cut nor produce a short circuit in this position, even though the current is left on. The distal end is open and is bevelled, making the upper portion of the sheath longer than the lower. In the original standard model, this bevel is long enough to form a protection over the loop in its extended position (Fig. 123). The maneuverability of the distal end is limited because of its impingement upon the bladder wall or vesical neck, but this very feature provides a safeguard against too deep resection and makes the instrument excellent for use by the novice. It does, however, prevent proper excavation of the prostatic bed, and is therefore not as satisfactory for the experienced resectoscopist as a shorter bevel sheath (Fig. 124) (JAHR),

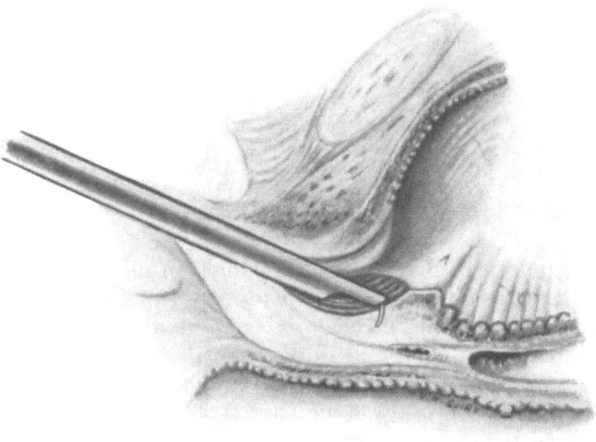

Fig. 124. Short bevel sheath can be depressed into prostatic urethra and does not infringe on bladder neck

which permits adequate excavation of the prostatic cavity. The proximal end of the sheath is fitted with a metal cone to receive the working element, and the water inlet tube and a key for holding the working element in place are mounted on the metal cone.

A metal covered sheath has been made as a modification of the standard sheath, to provide a smooth surface finish which facilitates passage of the instrument and eliminates its clinging to the urethra during operation. The authors have observed, however, that there is some tendency for the metal to heat —probably due to a partial short circuit to the urethral mucosa — and that stricture formation is more frequent following the use of this sheath.

A *thumb rest* has been designed by the author to increase leverage, producing pressure of the distal end of the electrotome against prostatic tissue. This attachment is applied to the locking device which is mounted on the metal cone at the proximal end of the sheath (Fig. 125).

The *working element* (Fig. 15, Chap. I) carries the telescope and also the cutting loop and its control mechanism; the outlet tube is mounted on the side. The telescope is inserted through the proximal end of the working element into

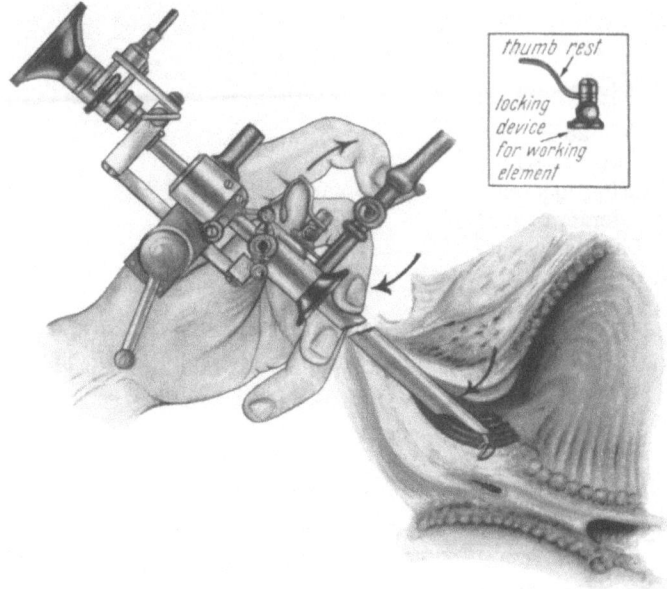

Fig. 125. Proper method of holding McCarthy electrotome. Arrow indicates direction in which pressure is maintained during cutting excursion of loop. Forefinger manipulates irrigating fluid intake. Inset shows author's thumb rest

a guide provided for it, and is held in place by a screw clamp. The loop is passed through a guide from the distal end, fits snugly into a socket and is held in place by a screw clamp. A rack and pinion apparatus with a handle for manipulation controls the loop and imparts to it an excursion of $2^1/_2$ cm.

b) Modifications of the McCarthy electrotome. These are listed on p. 6. The *McCarthy convertible* is designed to permit a closer view of the loop in its extended position. This is accomplished by the use of an adjustment which fastens the telescope to the loop carriage so that the loop and the telescope are moved together and are always in the same relative position. Therefore, when the loop is extended to begin the cut, the telescope is also advanced and perspective of the point of origin of the loop excursion is not lengthened.

The *McCarthy remote control* modification has the electronic control switch mounted on the rack and pinion handle. No foot switch is necessary. The cutting current may be activated by depressing the red button located at the top of the switch. The coagulating current is controlled by depressing the green button. If a blended current is desired, both buttons are depressed. Surgeons who use this instrument believe that a more immediate application of current or change of one current to the other can be applied by the switch on the handle than by the foot switch.

The *McCarthy of smaller calibre* in sizes 16 F. and 20 F. is standard length and is used for patients who have a small urethra, including children. The *McCarthy infant* is shorter and is useful in resecting the bladder neck in infants and small children. It is possible to remove only small bits of tissue with the small resectoscopes; therefore they are not used when more than a gram or two of tissue should be resected.

c) One hand operated resectoscopes. Resectoscopes which are operated with a single hand, such as the *Nesbit*, the original one hand operated instrument, leave the other hand free for digital rectal palpation during resection. Tissue can thus be pressed up toward the cutting loop. "Third dimensional perception" is also obtained; the operator can feel the proximity of the sheath to his finger in the rectum and thus estimate the amount of tissue which remains to be resected. The loop of the Nesbit is manipulated by the thumb which pushes the loop inward against a spring; the loop is then returned by the spring to its original position within the sheath.

The *Creevy*, the *Foley* and the *Miller* modifications have similar loop activating mechanisms, the first having two finger grips instead of the thumb rest and the latter two a disc surrounding the telescope for manipulating the loop. The *Ingleasias* has a leaflike spring to hold the loop in the retracted position.

Some surgeons who use the one hand resectoscope believe that the cutting loop can be moved back and forth more easily and naturally by the index finger than by the thumb. The *Baumrucher*, the *Gibson*, and the *Scott* modifications all pull the loop against a spring which returns it to the extended position. Of these, the GIBSON and the SCOTT are equipped with a pistol grip; with it the instrument can be held more steadily and the inner end pressed more forcefully against the tissue being resected.

d) Control of the cutting loop. The cutting loop of the McCarthy resectoscope can be controlled more accurately than that of the one hand operated instruments because: 1. the rack and pinion mechanism is a reduction gear — the handle moves a greater distance than the loop; 2. the loop does not stick or bind as it often does in the one hand operated resectoscope. Some of the modifications of the latter are designed to provide smoother operation of the loop; the leaf springs and the roller bearings are for this purpose.

e) Rotating modifications. The rotating modifications such as the *Foley* and the *Scott* allow the operator's hands and the inlet connections to remain stationary while the sheath, the telescope and the loop electrode rotate through 380°. Awkward positions of the surgeon's hands and entanglement of the inlet connections are thereby reduced to a minimum.

f) Loop electrodes. These are available in several wire diameters from No. 12 (0.012″ di.) to No. 20 (0.020″ di.). The size is stamped on the metal covering of the shaft near the loop of the electrode (Fig. 126). The larger size (No. 20) cuts more slowly, but coagulates better than the fine wire (No. 12). It is, therefore, preferable for use by the amateur. The experienced endoscopic surgeon usually prefers the finer wire because it cuts the tissue more easily and rapidly; it does not slide over the tissue when fibrous tags are being removed.

g) Resectoscopes for bladder tumors. Resectoscopes which are used for removing the prostate are also used for the endoscopic removal of bladder tumors. A short beaked sheath is preferable because it does not infringe against the bladder wall, especially when a tumor is being removed from the area opposite to the vesical outlet. The cutting loop electrode can be bent so that it projects straight out ahead of the sheath to resect tumors in this location; the inner end of the instru-

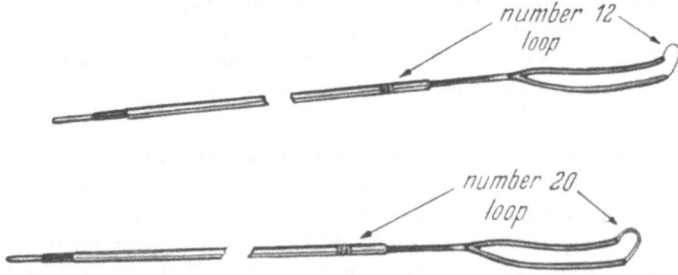

Fig. 126. Size of loop is stamped on metal shank and indicates gauge of wire used. No. 20 loop is heavy wire; No. 12 loop is light wire

ment is moved up and down or sideways like a paint brush. Special attachments to use with the resectoscope have been designed to facilitate removal of bladder tumors (KRAMER).

2. Electrosurgical units

These are described in Chapter XVI.

3. Table and stool

Flexibility of the table where the patient is positioned, and of the stool upon which the surgeon sits, is necessary to facilitate the performance of endoscopic

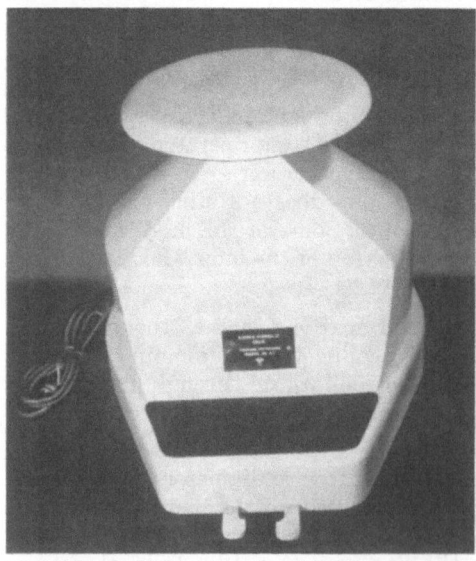

Fig. 127 A. Barnes hydraulic adjustable cystoscopic stool. Seat is raised and lowered by reversible electric motors

surgery. The position of the ocular end of the resectoscope must be changed frequently during endoscopic surgical procedures. A rapidly adjustable table to raise or lower the patient or a similarly mobile stool to change the height and position of the surgeon, reduces the amount of alternate back bending and stretching which is required of the operator. Some endoscopic surgeons prefer to operate in the standing position; the adjustable table is preferable for them. A rapidly adjustable stool is, however, moved more easily and quickly up and down and from side to side than is the operating table. The BARNES adjustable hydraulic stool (Fig. 127A)[1] is raised and lowered with reversible electric motors through a hydraulic system, and is mounted on casters for easy movement from side to side.

The BARNES rapidly adjustable spring activated stool (Fig. 127 B)[2] is lowered by the weight of the operator and is raised by coil springs when the weight is lifted. A key fits into notches in the shaft of the stool to hold the seat at

[1] Manufactured by American Cystoscope Makers Inc., New York.
[2] Manufactured by Medical Research Specialties, Loma Linda, Calif.

any desired height; the key is manipulated by a foot lever. The adjustment of the spring stool requires coordination of the operators leg muscles; when the foot lever is depressed, releasing the pin from the notch in the shaft, the seat will drop suddenly and then come to an abrupt stop unless the operator bears his weight on his legs and eases himself down as the seat lowers. Conversely, he must raise his weight with his legs when he wishes the seat of the stool to be

higher. The advantage of this stool is its rapid mobility; it can be raised or lowered in a fraction of a second and can be easily shifted from side to side on the large casters upon which it is mounted.

Most cystoscopic tables are adjustable; some by tipping the caudal end up or down and others by raising or lowering the entire table. There are several such tables such as the FOLEY which are especially designed for endoscopic surgery. The top of a routine cystoscopic table mounted upon the adjustable base of a dental chair is used by some surgeons.

4. Attachments to the table

Numerous attachments to the table used for endoscopic surgery are recommended by different urologists. These include an overhanging arm for hanging the inlet attachments for the resectoscope, a stand to hold the irrigating fluid, and special leg and foot rests. The most important is the support for the inlet attachments (CUMMINGS); when

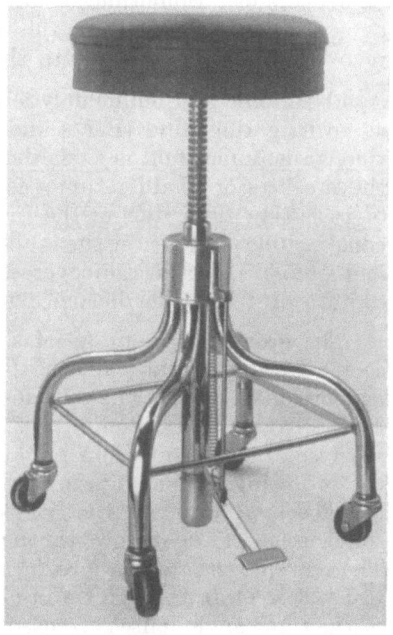

Fig. 127 B. Barnes adjustable spring stool.

these drag on the resectoscope, the easy mobility of the instrument and the fine sense of touch of the operator are disturbed. They should hang loosely from an overhanging arm in such a manner as to eliminate all drag on the instrument. A simple method of suspension is to attach them to the arm holding the x-ray tube above the table. The irrigating fluid supply source suspended from the ceiling is usually preferable to one attached to the table. Knee rests and foot rests which distribute the weight of the patient's legs are desirable.

5. Irrigating fluid

a) Sterile water. This was used as an irrigating or distending fluid during endoscopic surgery until Creevy and others advocated the use of an isotonic solution. During transurethral prostatic resection absorption of varying amounts of irrigating fluid takes place (GRIFFIN et al.). The quantity of fluid absorbed is increased by prolongation of operation time (HAGSTROM), by exposure of venous spaces, by perforation through the capsule and by increased inflow pressure of the fluid (NICOLAI and CORDONNIER). When water or other hypotonic fluids are used for irrigation during endoscopic surgery, absorption is sufficient to cause intravascular hemolysis resulting in hemoglobinemia. Heat generated by the electrosurgical unit might also be a factor in increasing free hemoglobin in the blood (FERGUSON and MILLER). More marked tissue reaction is produced by infusion of fluid which has a different osmotic pressure than that of the interstitial fluids.

Therefore, extravasation of water into the periprostatic and perivesical spaces causes tissue reaction; when this is extensive, shock is the result. Nonisotonic and nonisosmotic fluids may also diffuse through the bladder wall (MALUF). Intravascular hemolysis and tissue reaction due to the use of nonphysiological fluids predisposes to shock; when there are additional immediate complications such as excessive hemorrhage and prolonged anesthesia the state of shock may be irreversible. Sometimes acute renal failure (lower nephron nephrosis) develops as a result of these factors; this may lead to a fatality even though the patient recovers from the immediate shock.

b) Isotonic and nonhemolytic fluids. Although some urologists still use water as an irrigating fluid (PITTS and HINMAN) there is a preponderance of evidence that an isotonic fluid is safer than water, especially when adequate removal of the prostate or bladder tumor is being performed. There is minimal absorption of irrigating fluid during the first part of a prostatic resection and as long as considerable prostatic tissue is allowed to remain; but as the capsule is approached and venous spaces are uncovered, absorption of fluid is increased and the danger of untoward reactions becomes greater.

The great disadvantage of a nonhemolytic fluid for irrigation during endoscopic surgery is the cloudiness of the medium when it contains even a small proportion of blood. The red blood cells are not lysed by the fluid, and cloudiness results from a suspension of these cells. Conversely, when water is used the cells are lysed and the medium remains more transparent. By proper adjustment of the inflowing nonhemolytic fluid and by manipulation of the ocular end of the resectoscope, however, the field of vision can be kept clear. The nonhemolytic solution is clear as it flows through the resectoscope and across the field of vision. Therefore, if the flow is forceful and the ocular lens is placed close to the tissue, vision is as clear as when water is flowing in. Unfortunately, when the retrograde or the right angle lenses are used the view can not be as close as it is through the foroblique optical system. Therefore, it is not possible to obtain a clear field of vision through nonhemolytic fluid with these telescopes unless there is no bleeding. Inasmuch as vision while performing endoscopic surgery is always very close, the use of nonhemolytic irrigating fluid should not be a handicap.

c) Satisfactory irrigating fluids (Table III). These must be isotonic, isosmotic, nonhemolytic, nontoxic, and nonelectrolytic. They must also be clear, stable and easily sterilized. An electrolyte such as normal saline dissipates the electric current so that the active electrode will not cut tissue.

d) Glucose, 4 or 5 per cent solution, was one of the first isotonic solutions used for endoscopic surgery. It meets all the criteria for a satisfactory fluid. It is easily available almost any place in the world, and is inexpensive when prepared by the hospital. Some urologists have found it to be sticky, but in the authors' experience its adhesive qualities are no more noticeable than those of water. It is probably the best all round solution for use as an irrigating fluid during endoscopic surgery.

e) Glycene. 1.9 per cent solution, is used by many urologists. It also meets all the criteria for a satisfactory irrigating fluid. There is, however, some evidence of its being mildly toxic; moreover, it is not available in some areas.

f) Sorbitol, Mannitol. A mixture of sorbitol and mannitol with an antiseptic (Cytal-Cutter) is prepared in flasks ready for dilution with nine parts of water (SCHULTE et al.). Either sorbitol or mannitol 3 per cent may be used alone.

Table 3. *Comparison of irrigating fluids for use in endoscopic surgery*

Solution	Strength of solution	Advantages	Disadvantages	Cost per 20 litres (5 gallons)
Glucose	5% (3%:6%)	Isotonic, nontoxic available everywhere	Cloudy medium	$ 3.00 ($ 50.00 if intravenous solution used)
Glycine	1.9% (1.1%:2.1%)	Isotonic	Cloudy medium	$ 2.50
Mannitol	3%		Cloudy; not entirely isotonic	$ 2.50
Sorbitol	3%		Cloudy, medium; not entirely isotonic	$ 2.50
Urea	1%	Isosmotic	nonisotonic; causes hemolysis	$ 1.00
"Cytal" (Cutter)	Dilute 1:9 with water (3.2%)	Isotonic. No sterilization required. Easily mixed	Cloudy	$ 7.00

6. Miscellaneous armamentarium

a) **Aspiration apparatus.** One of these is sometimes required to remove pieces of resected tissue or fragments of stone. Most fragments, however, can be removed from the bladder by manipulation of the sheath while the irrigating fluid is flowing out (p. 205). The *Freyer* evacuator is effective in aspirating fragments of stone through an evacuating tube. A bulb provides the suction and the fragments drop into a bottle attached to the bulb. The *Ellik* and the *Hutch* are similar; the connecting tube fits into the outer end of A.C.M.I. adult cystoscope and resectoscope sheaths. The bulb is placed at right angles to the inlet; thus a whirling motion of the water is created which permits the pieces of tissue or stone to drop into the glass receptacle. The original McCarthy evacuator consists of a rubber bulb with a glass barrel attached to a two-way valve. The valve is fitted with a cone for attachment to A.C.M.I. cystoscope sheaths. To operate, the bulb is compressed before attaching the evacuator to the sheath. Water is allowed to run in through the sheath stopcock and fill the barrel. A quarter turn upward closes the portal into the sheath and opens the discharge portal below. Compression of the bulb in this position empties the barrel into the discharge pan.

The *Toomey* syringe evacuator also has a cone for attachment to cystoscope sheaths. *Suction* is applied with a piston; it is more powerful than the bulb evacuators. Some urologic surgeons prefer an attachment connected to the regular suction which is piped into most operating rooms.

b) **Alligator forceps.** These may be necessary to grasp and remove larger pieces of tissue which have not been cut up during resection. The piece may fill the fenestra of the sheath and suddenly obstruct the outflow of fluid from the bladder. When this occurs, the inner end of the sheath is depressed, pinning the piece of tissue to the floor of the bladder. The alligator forceps are then passed through the sheath, opened, pushed forcibly against the fragment of tissue, then closed over it; the sheath and the forceps with the piece of tissue in its grasp are all withdrawn through the urethra.

c) **Drapes.** For use during endoscopic surgery drapes are usually of cotton material, and should be large enough to cover the patient's legs and feet and the cystoscopic table pan. An opening through the drape at its lowest point in the pan, allows the irrigating fluid to run out. A screen to catch pieces of tissue and fragments of stone is either under the drape, or is sterilized to put over it. For surgeons who use the one hand resectoscope, a rubber finger cot surrounded by a shield is incorporated in the drape to allow intrarectal palpation during operation. A waterproof drape is more aseptic than one which becomes wet; bacteria may penetrate into the sterile field through the wet drape (GARSKE).

7. Lithotrites and lithotriptoscopes

These instruments are described in Chapters I and XIV.

Chapter XVI

Electrosurgical units

(C. J. BIRTCHER, Electrical Engineer, co-author)

I. Development of electrosurgical currents

The apparent ability of a certain type of high frequency electric current arc to divide animal tissue was reported by Dr. LEE DE FOREST in 1808 with the suggestion that this electric current might be useful in surgery. CERNY similarly reported in 1910. The history of electrosurgery combines the story of the physics of electric currents and of surgery. Although the many electrosurgical units built between 1920 and 1930 gave satisfactory cutting and/or coagulating service in most electrosurgical procedures, they failed in endoscopic surgery. This failure was found to be due to the fact the water used for bladder distention, irrigation and visibility dissipated so much of the electric current that only a relatively small portion of it reached the prostatic tissue which the surgeon was trying to cut with the wire loop. Since 1930, when demands for electrourosurgical equipment became insistent due to publicity and approval of endoscopic prostatic surgery, units have been improved and developed; now the best ones cut tissue rapidly and electrocoagulate bleeding vessels adequately even though the active electrode is immersed in water or a nonelectrolyte fluid.

II. Characteristics of electrosurgical currents

a) Requirements for surgery

Electrosurgical cutting and coagulating currents must be distinguished from electrocautery, which is simply an electrically heated wire applying sufficient heat externally to burn tissue. Electrosurgical currents are high frequency oscillating (several million cycles per second) currents which, when guided by a needle, dull blade or wire loop, will cut or coagulate animal tissue upon striking it according to the quality and quantity of the current. With the aid of properly designed surgical current generators and with technical skill on the part of the surgeon, the accuracy of cutting and coagulating approaches the control customarily accomplished with conventional scalpels and scissors. There is also the material advantage of immediate control of bleeding and of a mechanical approach and dexterity in surgical problems not otherwise obtainable. Both are of exceeding importance in endoscopic surgery.

b) Cutting current

For the cutting of tissue a high frequency current in which the peak voltage is sustained at a continuous even level is essential. This is known as an undamped HF (high frequency) current (Fig. 128). It is the kind of current emitted by a HF vacuum tube and oscillator circuit.

c) Coagulating current

For the coagulation of tissue and control of bleeding a high frequency current which will coagulate the tissue (bleeding points) without further cutting is essential. This is accomplished by a similar HF oscillating current in which, how-

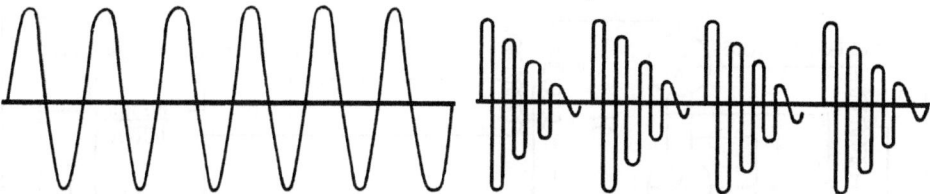

Fig. 128. Undamped high frequency current. The oscillations are all of c ual intensity. It cuts tissue with a minimum of coagulation

Fig. 129. Damped high frequency current. The oscillations are of unequal intensity. It coagulates and is used for controlling bleeding

ever, the peak voltage is not continuously sustained, but through a relatively short series of wave trains rapidly drops to near zero. It is known as a damped HF current (Fig. 129), and is the kind of current emitted by a spark gap oscillator circuit.

d) Combination currents

Over the years since 1910, makers of electrosurgical machines have designed equipment delivering either one or the other of these two basic HF currents. By various ingenious methods, several manufacturers attempted to modify one current to perform the function of the other. This led to production of equipment with considerable shortcomings; inadequate coagulation by the strictly tube oscillator machine and poor cutting on the part of spark gap oscillators. Early in the 1930's one of the authors (BARNES) had built for him a special machine containing both of these HF oscillator circuits with a means of selecting either with a twopedal foot switch. Since 1936,

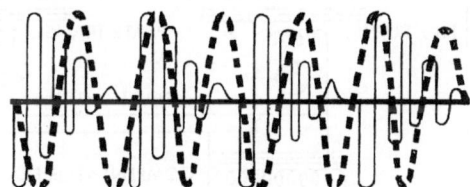

Fig. 130. Combined damped and undamped current superimposed. It is used for cutting when coagulation is also desired

however, the authors have employed equipment providing not only both basic types of HF current but having the ability to blend or mix at will, through the patient circuit, the two currents simultaneously. This provided a means of adding whatever amount of coagulation was needed according to the speed of the cutting stroke and the nature of the tissue. All major producers of electrosurgical machines in the United States (with one exception) have now come to this principle (Figs. 130 and 131). This basic idea of two completely different circuits emanating from the same machine, readily selectable by the operator, has been improved upon. At the recommendation of several surgeons throughout the country, electrosurgical units have been constructed equipped with a triple pedalled foot switch. At the operator's choice any one of three kinds

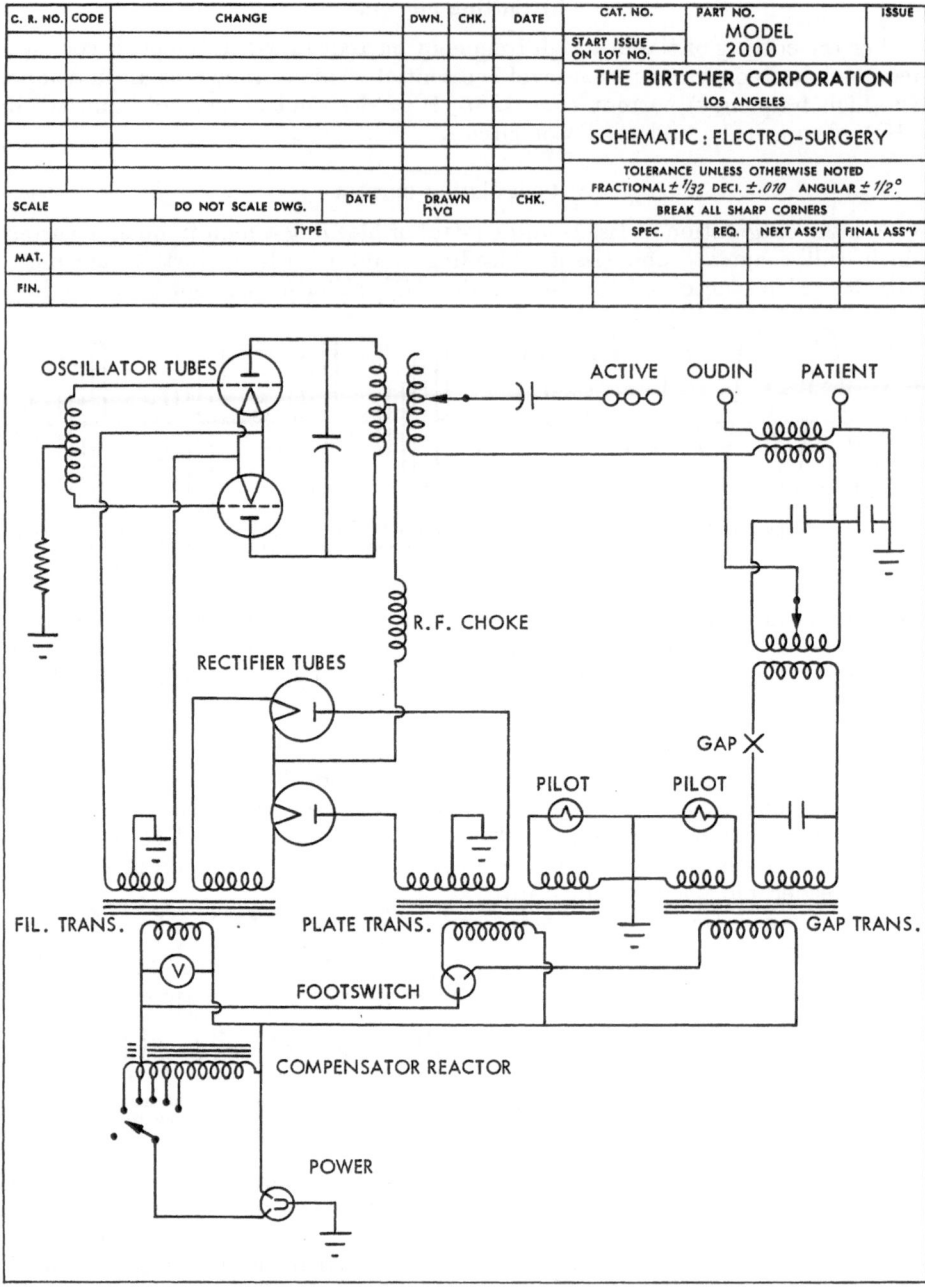

Fig. 131. Circuits of an electrosurgical unit which generates an undamped cutting current, a damped coagulating current and a combined or superimposed current. This machine requires a current input of 25 amps and 110 volts and has an output of 500 watts from the cutting current circuit and 275 watts from the coagulating current circuit

of current is instantly available: a true undamped cutting current from a tube circuit, a true damped spark gap coagulating current and a combined or superimposed cutting and coagulating current.

e) Modern electrosurgical units

Modern electrosurgical machines have had additional refinement to achieve a high quality electric current. They include a complete dual tube rectifier circuit preceding a heavy duty, dual tube oscillatory circuit, which definitely improves the speed and ease of cutting, particularly in neatly removing those elusive tags in the final steps of a prostatic resection. Both currents have been made more dependable also by building in means of continually measuring and quickly correcting supply line voltage deficiences. Such complicated equipment is of necessity both bulky and expensive, but is essential to the exquisite demands of urologic endoscopic surgery. This is the type of equipment which the authors now use exlusively and which they consider particularly essential to successful endoscopic prostatic surgery.

III. Effect of currents on tissue

a) Electrodes

Inasmuch as current density varies inversely with the surface area of the electrode, a large plate is used when the heat effect is to be kept at a minimum, whereas a small active or working electrode is employed when current density in the tissues in its immediate vicinity is to be increased. Obviously, an active electrode made of wire 10/1000″ in diameter requires less current and less time to produce reaction in adjacent tissue than one 18/1000″ in diameter. Active cutting loops employed in endoscopic prostatic surgery vary in diameter and may be selected according to need. Particularly tough, fibrous tissues sometimes do not yield readily; therefore, the operator selects a loop made of smaller diameter wire, 10/1000″ or 12/1000″. The larger diameter loops give longer service and control bleeding somewhat better, but cut slowly and tend to slide over the tissue rather than to cut clean.

b) Tissue change

A strong circuit flowing through the tissue creates such a high current density around the active electrode that desiccation is immediate and coagulation quick to follow; a still stronger current will cause an arc of sparks to pass between the operating electrode and tissue. The effect of a highly damped high frequency current is limited to heating and desiccation, but an undamped high frequency current of sufficient density produces a clean incision with a minimum of coagulation provided the excursion of the working electrode is executed with dispatch. If it is drawn through tissue slowly, more intense coagulation takes place. Considerable surgical and laboratory study has been conducted upon the effects of both the undamped and the damped high frequency current on tissue. Tissue change is minimal under the following conditions: 1. Powerful undamped tube rectified cutting current; 2. small diameter wire loop electrode; 3. rapid excursion of the loop through the tissue. Tissue removed under these conditions is suitable for biopsy study because the coagulating effect of the current penetrates into the tissue only 3 to 4 cells deep. Conversely when the cutting current is weak, necessitating slow excursion through the tissue, and/or is not completely undamped, the tissue change is deep; sections removed under these conditions are not suitable for biopsy study. There is also slower healing of the resected area when excessive coagulation occurs during the operation.

c) Faradism

Both cutting and coagulating currents should be completely free of faradism. This may be checked at any time by the operator by attaching two pieces of block

11*

tin, one to each side of the circuit, and pressing these snugly to a portion of the body. Each current may then be turned on. There should be no sensation of the passage of current through the body other than a mild heat. Any muscular twitching or excessive tickling sensation is an indication of faradism. This may be caused in a tube circuit by a loose connection in the machine, or by a leak in a condenser. In the spark gap coagulating circuit it may be due to a leak in a condenser, to a loose connection in the internal wiring circuit, or more often to pitted spark gap points in this section of the machine. It may also be an indication that the spark gap points are adjusted too far apart. On spark gap oscillators the gaps should be adjusted to 4/1000 to 5/1000 inch, should always be kept clean by polishing their surfaces with emery cloth every six months. At any time, an irregular or interrupted sound of the spark across the gap is an indication that the gaps need adjusting, perhaps cleaning.

Even though the electrosurgical unit is free of faradism, muscle jerking sometimes occurs. This is particularly hazardous during resection of tumors located on either lateral bladder wall. Jerking may be so violent that the loop electrode is thrust entirely through the bladder wall. The cause of this faradic effect is not definitely known, for it occurs even though no faradic current goes through the machine. It may be due to generation of faradism in the tissue by the reaction of the electrode on the tissue. It may be lessened by reducing the power of the electric current. Sometimes changing the plate electrode from the patient's back to the abdomen or to the thigh may help.

IV. Checking machine failures

At times the surgeon may be prepared to operate, and when the first stroke is made with the cutting loop he finds that there is inadequate current or no current whatsoever. The first thought is to blame it on the machine; but failure in first class equipment is relatively rare. Usually the trouble is due to some oversight on the part of an assistant in setting up the equipment for the operation. We have therefore listed, in the order of likelihood, mechanical items which should be checked. Generally these troubles can be discovered in a minute or two, and the operation continued without further mechanical difficulty.

1. Be sure the irrigating fluid is not an electrolyte such as saline solution.
2. Note that the cord which carries the current from machine to operating instrument is in good contact with the terminal of the operating cystoscope.
3. Check the contact of the indifferent electrode which is applied to the patient's back or buttocks. Sometimes when adjusting the patient to the proper operating position the indifferent electrode is permitted to slide upward to the small of his back, or his gown or the sheet has fallen between the electrode and his skin, resulting in poor contact. Remember that as large a portion as possible of the entire surface of the indifferent electrode should be in good, smooth contact with the patient's buttocks, for if such contact is not made there is great danger of skin burns, as well as of poorly functioning electric current. Observe also that no portion of the metallic part of the indifferent electrode is in contact with the metal of the operating table.
4. The bare skin of the patient should never be in contact with the metal of any part of the operating table; there is danger of severe burns where skin and metal are in contact. Note that the cord from the indifferent electrode is in contact with the terminal of the machine and is in the proper terminal.
5. Observe that the line cord from the wall socket to the machine is connected. If the machine is made to operate on the 110 volt alternating current, it must be attached to a plug in the wall providing this current, and not to a wall socket providing 110 volt direct current or 220 volt alternating current. If the machine has been designed for some type of current other than 110 volt 50 to 60 cycle, it will be so labeled specifically.
6. Observe that when the foot switch pedals are depressed, the tubes in the tube circuit glow brightly, and a spark passes across the gap in the spark gap circuit.

7. Some tube machines do not preheat the filament of the tubes, and the tubes merely turn on when the foot switch is depressed. Better machines provide a switch on the operating panel which must be turned on in advance of the operation and left on during it, to preheat the filament of the tubes. On this type of machine, note that the Filament Switch is turned to the "On" position.

8. Remove the cutting loop from the operating instrument and ascertain that it is intact, and that insulation which covers a portion of it is also intact and not "dog eared". Note that the shaft of the cutting loop is fully inserted into the socket of the cystoscope designed to hold it. Heat generated in the base of the cystoscope where it is held by the operator is an indication that the contacting point at the base of the loop, or lead cord carrying the current to the cystoscope, is failing to make good contact and creating a resistance which causes the heat.

9. If all aforementioned parts seem to be in order and there appears to be current at the active electrode although insufficient to cut or coagulate properly, check the possibility of a lowered supply of current at the proper voltage from the line socket. A large infra red lamp, fluoroscopic unit, electric sterilizer or large operating room lamp in the same line as the surgical unit may so lower the voltage that inadequate current is being provided the machine. The only correction is to attach the machine to a separate circuit or to have the other devices that are using the current turned off during the period of operation. A meter connected to the terminals of the line circuit should show a minimum of 110 volts and 15 amperes, and must not drop below this amount while the current is being used.

10. When current is turned on with the foot switch depressed, should one or both tubes fail to light, appear milky white inside the glass or completely black throughout, it is an indication that the tube is burned out and a new one is required.

11. If all other aforementioned items seem to be in order, a loose connection in the machine may be suspected. Often even the inexperienced may merely look into the machine and by jiggling various wires discover a loose connection. Usually it can be readily reconnected to the proper point, as ordinarily the wiring circuits are held in such position that a loose connection will reach one point only. If a coil of wire (resistors) or other electric device seems to be charred, it is an indication that the condenser or resistors have burned out and need replacing.

12. If the cystoscope light burns out when current from the machine is applied through the loop, it is an indication of a short either in the electrotome or from the light connection wires to the table.

V. Care of the machine

Once a year electrosurgical units should receive a general inspection, servicing and tuning. Do not permit inexperienced men to do this. Most manufacturers, being proud of their eqipment, will be glad to perform this service without charge provided shipping costs have been paid. The nominal cost of shipping represents little in comparison to keeping equipment ready for maximum service at all times.

When not in use the machine should be kept covered, so that it does not accumulate dust inside or out. Insist upon nurses keeping the surface clean and polished. Never replace a burned out tube with any other than the exact type and make that came with the machine.

Chapter XVII

Indications for endoscopic surgery

I. Training, ability and experience of the surgeon

The indications for endoscopic surgery vary within wide limits, depending chiefly upon the surgeon who is to perform the operation. If he has had adequate training and extensive experience in endoscopic surgery, the indications for him will include the removal of prostates weighing up to 150 to 200 grams, the resection of bladder tumors as large as 4 to 5 cm. in diameter, the crushing of stones up to 3 to 4 cm. in diameter and other procedures which are technically difficult.

Surgeons who have not had this extensive training and experience, or who are unequipped in other ways to perform the more difficult endoscopic procedures, will limit the use of this approach to the small sized lesions. If a surgeon has had no training or experience in endoscopic surgery, there is no indication for its use by him; it is better for him not to attempt it. For the average urologist, endoscopic surgery is indicated for relief of bladder neck obstruction due to prostatic carcinoma which does not respond to hormone therapy and for removal of bladder neck contractures, median bars and the smaller prostatic adenomas. Small bladder tumors, small stones and sometimes Hunner ulcers and other intravesical lesions can usually be removed more successfully by the average urologist through the endoscopic approach than by open surgery. The ultimate factor determining the approach best suited to a given case is the operator himself. If he has established proficiency at removing large prostates and other lesions endoscopically, there is no contraindication to continuance of this method in his hands. But if his series of endoscopic operations is comparatively limited, he should restrict utilization of this approach to the smaller lesions and not attempt more extensive procedures until he has worked up to them. Should he display no aptitude for this type of surgical performance, he should rule it out of his technical armamentarium entirely, and continue to give his patients the benefit of his skill in open surgery.

II. Differential diagnosis

1. Indefinite symptoms

An accurate preoperative diagnosis is one of the components of successful endoscopic prostatic surgery. It is extremely important to know whether or not the symptoms are actually due to the lesion at the vesical outlet. There is a greater tendency to carry out an endoscopic procedure on a small prostate which might not be causing symptoms than there is to remove a small gland through the suprapubic or perineal approach. The knowledge of the difficulty of enucleation of the small prostate deters the surgeon from attempting to remove it through the open approach; but he often decides to resect a few pieces from the bladder neck, hoping it will bring relief of symptoms. Those who follow this reasoning are disappointed in the results they obtain from the endoscopic operation.

The patient who complains of vague "pain in the prostate", discomfort in the perineum, rectum and external genitalia, burning pain on urination and urinary frequency — especially during the daytime — requires a critical evaluation. If examination reveals no residual urine, little or no prostatic enlargement and no trabeculation in the fundus of the bladder, surgery will not benefit him. Even though some elevation of the dorsal lip of the bladder neck or of the interureteric ridge exists, to remove this questionable obstruction will probably not relieve his symptoms.

2. Residual urine

The presence of residual urine is one of the most important criteria for determining whether or not endoscopic prostatic surgery is indicated (KIRWIN et al.). In the absence of residual urine, the probabilities are that endoscopic resection of the bladder neck or prostate will be of no benefit. There are two exceptions to this rule: 1. an exceptionally strong bladder muscle which forces the urine out through the obstruction (trabeculation in the fundus of the bladder is present in these cases), 2. sufficient intravesical protrusion of the prostate for it to act as a foreign body causing irritation; the bladder tries to expel the foreign body.

Endoscopic surgery for removal of vesical neck obstruction is not often indicated when residual urine measures less than 50 cc. There are exceptions to this rule, as noted above, but nine times out of ten the results of surgery will be disappointing unless patients presenting less than 50 cc. of residual have been carefully evaluated and selected.

3. Bladder tone

An estimate of bladder tone (p. 58) may be helpful in making a differential diagnosis and in deciding if surgery is indicated. Poor tone may be myogenic, due to a large amount of residual urine which has persisted over a long period of time. In these cases residual urine may persist following surgery, but if there is a definite obstructive lesion at the outlet from the bladder, surgical removal of the lesion is indicated regardless of hypotonia. Poor tone is often due to a central nerve lesion. Wide resection of the bladder neck and prostatic urethra will help these patients to empty the bladder unless there is already a wide open funnel shaped vesical outlet. Hypotonia may also be due to a large vesical diverticulum. The prostate is not greatly enlarged in most of these cases; removal of the obstruction without extirpation of the diverticulum does not eliminate the status of residual urine unless the diverticulum is small and empties readily.

4. Cystoscopic examination

This should be carried out for the purpose of differential diagnosis unless symptoms and findings are diagnostic. When symptoms are vague, when the prostate is normal by rectal palpation, when there is no residual urine and when bladder tone is poor, endoscopic examination of the bladder is indicated.

5. Cystogram

This is helpful in differentiating the causes of hypotonia. The bladder outline in neurogenic hypotonia is small and regular and the dome is rounded. Myogenic hypotonia is more likely to cause a slightly irregular bladder outline and a more or less pointed dome. A diverticulum is easily demonstrated by the cystogram.

III. Size of the lesion

1. Duration of operation

The size of the lesion to be removed is an important criterion in deciding whether the endoscopic or the open approach is indicated. The speed and accuracy with which an operator can remove a prostate endoscopically play the decisive role in determining the size of gland he should undertake to operate upon in this way. If he is unable to remove more than one-half gram of tissue per minute, he should not attempt to resect prostates which will yield more than 30 grams, but should limit his work to Grade I and smaller Grade II enlargements. If, however, he is capable of excising one gram per minute, he may safely engage in extirpation of all Grade II and the smaller Grade III hypertrophies, for his speed will be equivalent to 60 grams an hour. When the resectoscopist has become especially proficient and can take out two or more grams per minute, it is evident that this ability will enable him to remove practically all Grade III hypertrophies within one hour and a half (O'BRIEN et al.). Bleeding from a large gland retards the operation considerably, and sometimes it is difficult to determine beforehand just how much bleeding will be encountered. As a general rule it may be stated that the softer and larger the prostate and the longer the operating time the greater the hemorrhage. If preoperative cystoscopic examination had been done, the amount of bleeding caused by this instrumentation and the degree of dilatation of the vessels in the prostatic mucosa and about the bladder neck provide a rough index of the volume which may be expected during operation. Most carcinomatous glands bleed very little, and the rigid malignant tissue does not fall into the prostatic urethra as this region is being made concave, but remains in a stationary position. In benign hypertrophy, especially in the softer glands, there is increased tendency to bleeding and considerable falling in of the prostate as it is being resected, necessitating removal of a greater amount of tissue.

2. Estimate of size and consistency of the prostate

a) Digital palpation through rectum

This is an examination which must be done carefully, thoroughly and intelligently. Rough or forceful manipulation traumatizes the prostate and may result in dissemination of infection into the blood stream, producing pyelonephritis which is sometimes called "urethral fever". It is therefore essential that palpation be very gentle and by all means not repeated by more than two consecutive examiners. The entire surface of the gland is covered with the palpating finger, and especially is the depth of the groove between the lateral surface of the prostate and the lateral rectal wall noted. The size, consistency, mobility, tenderness and regularity of the gland are remarked.

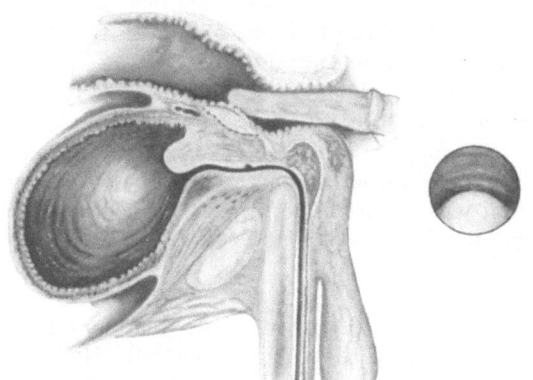

Fig. 132. Grade I enlargement of prostate as palpated through anterior rectal wall. Prostate encroaches into rectal lumen from 1 to 2 cm

An accurate estimate of the size of the prostate by rectal palpation is sometimes difficult to make, but experience increases the examiner's acuity. It is important to distinguish its approximate volume, as this is often a decisive factor in selecting the proper method of surgical treatment. For convenience prostates are divided into four grades according to size, and although this classification is more or less empirical, it provides a satisfactory means of recognizing group measurements. To assist in standardizing the grading of prostates, Table IV is given (Fig. 132).

During palpation of very large prostates, the finger is stopped by the protuberant gland which extends into the rectal lumen at almost right angles to the anterior rectal wall. It is impossible to proceed up over the enlarged prostate unless the finger is rotated and flexed so that the back of it passes upward and palpates the bulging prostatic mass; this size prostate is Grade IV.

The fibrotic prostate, which is an almost constant finding in patients with bladder neck contracture or median bar, is no larger than a normal prostate and may be somewhat smaller than average. When palpated through the anterior rectal wall, it is found to be harder than normal, and smooth and movable; if a cystoscope is in the urethra, a transverse ridge is felt to extend across the cystoscope at the bladder neck.

Table 4. *Summary of methods of estimating size of prostatic enlargement.*

Size of gland	Method of examination				Approximate amount of tissue to be removed
	Rectal palpation	Cystoscopy			
		Intraurethral lateral	Intraurethral middle or dorsal portion of laterals	Intravesical middle or dorsal portions of laterals	
Normal	Encroaches 0—1 cm. into rectal lumen	Concave lateral prostatic urethral walls	1—2 cm. between veru and prostatic border	Does not cover trigone	Up to 10 gm. (average normal weight 8 to 10 gm.) excluding capsule
Grade I	Encroaches 1—2 cm. into rectal lumen	Lateral lobes bulge inward but do not touch in midline	2—3 cm. between veru and prostatic border	Covers up to ½ of trigone	Up to 20 gm.
Grade II	Encroaches 2—3 cm. into rectal lumen	Lateral lobes touch in midline	3—4 cm. between veru and prostatic border	Covers from ½ to all of trigone	20—50 gm.
Grade III	Encroaches 3—4 cm. into rectal lumen	Lateral lobes touch in midline for 2 to 3 cm.	4—5 cm. between veru and prostatic border	Covers more than trigone	50—125 gm.
Grade IV	Encroaches more than 4 cm. into rectal lumen	Lateral lobes touch in midline more than 3 cm.	More than 5 cm. between veru and prostatic border	Extends up into fundus	More than 125 gm.

The carcinomatous prostate is always harder than the normal gland; if only a small area is involved, this section alone is indurated and the remaining portion is normal in consistency. The malignant prostate is also more fixed than the normal gland, and if the new growth is extensive it may have the consistency of bone. Its contour is not well defined, the edges extending laterally beyond the side of the rectal wall. It is sometimes difficult, or even impossible, to differentiate carcinoma of the prostate from subacute or chronic prostatitis. Sometimes the latter condition is complicated by periprostatitis which results in marked inflammatory induration extending beyond the confines of the prostatic capsule, so that by palpation it cannot be distinguished from malignancy.

b) Cystograms and urethrograms

These may aid in determining the size of the prostate. A smooth rounded negative shadow projecting from the base of the bladder up into the cystographic outline is suggestive of intravesical enlargement; when a corresponding shadow is positive in the contrast cystogram the diagnosis is confirmed. Intraurethral hypertrophy is indicated by an elevation of the base or lower margin of the cystogram above the pubis. Normally the base is not more than 1 cm. above the upper margin of the pubic bone. A urethrogram taken in the oblique position shows an elongated prostatic urethra and an elevation of the vesical orifice when the prostate is enlarged (Fig. 133). Determination of the size of the prostate by cystogram and urethrogram is, however, not always accurate. A space filling lesion may cast a shadow 2 or 3 cm. up into the bladder, or the bladder may be elevated 2 or 3 cm. above the pubis, whereas subsequent cystoscopic examination may reveal a prostatic enlargement not in excess of Grade I. Conversely, very little or no evidence of prostatic enlargement may be apparent on the cystogram, but at the time of endoscopic resection 50 or 60 gm. of tissue must be resected. An estimate of the size of the prostate by measuring the length of the urethra on the urethrogram is somewhat more accurate in determining its size.

c) Cystoscopic examination

Endoscope used. Whenever uncertainty exists regarding whether intravesical hypertrophy is too extensive for endoscopic removal or whether other pathology is present in the bladder, cystoscopic examination should be carried out (SHIVERS). Different types of instruments are recommended for this procedure, but the author has found the No. 16 F. BROWN-BUERGER with a convex sheath to be satisfactory. It passes easily into the bladder, and the convex sheath can be drawn out into the prostatic urethra so that the intraurethral portion of the prostate may be thoroughly examined. When water is run through the cystoscope, the lateral lobes of the prostate separate, and the size of these lobes, the distance between the verumontanum and the internal prostatic border, and the extent of intravesical encroachment of the middle lobe can be estimated. The foroblique telescope is preferred by many urologists.

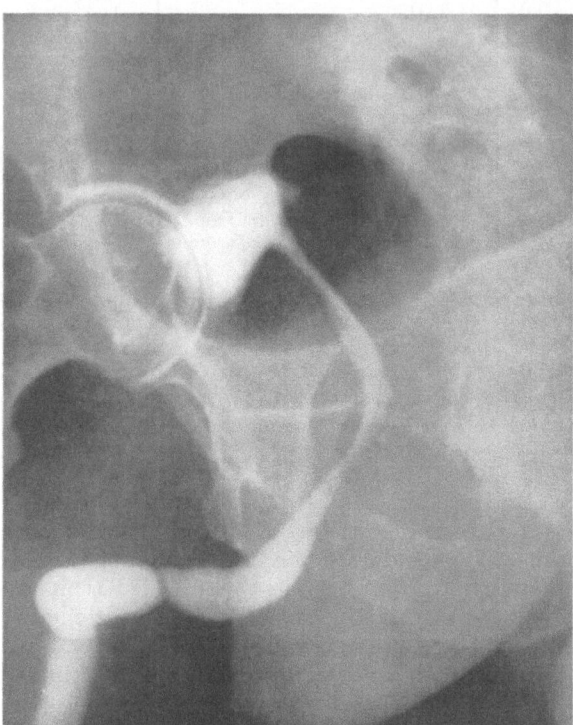

Fig. 133. A cystourethrogram showing elevation of vesical outlet and intravesical protrusion of prostatic hypertrophy. Forty grams of tissue were removed from this patient by transurethral prostatic resection. Courtesy Hector Le Duc, M.D. Note length of compressed prostatic urethra

Lateral lobes. Cystoscopic grading of prostates into size groups is empirical, like rectal grading. The intraurethral lateral lobes are classified as follows: when their intrusion into the prostatic urethra is marked by only slight inward bulging and they do not touch in the midline, they are Grade I. When they meet in the midline, they are Grade II. When they join for a distance of from 2 to 3 cm. they are Grade III. Grade IV intraurethral enlargement is rare, and is found in cases in which the prostatic urethra is more than 5 cm. long.

Length of the prostatic urethra. This is a good indication of the size of the gland. The distance between the verumontanum and the intravesical prostatic border can be determined during cystoscopic examination by measuring the excursion of the ocular end of the cystoscope as it is passed inward from the verumontanum to the prostatic border. This distance is from 1 to 2 cm. in the normal prostate (Fig. 134); from 2 to 3 cm. in Grade I intraurethral hypertrophy (Fig. 135); 3 to 4 cm. in Grade II; 4 to 5 cm. in Grade III, and more than 5 cm. in Grade IV.

Often there is minimal, and occasionally marked, hypertrophy between the external sphincter and the verumontanum, but this does not occur often enough to constitute a prominent factor in determining the size of the gland. In most

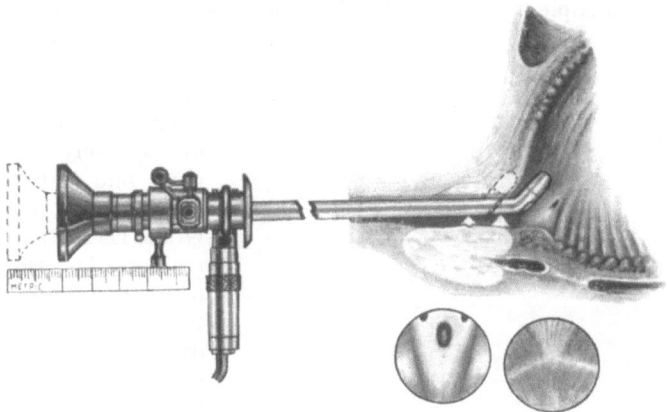

Fig. 134. Normal prostate. Distance between verumontanum and bladder neck is 1 to 2 cm as measured with cystoscope. There is no elevation between the apex of the trigone and the dorsal lip of the bladder neck. Insert shows cystoscopic appearance of verumontanum and bladder neck

cases it can be disregarded when calculating the amount of tissue to be removed during resection.

Intravesical protrusion. The dorsal portion of the normal bladder neck or prostatic border is elevated very little above the apex of the trigone, the floor of

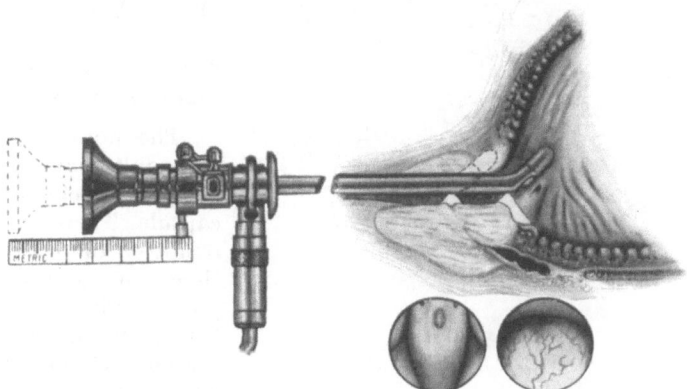

Fig. 135. Grade I hypertrophy. Distance between verumontanum and prostatic border is 2 to 3 cm

the prostatic urethra being on practically the same plane as the trigone (Fig. 134). The volume of the intravesical enlargement is estimated by the proportion of the trigone or fundus covered by the gland as it projects upward into the bladder. When encroachment of the middle lobe over the trigone is slight and does not cover more than half of the trigone as viewed through the right angle lens, it is graded I (Fig. 136). When the middle lobe covers the trigone up to the ureteral orifices, the grading is II (Fig. 137). When it extends above the ureteral orifice, it is Grade III. When the entire lower protion of the bladder is filled, it is Grade IV.

d) Correlation of all examinations

So many ways of defining the size of a prostate may seem confusing; however, a correlation of all these examinations is helpful in determining which glands are suitable for endoscopic removal. For instance, rectal palpation may lead to a diagnosis of Grade I enlargement, but cystoscopic appraisal of the same gland may reveal that the middle lobe entirely covers the trigone and extends up into the fundus, making it a Grade III intravesical hypertrophy. Or, there may be no intravesical encroachment to cover the trigone, but the distance between the verumontanum and the prostatic border may be $3^1/_2$ cm. — which would make it a Grade II enlargement. This combined method of grading prostates as to size provides the surgeon with a practical estimate of the amount of tissue which must be removed from any given gland in order to assure complete resection (Table IV, p. 69).

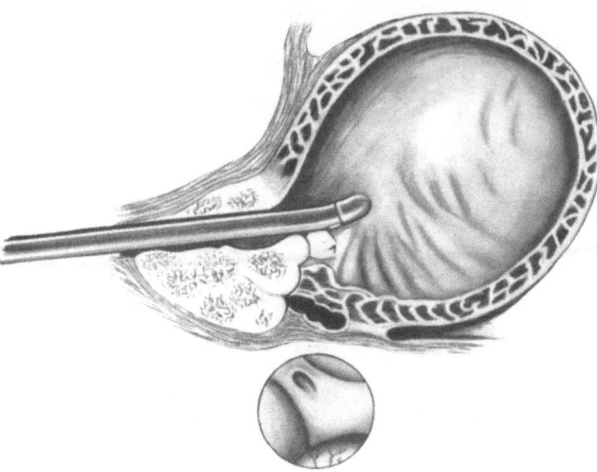

Fig. 136. Grade I intravesical hypertrophy. Dorsal prostatic border covers apex of trigone as viewed through right angle lens. Insert shows cystoscopic view

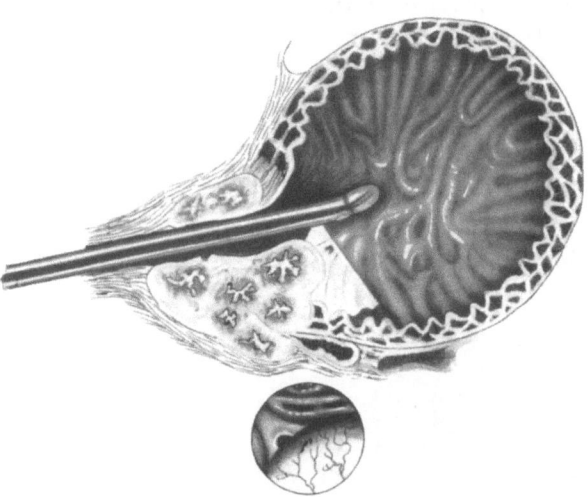

Fig. 137. Grade II intravesical hypertrophy. Prostate covers trigone. Ureteral orifice is barely visible above prostatic border. Insert shows cystoscopic view through right angle lens

3. Correlation of size with amount of tissue removed

The normal prostate in the adult male weighs 8 to 10 gm. exclusive of the capsule. Therefore, complete removal of the glandular tissue of a normal sized prostate yields 8 to 10 grams. When the prostate is calibrated as Grade I, from 10 to 20 gm. of tissue are removed when the entire adenoma is extirpated. A Grade II gland yields 20 to 50 gm., a Grade III, 50 to 125 gm., and a Grade IV size weighs more than 125 gm. These weights are obviously approximate, and sometimes even after careful evaluation of the size of the gland, more or less tissue is removed by endoscopic means than is estimated before surgery. When, however, the amount of tissue removed consistently weighs less than the preoperative estimate, it is probable that adequate removal of the gland is not being accomplished.

4. Estimate of size of vesical tumors and stones

a) Cystoscopic examination

The size of tumors and stones in the bladder can be estimated through the cystoscope by measuring the distance from one edge of the lesion to the other just as the length of the prostatic urethra is measured (p. 170). The objective lens of the cystoscope is placed as close to the lower edge of the lesion as possible, and the position of the eyepiece noted. The instrument is then advanced until the upper edge of the lesion comes into view, and the distance measured. The width of the lesion cannot be measured in this way but the experienced cystoscopist can estimate the size fairly accurately (MARSHALL).

b) X-ray examination

Stone. X-ray examination is the best method of measuring the size of a vesical calculus. The opacity shown in the film is only very slightly larger than the actual size of the stone. If the long diameter of the calculus is in the anteroposterior plane, the opacity may not be as large as the stone. However, at the time of litholapaxy, it is possible to engage the narrow diameter of the stone with the lithotrite jaws. Cystograms show a filling defect corresponding to a nonopaque bladder stone.

Tumor. The space filling lesion of a bladder tumor is also evident in most cystograms (BRUNKOW). Estimation of the size of a tumor, however, cannot be as accurate as that of a stone. The tumor may not project out into the bladder enough to produce a negative shadow, or the nonopaque space may not contain the entire tumor. Conversely, bladder spasm or neoplastic or inflammatory induration may produce an encroachment which is larger than the actual size of the tumor.

5. Indications based on size

a) Stone

The indications for endoscopic removal of a bladder stone or tumor depend to a large extent upon its size. It is obvious that a calculus which is too large to be grasped by a lithotrite cannot be crushed. The largest size which will not slip from the jaws of the instrument measures about 3 cm. in diameter. When the stone is soft and/or rough it will stick in the jaws, whereas a hard and/or smooth one of the same size will more likely slip out. The surgeon who has had training and experience in litholapaxy can crush stones up to 3 cm. in diameter easily and atraumatically (MIDDLETON).

b) Tumor

The size of a bladder tumor which can be removed endoscopically depends largely upon the training and skill of the surgeon. Criteria are similar to those pertaining to prostatic resection (p. 165). Small tumors up to 1 cm. in diameter can be electrocoagulated, but removal with a resectoscope loop is preferable because by so doing a specimen can be obtained for biopsy study. The skilled endoscopic surgeon can safely and adequately remove tumors up to 6 or 7 cm. in diameter unless they have invaded through the bladder wall (THOMPSON and KAPLAN). The surgeon who is unskilled in endoscopic procedures should not attempt to remove tumors larger than 2 cm. in diameter by this approach (HECKENBACH). The same criteria apply to the surgical management of multiple tumors.

IV. Invasion and malignancy of bladder tumors

1. Invasion

Invasion of a tumor through the bladder wall is a contraindication to its removal by endoscopic means (MELICOW). It is not always possible to make a preoperative evaluation of the depth of invasion. Tumors which are flat and ulcerative are nearly always deeply invasive; when this type measures more than 3 cm. in diameter, it is more amenable to extirpation by the open approach (BAKER). Some tumors of bilharzial origin are infiltrative but others can be removed adequately by endoscopic resection (NEWMAN).

2. Malignancy

Highly malignant tumors, as determined by cell structure, can be as successfully removed by endoscopic surgery as by an open operation unless they are invasive through the bladder wall or are too large for extirpation through the transurethral approach. Most of the rare benign bladder tumors except endometrioma (MOORE et al.) can be treated successfully by endoscopic surgery (GRAHAM and BUCKLEY). They are almost never too large for extirpation by this approach.

V. Position of the lesion

1. Bladder tumors

a) In the dome

Tumors located in the dome of the bladder are difficult to remove by endoscopic means. When they are close to the vesical orifice on the ventral wall, it is almost impossible to reach them with the resectoscope loop. Heavy suprapubic pressure by an assistant's hand depresses the ventral wall, making it easier to reach with the instrument. Women patients may be placed on the table face down; in this position the lesion is on the floor of the operating field and the pressure of the table pushes the ventral bladder wall inward. Most bladder tumors larger than 1 cm. in diameter, located in the dome and ventral wall of the bladder are better removed through a suprapubic incision; they are easily accessible by open surgery and they may be urachal in origin and extend entirely through the bladder wall (WESSELL et al.).

b) On the floor

Tumors located on the floor of the bladder are more accessible to the resectoscope. If they cover a ureteral orifice they are removed with the cutting current applied rapidly; excessive coagulation predisposes to fibrosis. Sclerotic stenosis of the ureteral orifice is more likely to occur when the area is electrocoagulated. The ureter is usually dilated above the tumor in these cases, and when the growth is resected the ureter is seen to be wide open, and often the cut end of the ureteral mucosa projects outward for one or two millimeters; when this occurs, fibrotic stenosis is not a sequela.

c) In the fundus

Most tumors located in the fundus, including the lateral walls of the bladder, are also accessible to the resectoscope and can be adequately treated endoscopically unless they are too large or invade through the bladder wall.

2. In a diverticulum

Bladder stones and tumors located in a diverticulum are usually not accessible to endoscopic removal. When the diverticulum is small its orifice may be incised with the resectoscope knife and the end of the instrument advanced into its cavity. The stone can then be manipulated out into the bladder where it is crushed. A small tumor in a small diverticulum may be electrocoagulated or resected. Most tumors in diverticula, however, can be removed better by open surgery.

3. Beneath an overhanging prostate

A stone or tumor located beneath an overhanging middle lobe usually cannot be removed transurethrally if the intravesical prostate protrudes above the trigone. It is necessary to resect the prostate first, then the lesion beneath it is accessible to endoscopic surgery. However, by careful manipulation of the lithotrite with the jaws pointed dorsalward, a stone in this position may sometimes be crushed before removal of the prostate.

VI. Prostatic carcinoma

1. Use hormone therapy first

Urinary obstruction due to carcinoma of the prostate is relieved in most cases by hormone therapy. After three or four weeks of estrogen therapy the patient can usually void freely. Endoscopic removal of the obstruction is therefore not indicated. If the obstructive symptoms are not relieved by hormone therapy, which is true in one case out of ten, endoscopic removal of the obstruction becomes necessary. Sometimes the obstruction from prostatic carcinoma is relieved for several years by endocrine therapy, then recurrence takes place which this time does not respond to medical treatment. In these instances the transurethral approach is far superior to open operation. The obstructing tissue can be easily removed by the resectoscope, whereas removal by enucleation through either a suprapubic or a perineal exposure is difficult; the entire area is indurated, there is no line of cleavage to follow and it becomes necessary to tear or cut out sufficient tissue to open the channel.

2. Occult carcinoma (de Albuquerque)

Small areas of malignancy in the prostate can be removed by adequate endoscopic surgery as well as, if not better than, by enucleation through any open approach. Even the surgical or false capsule of the gland can be removed by endoscopic resection, whereas this is always left in place when enucleation is employed. Therefore, unless the extent of malignant involvement is small enough to be amenable to complete extirpation by total prostatectomy, there is no contraindication to removing it by endoscopic means. There is no evidence that electroresection of the neoplasm tends to spread it.

VII. Bladder neck contracture and median bar

1. Suitable for endoscopic surgery

Some urologists limit their endoscopic prostatic surgery to the removal of obstruction due to bladder neck contractures and median bars. These lesions are the most suitable of any for endoscopic extirpation. When they occur in adults, either men or women, the standard electrotome is used; when they exist in children, the instruments designed for use in infants or children are indicated.

2. Difficult to evaluate

Bladder neck contracture and median bar are the most difficult conditions to evaluate in attempting to determine whether or not patients will benefit by endoscopic resection. The criteria given on p. 166 (differential diagnosis) should be followed when making this decision. The end results following operation on these cases are not nearly so good as those after endoscopic removal of prostatic adenoma; the contracture often recurs, bladder neck irritation may persist and the patient is sometimes worse off than he was before surgery. When cases are carefully selected, however, and when the operation is performed skillfully, results on the whole are gratifying (PEYTON).

3. Contracture in women

The indications for endoscopic resection of the bladder neck in women are discussed in Chapter XXI.

4. Contraindications

There are three situations in which the open approach is preferable to the endoscopic route: 1. recurrent bladder neck contracture with hard extensive fibrosis; endoscopic resection tends to increase the fibrosis and stenosis recurs; 2. contracture of the vesical orifice in male infants, and 3. in children when no suitable resectoscope is available. In these situations a plastic procedure on the bladder neck performed through a suprapubic approach is indicated.

VIII. Chronic prostatitis and prostatic calculi

1. Intractable prostatitis

Chronic prostatitis may be intractable even after long continued nonsurgical treatment. When symptoms persist from such a condition they may be relieved by endoscopic resection. It is rarely necessary to resort to surgery, however, unless there is obstruction from fibrosis or hyperplasia of the infected gland. If the surgeon is unable to remove practically the entire gland by endoscopic means he should not attempt to treat chronic prostatitis per se in this way; for if more than a small fraction of the infected prostatic tissue is allowed to remain, symptoms will probably be more marked following the operation than they were before.

2. Prostatic abscess

A small abscess situated in the portion of the prostate near the urethra may be successfully treated by endoscopic means, but larger abscesses which are easily palpable by rectum and are fluctuant, should be opened by perineal incision.

3. Tuberculous prostatitis

The tuberculous prostate which is causing symptoms can, in most cases, be removed successfully by endoscopic resection (GREENBERGER and WINER). Healing is considerably slower than in nontuberculous prostatitis. Antibiotic treatment hastens healing and helps to prevent dissemination of the tuberculous infection.

4. Prostatic calculi

The presence of calculi in the prostate does not contraindicate endoscopic removal of the gland. Inasmuch as the concretions are most numerous near the capsule, and inasmuch as the calculous prostate is always infected, it is necessary that complete endoscopic removal be effected in order to provide complete relief of symptoms. Any rests of tissue will probably contain stones and harbor infection which will result in continued irritation even though the obstruction has been obliterated. Endoscopic removal of the entire prostate down to the true capsule provides more complete relief of symptoms than can be obtained by any other means except total prostatectomy (removal of capsule) through an open approach. Suprapubic and also perineal enucleation both are unsatisfactory in most of these cases, for frequently the prostate is fibrotic and sometimes the calculi reside in the surgical capsule. These contingencies make enucleation of the gland by the above techniques both difficult and incomplete.

IX. Neurogenic bladder dysfunction

1. Difference of opinion

Considerable difference of opinion exists concerning indications for endoscopic resection of the bladder neck associated with neurogenic bladder dysfunction. This varies between urologic surgeons who operate as soon as urinary retention occurs due to a neurological lesion (EMMETT and BEARE) and those who never attempt to improve micturition by endoscopic surgery.

2. Positive indications

The most positive indication for endoscopic resection of the bladder neck is urinary retention and/or voluminous residual urine of six months' duration or longer, associated with good bladder tone, and a definite elevation of the dorsal lip of the bladder neck or a collar contracture of the vesical outlet (MALAMENT and BUNTS). In the absence of residual urine the operation is of no benefit, unless the bladder muscle is hypertonic and forces the urine out through a definitely obstructive lesion at the bladder neck. When bladder tone is poor and the vesical orifice is wide open and funnel shaped, the operation is seldom indicated even though residual urine is present. It is, however, indicated in most cases which display residual urine and a vesical orifice which is not widely dilated.

3. Acute neurological lesions

Acute neurological lesions, such as injury to the spinal cord or disturbance of the nerve supply to the bladder during abdomino perineal resection of the rectum, often cause urinary retention. In many cases normal bladder function becomes restored within several months. It is therefore better to postpone endoscopic resection of the bladder neck for at least three or four months following the onset of symptoms; endoscopic resection is indicated when urinary retention persists after this length of time.

X. Multiple lesions

1. Obstruction and stone

Multiple lesions in the bladder can be removed endoscopically if the other criteria indicating the transurethral approach are met. When a bladder stone coexists with a bladder neck obstructive lesion, it is best to crush the stone first, then resect the prostate. By so doing the fragments of the calculus are evacuated with the pieces of prostate. When the bladder neck surgery is done first, there is more danger of injury and hemorrhage from the litholapaxy. However, if the stone is located under an overhanging intravesical protrusion of the prostate it may be necessary to remove the latter before the stone can be grasped with the lithotrite.

2. Obstruction and tumor

Bladder tumors and the obstructive prostate can be removed at the same time. Although cases of bladder tumors recurring in the urethra have been reported (KIEFER), there is no positive evidence that tumor implants occur in the prostatic urethra as a result of this combined surgery. It is usually best to resect the prostate first because removal of the tumor often leaves the bladder wall thin; when the tumor is removed before the prostate, the danger of rupture of the bladder wall is increased.

3. Obstruction and diverticulum

Vesical diverticula which are small and drain well as shown by postvoiding cystogram, do not require treatment (ADAMS). Larger ones which do not drain require surgical intervention. A few of these can be treated successfully by transurethral diverticulotomy at the time of endoscopic prostatic resection (BARNES and BERGMAN). Most of them, however, require suprapubic diverticulectomy which may be performed either before or after endoscopic removal of the bladder neck obstruction. Some surgeons prefer to carry out both procedures at the same session through a suprapubic approach. The prostate in these cases, however, is often small and difficult to enucleate; therefore the endoscopic approach is usually preferable.

XI. Multiple stage operations

Multiple stage endoscopic operations are, in the author's opinion, seldom indicated. If the prostate or a bladder tumor is too large to resect during one operation of one hour and a half, it is better in most cases to use the open approach rather than to plan two or more stages. An exception to this rule is the very poor risk patient who presents a large lesion and who would probably not tolerate open surgery. In these cases a planned multiple stage operation is sometimes indicated. When removal of the prostate in two stages is contemplated, or when for some unpredictable reason it is not possible to remove it all during one operation, it is preferable to resect one side completely and leave the other intact; when a portion of each side is removed, there is more postoperative bleeding, infection, sloughing, pain and irritation.

XII. Surgical risk

1. Tolerance to endoscopic surgery

Endoscopic surgery, when performed wisely and skillfully, is less shocking to the patient than is open surgery. Many patients whose general condition is too poor to tolerate an open surgical procedure suffer no ill effects from expert endoscopic removal of the lesion. Conversely, endoscopic surgery when performed unwisely and without skill is potentially more dangerous than open surgery.

2. Evaluation of surgical risk

An evaluation of surgical risk includes a general appraisal of the patient's strength, vitality and attitude as well as a knowledge of the results of laboratory examinations and tests. In fact his general condition is more important than almost any tests which might be made. If the patient is able to do a bit of work in his garden, if he likes to get up in the morning rather than remain in bed most of the day, and if his appetite is good, he is nearly always a good surgical risk — even though tests reveal renal function to be low and blood urea nitrogen elevated, and the electrocardiogram shows a cardiac infarct. Conversely, if the patient is lethargic, has a poor appetite, and is afraid of a surgical operation, he is a poor surgical risk even though all laboratory tests are within normal limits.

3. Improving the risk

It is seldom wise to delay performing endoscopic surgery more than two or three days in order to improve the risk. During this time of observation and treatment of the poor risk patient, diabetes can be controlled, digitalization started if necessary, most acute infections brought under control and anemia eliminated with transfusion of whole blood or red cells.

4. Poor renal function

Poor renal function and high blood urea nitrogen in themselves seldom contraindicate proceeding immediately with endoscopic surgery. If the etiology of renal damage is obstruction to urinary outflow, the sooner the obstruction is removed the better. Progressive kidney damage is reversed more quickly and safely by removing the obstructive lesion than by temporary means such as indwelling urethral catheterization. Therefore, when the general condition of the patient is good, he is a good surgical risk regardless of abnormal routine laboratory findings.

5. Poor risks

A few candidates for endoscopic surgery are bedfast due to disabling cardiac decompensation or recent infarct, or are so old and weak or disabled from other cause that they must remain in bed. These represent poor surgical risks, and even endoscopic surgery should be performed only after careful calculation of the risk involved. By expert handling, however, most of them can be brought through the endoscopic procedure without a high mortality rate (ANTONIO).

Chapter XVIII

Examination, preoperative care and selection of the anesthetic

I. Preoperative examination

This is similar to the examination required for open prostatic surgery (Vol.XIII). Examinations to determine the indications for endoscopic surgery are given in Chapter XVII.

II. Preoperative care

1. Bladder drainage

Advanced age and the related general condition of patients requiring endoscopic prostatic surgery sometimes necessitate an extended interval of preoperative care. On the other hand, preliminary treatment may be overdone, and long continued indwelling catheterization or even frequent catheterization may result in creating a poorer surgical risk rather than improving the condition. BUMPUS

12*

and MASSEY and others have shown that prolonged bladder drainage before endoscopic surgery is not necessary unless marked infection or impairment of renal function is present. Even these complications often clear up better after surgery than before. In preparation for suprapubic or perineal prostatectomy, it has been considered essential, by some surgeons, either to catheterize the patient several times daily or to leave a retention urethral catheter in place for a week before surgery. This, however, does not apply to preparation for endoscopic prostatic resection. In the past, many patients presenting obstruction associated with little or no infection in the urinary tract and no impairment of renal function were treated by indwelling urethral catheterization — a method which resulted in sepsis and impairment of kidney function. This practice is still frequently seen on a charity service which provides only one operation day a week. The patient may be admitted on the allotted operating day or the next day, and therefore is obligated to wait a week before he is taken to surgery. During this interval he may change from a good surgical risk to a poor one, and sometimes it becomes necessary to postpone operation until improvement takes place. It is safer to proceed with prostatic resection at once in all cases in which the general condition is good.

2. Poor surgical risk

When it is evident that a patient is not in proper condition to undergo surgery because of cardiac decompensation, general debility or severe acute urinary tract infection indicated by septic symptoms, retention urethral catheterization or frequent catheterization may be necessary. Most of these patients will improve sufficiently under this treatment so that endoscopic resection can be done at a later date, but some are so far gone that continuation of the above palliative measures provides the only means of relief for them. Bladder irrigation should be carried out once or twice daily with a suitable solution such as potassium permanganate 1:6,000. The retention catheter should be placed so that the tip does not extend more than 2 cm. into the bladder, for when it penetrates farther it does not drain satisfactorily. A bag catheter of 5 cc. capacity and No. 14 or 16 F. calibre is recommended, for the bag rests against the bladder neck and the tip thus assumes the proper position. It should be changed about once a week, or more often if calcareous material tends to collect on it or if excessive urethritis is produced.

3. Decompression of the bladder

Some urologists still advocate slow decompression of the bladder in certain cases (LAWSON et al.). From a practical standpoint, however, it has been demonstrated that rapid decompression does not lead to any more complications than gradual decompression. The majority of urologists now do not adhere to the practice of gradual decompression. There is a hazard in allowing the bladder to overdistend again after having been emptied; this is the time when infection, hemorrhage and acute renal failure may occur.

4. Suprapubic cystostomy

As a preparatory measure for endoscopic prostatic surgery this is indicated in only a few cases. The patient whose general condition is poor and whose prostate is large should receive this preliminary operative treatment. The drainage obtained by cystoscopy reduces the size of the prostate so that subsequent resection is more easily accomplished.

a) Trocar cystostomy

The present tendency is to substitute trocar cystostomy for the conventional operation. Ferrier performed bladder paracentesis on 50 autopsy bodies after filling the bladder with 500 cc. of fluid, and in none was the peritoneum touched. KREUTZMANN has devised a trocar which is guided into the bladder by a long needle previously inserted immediately above the pubis. When the end of the needle enters the distended bladder, fluid runs out through it; then the trocar is passed along the needle which guides it into the bladder. After the trocar is in place, the obturator is withdrawn and a No. 14 or No. 16 F. Wishard (wing tip) or a straight catheter is passed through the trocar and held in the bladder while the trocar is removed. Trocar cystostomy can be done easily under local anesthetic without moving the patient from his bed. It must never be attempted unless at least 500 cc. of fluid are in the bladder.

b) Permanent cystostomy

After cystostomy has been performed, it is wise to postpone further surgery until the patient is in good enough general condition to be up and about; at this time endoscopic resection is indicated, and is preferable to a second stage suprapubic prostatectomy because it is attended by less shock. If improvement has not taken place it is better to allow the cystostomy tube to remain permanently than to subject the patient to another operation.

c) Resection mortality

Suprapubic cystostomy preliminary to endoscopic prostatic surgery in the poor risk patient will reduce the operator's resection mortality but probably not his total mortality, for some of these patients do not survive the first procedure. However, the judgment of the conscientious surgeon will be influenced by his wish to give the patient the best possible treatment rather than by his desire to keep his resection mortality statistics favorable. If the prostate is not greatly enlarged, and resection can be done in an hour or less, it is preferable to proceed with the major operation rather than to carry out cystostomy first even if the patient is a poor surgical risk. A properly performed endoscopic prostatic resection which requires no longer than 45 minutes to complete produces no more surgical shock than does suprapubic cystostomy; so the choice of procedures is obvious.

5. Bed rest

a) Cardiac decompensation and extreme hypertension

Bed rest as preparatory treatment to endoscopic prostatic surgery is indicated in the presence of cardiac decompensation and extreme hypertension. In the former, digitalized action is instituted and bed rest adhered to until the heart compensates. If the blood pressure is over 200, bed rest should be maintained for a few days to stabilize pressure, although it is not essential that this be below 200 to make endoscopic surgery a safe procedure.

b) Avoid bed rest whenever possible

With these exceptions, patients requiring resection should not be kept in bed more than one day preoperatively. It is well known that they develop hypostatic pulmonary congestion very easily, and this predisposes to pneumonia. An old man's strength rapidly fades when he is bedridden, and unless he presents the above complications, allowing him to be up and about until the very day of surgery makes him a better risk. Even if it is considered advisable to prepare the patient with an indwelling urethral catheter, frequent catheterization or suprapubic cystostomy, he should be ambulant while this treatment is being carried out.

6. Cardiac care

Inasmuch as many patients who require endoscopic prostatic surgery have cardiac disease, thought must be given to the care of this condition. As noted above, bed rest is essential for management of decompensation, and digitalization is usually indicated. If it is possible to obtain the consultation of a cardiologist, the endoscopic surgeon is afforded a much better chance of pulling the patient through successfully, as well as considerable peace of mind for himself. Many general practitioners are experienced in care of the cardiac case, and can also be of great help in its management.

7. Infection

Preoperative infection, when severe or acute, is treated by appropriate chemotherapy, indwelling urethral catheterization or in rare cases cystostomy.

a) Chemotherapy

It has been reported that preoperative chemotherapy in the absence of definite urinary tract infection does not prevent postoperative infection (GAUDIN et al.), but later experience and reports have reversed this opinion (CREEVY and FENNEY).

b) Ureteral catheterization

Whether for treatment or for diagnosis this is rarely resorted to in patients with prostatic disease causing residual urine. If such manipulative procedures are required, they should not be applied until after the bladder neck obstruction has been removed.

8. Vasligation

Vasligation or vasectomy preceding endoscopic prostatic resection will prevent postoperative epididymitis in many cases (BERGMAN et al.). Occasionally this complication occurs, however, regardless of precautionary measures, in which case it is due to extention of infection down the vas producing funiculitis rather than true epididymitis. In some cases a hematogenous infection of the epididymis occurs. If the patient is to be prepared by frequent catheterization or inlying urethral catheterization, the vasa are ligated before this treatment is started. If catheterization is not necessary, the patient is spared some anxiety as well as pain and inconvenience by a combined procedure performed at the time of prostatic surgery.

9. Dilatation of urethral stricture

Preoperative care of the patient presenting urethral stricture as a complication of an enlarged prostate consists of indwelling urethral catheterization, beginning with as large a size as can be passed over a stilet, or with a filiform bougie if a catheter cannot be passed. This is changed every day, or every other day, a larger size being inserted each time until a No. 26 F. is passed. In this way the stricture is dilated more rapidly than can be accomplished with sounds passed once or twice a week, and there is also less danger of febrile reaction.

10. Fluids

Copious fluid intake is indicated in all candidates for endoscopic surgery except those suffering from cardiac decompensation or nephritis with edema. During each 24 hours 3,000 cc. of fluid should be consumed; if this is impracticable, 5 per cent solution of dextrose in sterile water is given intravenously to compensate the amount. Forcing of fluids is continued until the time of surgery, but breakfast is omitted on the morning of operation.

III. Selection of the anesthetic

1. General considerations

Criteria for selection of the anesthetic for endoscopic surgery are similar to those used for open operations upon the bladder and prostate (Vol. XIII). In most cases, however, the endoscopic procedure does not require as complete relaxation as do open operations. Therefore, a light anesthetic is usually sufficient. Sometimes the endoscopic procedure, performed by unskilled hands, is prolonged unduly, and in these cases continuous application of the anesthetic agent is necessary (RAINES et al.).

2. Intraprostatic

Intraprostatic local anesthesia may be used when tolerance to spinal or general anesthetic is low (O'HEERON et al.). A long needle (p. 24) is passed through the panendoscope and inserted into prostatic tissue and the local anesthetic agent injected through it (WISHARD et al.).

3. Intradural spinal

For the majority of endoscopic procedures a low intradural spinal anesthetic is the most satisfactory (LUNDY; GRAVES et al.).

4. Miscellaneous

Extradural caudal block does not produce adequate anesthesia consistently enough to make it entirely satisfactory. The anesthetic agent for inhalation anesthesia must be noninflammable, and that for intravenous injection, nondepressing.

5. Preoperative Sedation

Preoperative medication is an integral part of the anesthetic and is ordered by the anesthesiologist. In the old and debilitated patient it is better judgment to omit it unless he is especially fearful or unless it is of a character which will reduce the quantity of the anesthetic agent to be employed. The dosage given should never be sufficient to put him to sleep entirely, for the anesthetic superimposed on such extensive sedation will frequently prove more than he can withstand.

Chapter XIX

Technique with the Stern-McCarthy electrotome

I. Difficulties in mastering the technique

The purpose of setting forth in detail the technique of endoscopic prostatic surgery is not to teach the embryo urologist or the occasional urologist how to perform this operation; nor is it to displace or replace any hospital training of the urologic surgeon. Instead, it is offered to supplement such instruction, and to give the house surgeon a didactic knowledge of the procedure before he starts using it — as well as to save some of the time and energy of the attending urologist who patiently or otherwise endeavors to explain and demonstrate the fine points. The portion dealing with the more advanced technique might also help those who have had their basic training, and who are trying to perfect and adapt themselves to the more difficult operation of removing larger prostates by the endoscopic method. If, perchance, one of the many urologic surgeons who are expert resectoscopists should read this volume, he will undoubtedly find that the details of technique given here do not correspond with those used by him. It is hoped he will understand that the author makes no claim that his is the one and only method in use, and that the following exposé is not intended to convey the impression that this is the last word in technical knowledge. It is merely the description of a procedure which in our hands has proved workable, and is presented as such and no more. It should be pointed out that the basic principles which apply to endoscopic prostatic surgery also apply to other types of endoscopic surgery.

This phase of the first steps in learning the technique of endoscopic surgery is described in Chapter XV.

After the student has learned to recognize landmarks through a foroblique lens, is able to keep the field of vision sufficiently clear of indigocarmine to allow him to catheterize ureters, and has learned to coordinate eye, hand and foot in a rhythmic manner by practice on a beef heart, he is ready to launch on his clinical experience. The first few cases must be done slowly and carefully. The student should resect only a small amount of the prostatic border, and the instructor should finish the operation. When the beginner attempts to resect even the smaller hypertrophied gland, he almost invariably fails to identify the vesical and prostatic urethral landmarks, and is very likely to remove some of the trigone or even the ureteral orifices — or to take a section through the external sphincter. In either case he will have produced a result which will not soon be forgotten. If the trigone has been removed, the patient will return week after week complaining of continuation of frequency, urgency and bladder pain. The case presenting the injured external sphincter will give the surgeon the nightmare of a man wearing a rubber urinal and pointing a finger of accusation which will haunt him for the remainder of his days.

II. Importance and checking of the armamentarium

Endoscopic surgery is a highly technical procedure. The armamentarium must function perfectly; otherwise there is likely to be a breakdown in technique. Each surgeon becomes accustomed to certain kinds of instruments and special supplies; any variation in these may create a hazard to smooth performance of endoscopic procedures. Therefore, meticulous care must be exercised to ascertain that the entire armamentarium is set up and in perfect working order before the operation is started.

The description of armamentarium is given in Chapters I, XV and XVI. A few points, however, should be checked immediately before starting the surgery.

1. Loop electrode

The cutting loop electrode is placed and fastened into the working element of the resectoscope. Its excursion is tested by manipulating the ratchet; the loop is drawn backward into the sheath until it is entirely covered (Fig. 138). If the loop does not disappear completely within the sheath it will not sever the pieces cleanly, and they will cling to the unresected tissue. It is necessary that the loop be covered by the sheath for a distance of approximately 2 mm. in order to cut the pieces free. If the uninsulated portion of the loop contacts the metal of the telescope, the loop should be bent slightly toward the unaffected side to keep it clear of the telescope. The exposed portion of the loop should be bent to an angle slightly greater than 90 degrees from the shaft; when the angle is less than this it is more likely to catch on the edge of the sheath during the inward excursion of the loop. In large prostates, larger pieces of tissue can be removed if the distal end of the loop is bent downward, making it project one or two millimeters below the edge of the sheath.

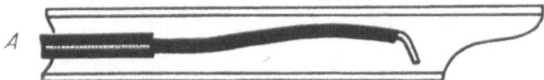

2. Illumination

Before the resectoscope is passed the light cord is connected to the post and the bulb on the resectoscope checked to make certain the light is on and of sufficient brilliance (p. 29).

Fig. 138 A and B. A Proper position of loop at end of excursion: well back into sheath. B Improper position of loop at end of excursion: comes only to edge of sheath and does not cut off tissue cleanly

3. Electrosurgical unit

The foot switch of the electrosurgical unit is depressed to determine whether the cutting and coagulating currents both are available through the unit. During this testing, the active electrode terminal must be removed from proximity to the patient to prevent burning. See Chapter XVI for procedure if the electrical unit fails to function.

4. Indifferent electrode

The indifferent plate electrode is placed under the patient's back. It is very important that the entire plate be firmly and evenly in contact with the patient's skin. If his gown or part of the sheet covers a portion of the electrode or if there is some air space or spotty dampness between the surface of the plate and the patient, an electrical burn is almost sure to result. These burns are very slow to heal and the surgeon or his malpractice insurance carrier may have to pay dearly for them.

III. Position of the patient

The position of the patient is that used for routine cystoscopy, except that knees are spread farther apart and the thighs are at a less acute angle with the body (Fig. 139). Sharply flexed thighs displace the prostate downward into the perineum, making it less accessible to the instrument in the bladder neck and prostatic urethra (Fig. 140). The wide spread of the knees is to allow adequate space between them for manipulation of the ocular end of the resectoscope.

IV. Position of the operator

The tendency of most operators who are performing their first cases of prostatic surgery is to have the patient too low and the stool upon which they are sitting too high. This results in a bent, cramped position for the resectoscopist, and increases the difficulty of operation. At the beginning of the resection the electrotome should be horizontal, and the operator sitting on the stool in an erect position. As the operation progresses this position is changed, the direction of the shift depending upon the section of the prostate being removed. When the dorsal portion is being cut, the ocular end of the resectoscope — and also the operator —

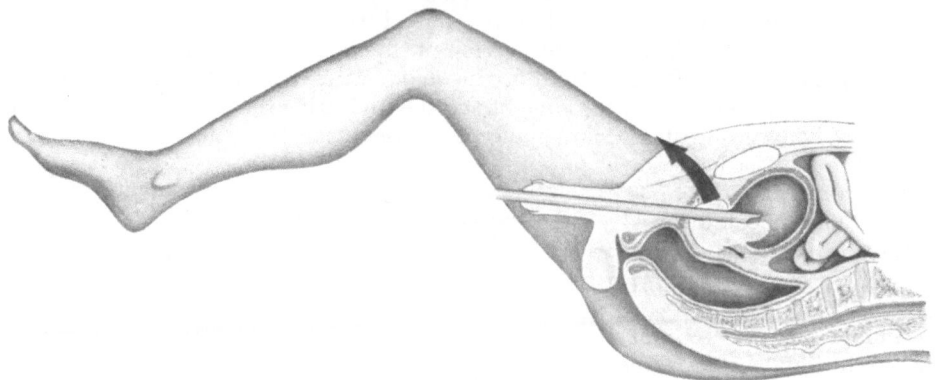

Fig. 139. Proper position of thighs for prostatic resection. Prostate is displaced upward into bladder and toward urethra as indicated by arrow

are elevated, thus depressing the distal end of the sheath (Fig. 141). When the ventral portion is being worked on, the ocular end of the sheath is depressed, thus

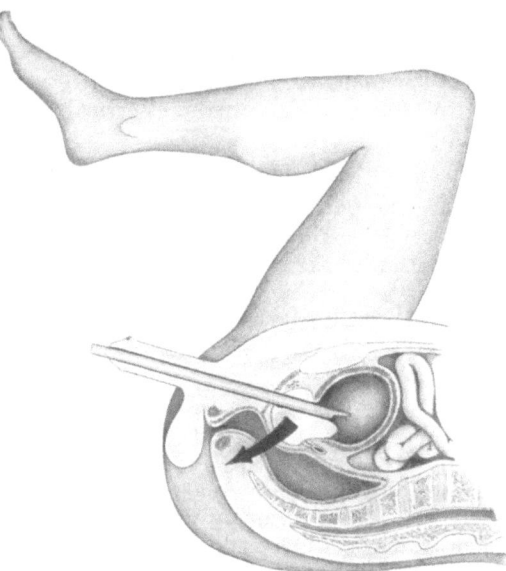

Fig. 140. Improper position of patient: thighs are flexed too much, throwing prostate outward toward perineum as for perineal prostatectomy. Arrow indicates direction in which the prostate is displaced when thighs are in this position

elevating the distal end, placing it closer to the desired section of the prostate above the 9 and 3 o'clock positions (Fig. 142). When the lateral portions of the gland are being removed, the ocular end of the electrotome is placed near the opposite thigh. The author's rapidly adjustable cystoscopic stool is a great help in carrying out the maneuvers. The seat can be quickly raised or lowered by the weight of the operator, and is held in position or loosened by means of a foot lever. The hydraulic model is adjusted by electric motors (p. 156). Proper height of the stool at all times is a great boon to the surgeon, as it eases back strain due to the continual bending and stretching necessary to reach the different parts of the prostate during resection.

V. Introducing the resectoscope

1. Preliminary dilatation

Introducing the resectoscope into the bladder is facilitated by first passing sounds, preferably No. 26 to No. 30 F., and by performing urethral meatotomy if the meatus is small. If this preparatory measure is not adhered to in patients who have a small calibre urethra, the slightly rough edges of the fenestra are likely to traumatize the urethral mucosa or even to strip some of it away, predisposing to later stricture formation.

2. Hinged obturator to follow urethral roof

A hinged obturator such as the Timberlake or Timberlake-Alcock (Fig. 13, p. 21) is preferable to the straight obturator, especially if the gland is very large, because the curve of the hypertrophied prostatic urethra is exaggerated and this obturator follows the curve more satisfactorily. Gentleness and avoidance of all force are even more essential while introducing the resecto-scope than while passing sounds or a cystoscope. The introduction of the in-

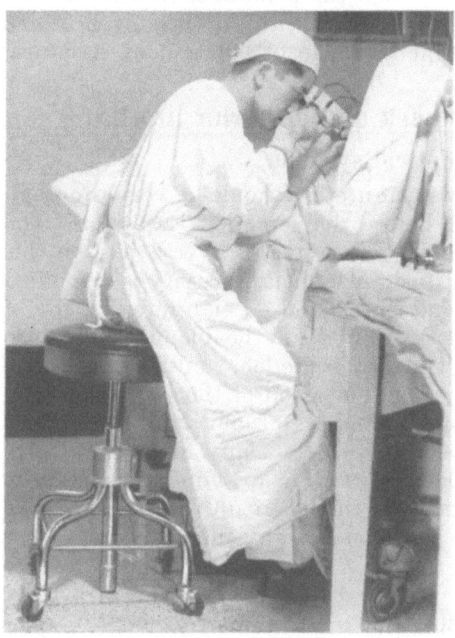 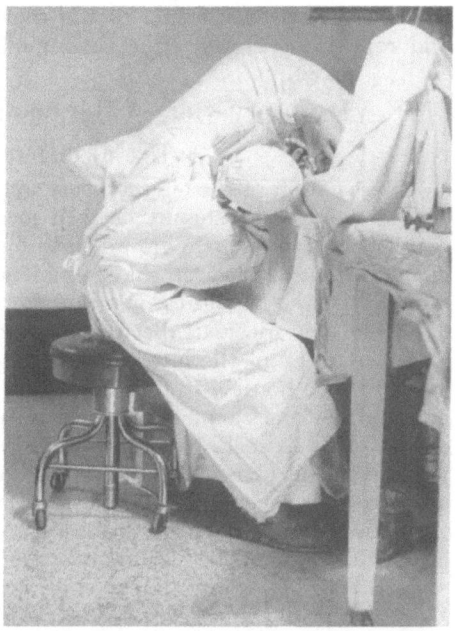

Fig. 141. Position of operator while resecting dorsal portion of prostate. Operator and ocular end of electrotome elevated, thus depressing distal end of instrument against floor of prostatic urethra. Note elevated position of stool

Fig. 142. Position of operator while resecting ventral portion of prostate. Operator and ocular end of electrotome lowered, thus elevating distal end of instrument against prostate situated above 9 and 3 o'clock positions. Note lowered position of stool

strument is usually facilitated by elevating its inner end so that the tip of the obturator is against the roof of the urethra. This is accomplished by a leverage action on the outside of the sheath; the extreme outer end is depressed while the portion of the sheath immediately outside the meatus is elevated. It may also be accomplished by inserting a finger in the rectum and pressing the anterior rectal wall forward.

3. Bypassing a false passage

Sometimes the tip of the resectoscope enters a false passage in the prostatic urethra and it is not possible to manipulate the instrument by it into the bladder. Sometimes a No. 18 or No. 20 F. semisoft rubber coudé tip catheter succeeds in entering the bladder; the sheath of the resectoscope is then passed over the catheter, which guides the sheath past the false pocket.

4. Perineal urethrotomy

Perineal urethrotomy is indicated when the urethral lumen is narrow (NESBIT); divulsing or forcibly dilating the small calibre urethra in order to pass the instrument traumatizes and usually results in stricture formation. Perineal urethrotomy is performed by incising through the perineum on to the tip of a metal sound in the urethra; the tip is pressed out toward the perineum and is easily palpable through the skin. When the incision has penetrated the urethra a suture is passed through each edge; lateral tension on these opens the urethra and the resectoscope is passed into the bladder between them.

5. Internal urethrotomy

Internal urethrotomy may be performed when the pendulous urethra is constricted. MacDonald incises the roof of the urethra with an Otis urethrotome and reports no trouble due to immediate hemorrhage or subsequent stricture.

VI. Observation of the bladder neck and posterior urethra

1. Use of different optical systems

Observation of the bladder neck and prostatic urethra should be very thorough and all landmarks should be well located before the resection is started. The retrograde, the right angle and the foroblique telescopes all have a part in this examination. Each ureteral orifice is identified. The bladder cavity is examined minutely with the right angle and foroblique lenses, and special note is made of diverticular orifices, trabeculation, stones or tumors in the bladder, and also of hypertrophy of the interureteric ridge. The retrograde lens is used to examine the prostatic border, to determine the degree of elevation of the vesical neck or of glandular encroachment upon the trigone[1]. The lateral and anterior portions of the bladder neck are also observed carefully to detect any intravesical projection of the prostate. The foroblique lens is used to examine the prostatic urethra, and reveals the amount of lateral lobe intrusion into its lumen. The verumontanum is identified, and observation made of any infringement of the lateral lobes over the verumontanum.

2. Composite view

The many fields of vision which are seen through all the lenses are blended into one composite picture, and the operator must retain this picture in his mind throughout the entire resection. In order to keep the operation within the prostatic urethra, its length as noted by means of the foroblique lens must be remembered. The amount of intravesical bulging of the middle and lateral lobes as seen through the retrograde lens must be recalled during operation, so that this amount and no more is removed. If the surgeon will train himself to visualize the prostatic urethra as a conglomerate whole, rather than limiting his perception to one field at a time, he will find that this is a valuable aid to orientation throughout the operative procedure. His pictorial memory must encompass not only the original picture of the prostatic border and prostatic urethra, but also the changes which are being made as tissue is removed during the progress of the operation.

VII. Holding the resectoscope

After the bladder neck and prostatic urethra have been thoroughly examined, the working element is inserted and the electrotome grasped with the left hand so that the thumb lies against the lock on the upper side of the sheath. This prevents the working element from slipping out when the resection is in progress. The author's thumb rest aids in keeping the thumb in this position, as well as in

---

[1] See p. 15 for orientation with the retrograde lens.

allowing upward pressure on the ocular end of the sheath. The first finger of the left hand is free to manipulate the water inflow, and the handle of the pet-cock is placed so that it can be easily manipulated with this finger (Fig. 143). The second and third fingers of the left hand are placed on the upper side of the sheath near the glans penis so that downward pressure can be exerted there, thus

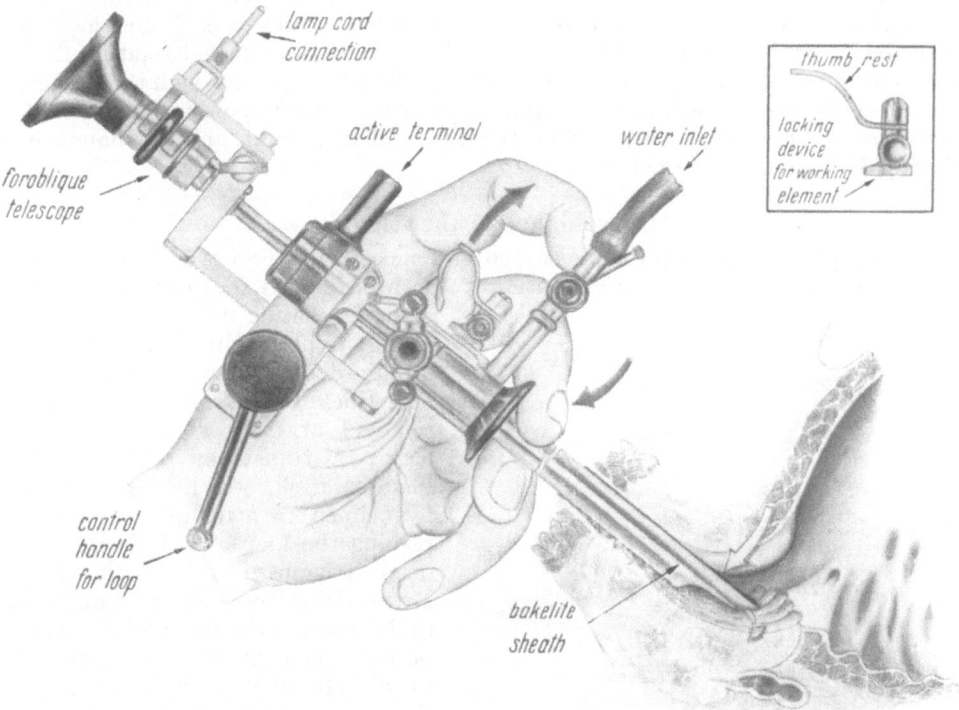

Fig. 143. Proper method of holding electrotome. Arrows indicate direction pressure is maintained during cutting excursion of loop. Forefinger manipulates water intake. Inset shows author's thumb rest

forcing the bevel of the beak of the sheath down into the prostatic tissue. The left elbow is rested against a stationary object, such as the pan or edge of the table. This support enables the surgeon to hold his instrument steadier with less likelihood of unconsciously slipping it farther in or out as he operates. The right hand is used to manipulate the ratchet.

VIII. Starting the resection

1. Removal of first pieces

The resection is started by placing the loop barely over the prostatic border at the apex of the trigone. The ocular end of the resectoscope is now elevated, keeping the eye against the eyepiece, and the inner end of the sheath is depressed as described above, while the thumb exerts upward pressure at the extreme outer end. This manipulation is facilitated by the use of the author's thumb rest. By creating pressure in this way, a larger section of tissue can be removed from the prostate. The cutting current is then applied by depressing the proper foot switch, and the loop is slowly drawn through the tissues and entirely back into the sheath. When the resectoscopist becomes more proficient, he may draw the loop rapidly, but the initiate will control bleeding better and will be less apt to lose sight of his landmarks if he withdraws slowly. The moment the loop is entirely encased within the sheath, the electric current is discontinued to avoid burning the edge of the sheath. This is particularly important following the use of the coagulating current. At first it may be preferable to remove each section as it is cut off, for the new resectoscopist has difficulty in obtaining very large pieces. After the first few pieces are removed, however, each one, as it is resected, is washed into the bladder and all are evacuated as the irrigating fluid runs out of the bladder (p. 205). The slower the loop is drawn through the tissues, the more the pieces stick to the loop and obscure the field of vision. After each bite, the resectoscope is rotated slightly either to left or right, and the maneuver repeated to remove the next bite.

2. Avoidance of the trigone

When placing the loop for the second and successive sections, it should never be advanced higher on the trigone than its position for the first cut, because in cases presenting median bar or low grade hypertrophy the prostatic margin does not project more than the thickness of one loop. The removal of one or more sections leaves a transverse ridge at the bladder neck, which the operator is inclined to attempt to level by placing the loop a little farther up on the trigone as each bite is taken. By so doing the ridge retreats and in this manner the entire trigone may be resected (Fig. 144). In order to avoid "climbing up on the trigone", the upper limit of the previous section is visualized, and this transverse boundary line is respected when succeeding sections are made. Even though an apparent definite elevation is still present at the bladder neck, it should be allowed

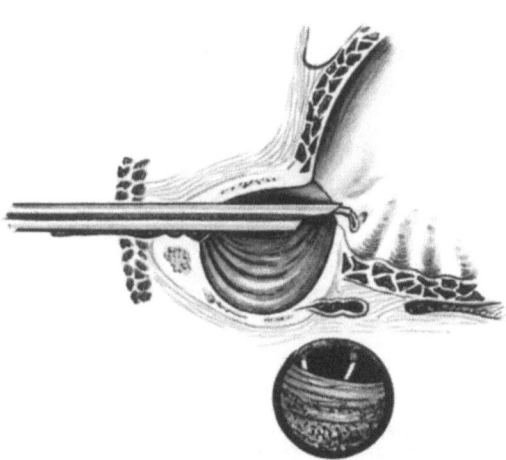

Fig. 144. Elevated bladder neck and trigone may be near ureteral orifice which is invisible from this position through foroblique lens. By removing this elevation with loop, a ureteral orifice may be removed. Inset shows cystoscopic view

to remain until the ureteral orifices have been identified with the foroblique lens or until adequate hemostasis permits a satisfactory view with the retrograde lens. Then only will it be apparent whether there is still tissue to be resected in that location. The beginner should always coagulate each bleeding point as it occurs, for when several are active at once they are difficult to identify and to fulgurate well. A more experienced operator leaves most of the bleeding points for coagulation until the section of the prostate which is being removed at the time is completed; but this practice is dangerous for the unskilled.

IX. Orientation

The surgeon becomes orientated by identifying landmarks before starting the resection of tissue. Occasional identification of the ureteral orifices and frequent observation of the characteristic bladder neck fibers helps orientation during resection of the intravesical section. The verumontanum must be identified previous to the removal of each series of pieces from the intraurethral portion. The resectoscope must be kept in the same in and out position to avoid resecting too far out into the membranous urethra or too far in onto the trigone. During resection of bladder tumors, an occasional observation of the edges of the tumor helps the operator to remain oriented. Good visualization is necessary to maintain proper orientation.

X. Method and rhythm

1. Planned approach

A planned approach to the endoscopic removal of tissue is even more important than an orderly sequence of technique for open surgery (BARNES 1951). Alternate

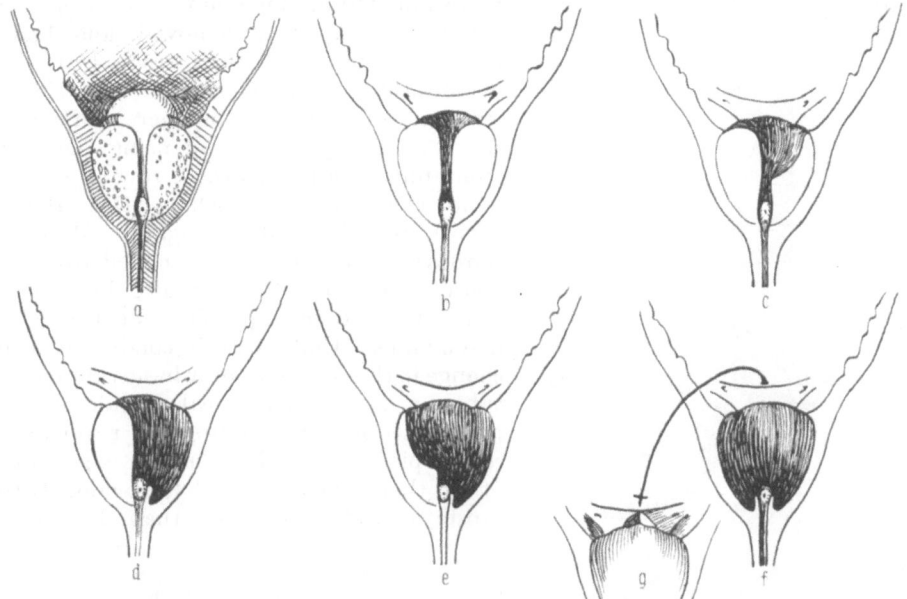

Fig. 145. One section of prostate is completely removed before another is started. *a* Horizontal view before operation is begun. *b* Middle lobe removed. *c* Middle and intravesical left lateral lobes removed. *d* Middle and entire left lateral lobes removed. *e* Intra-urethral right lateral lobe only is left. *f* All of prostate has been removed. *g* Trigone incised

resection of a piece of tissue from one side and then the other of the prostate or a bladder tumor, is an inefficient method of performing endoscopic surgery. Several efficient plans have been advocated.

a) Superficial to deep

Probably the best plan for removal of any tissue by the endoscopic route is to begin at the most superficial portion of the tissue mass, either prostate or bladder tumor, and remove it symmetrically down to the base or deepest portion. When the prostate is being removed endoscopically the intravesical protrusion and the tissue at the bladder neck are first removed from one side (Fig. 145).

b) 6 to 12 o'clock position

The resection is started at the 6 o'clock position and a row of pieces removed from this point up to the 12 o'clock position as the resectoscope is rotated back down again to the 6 o'clock position. When the adenomatous prostate is small, the removal of one row of pieces exposes most of the bladder neck fibers. When it is larger, several rows — each deeper than the previous one — must be removed before the bladder neck fibers are exposed (Fig. 146).

c) Removal by sections

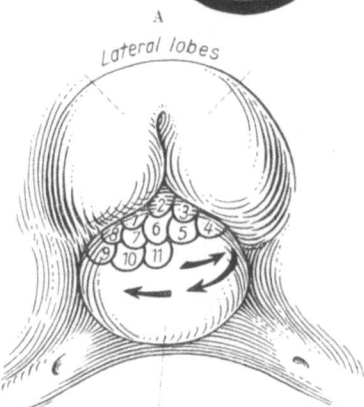

After the intravesical portion of the gland has been removed on one side, and the bladder neck fibers are exposed, bleeding is controlled in that section. If for any reason a two stage operation is necessary, the first stage should be terminated at this point — after one side has been completely removed, and before the other side has been started. Next, the intraurethral portion on the same side is removed in a similar manner, starting at the 6 o'clock position and using the verumontanum as a landmark; a row of pieces is removed to the 12 o'clock position, then a deeper row back down again. After the prostatic tissue has been removed from this section exposing the capsule, bleeding is controlled and the opposite side treated in like manner. Under certain conditions, some change in this sequence may be advantageous. When a large intravesical middle lobe is present, it is resected first by removing a row of pieces across it from one edge to the other, then a deeper row back again. If the prostatic urethra is more than 3 to 4 cm.

A

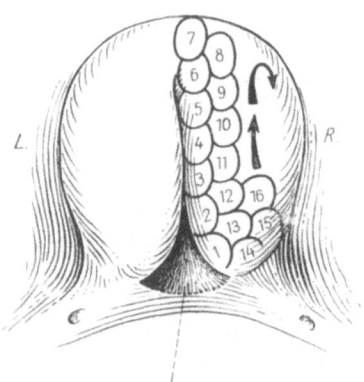

a *Middle lobe*   b *Middle-lobe removed*

Fig. 146 A and B. A planned approach to endoscopic removal of prostate. Resection is started at 6 o'clock position and a row of pieces is removed from prostatic border up to 12 o'clock position, as resectoscope is rotated through 180 degrees. Then a deeper row is removed as resectoscope is rotated back down again to 6 o'clock position. A Composite of six visual fields. B View from within the bladder

long, three sections may be required on each side instead of two — one intravesical and two intraurethral — the most distal one at the level of the verumontanum.

d) Advantages of starting the resection at the 6'clock position

The chief advantage of beginning the series of loop excursion at the dorsal portion of the bladder neck rather than at its ventral portion is the greater number of easily identifiable landmarks existing in that area. The trigone and ureteral orifices in the bladder and the verumontanum in the urethra are aids to orientation and to verification of the inner and outer limits of the resection. After the operator becomes orientated and begins removing the consecutive series of pieces from the 6 to the 12 o'clock positions, it is very important to keep the resectoscope in the same relative in and out position. If this is done, orientation will not be necessary again until after the irrigating fluid has been emptied from the bladder and the removal of another series of pieces has begun.

2. Coordination of movements

The endoscopic surgeon must develop coordination of movements and rhythm of motion before he is able to remove large prostates rapidly.

a) Foot and eye

The operator's foot must depress the foot switch the instant the loop touches the tissue for the beginning of its excursion, and release it at the exact time the loop enters the sheath at the end of its excursion. His eye must be kept against the ocular end of the resectoscope continuously while pieces are being resected.

b) Fenestra alternately against and removed from tissue.

An important maneuver is the alternate pressure and release of the inner end of the resectoscope against, then away from, the tissue being resected. When the loop is in the advanced position ready to begin removing a piece of tissue, the inner end of the resectoscope is pressed against the tissue to be removed by swinging the ocular end of the instrument in the opposite direction, and by depressing the beak with a leverage pressure of the left hand which is holding the instrument. This enables a deeper bite to be engaged by the loop. At the completion of the excursion, when the loop is again within the sheath, the opposite maneuver is performed. The inner end of the resectoscope is swung away from the tissue so that the loop can be extended once more for the next bite (Fig. 147). If this latter maneuver is not carried out, the advancing loop may impinge on unresected tissue. Moreover, if the inner end of the sheath is not swung away from the unresected tissue, the last piece removed may be held against the side of the prostatic urethra, by the beak of the sheath. This would prevent its being washed back into the bladder and would result in obstruction of view through the lens. This swing into, and then away from, the tissue is developed into a rhythmic series of movements and is coordinated with the action of the foot switch, with the excursion of the loop, and with the eye against the ocular end of the lens. It should be practiced and learned by the trainee before he begins operating on patients.

c) Manipulation of water inflow

Another maneuver which saves time is the proper manipulation of water inflow. If this is reduced or turned off whenever a forceful flow is not needed for clear visualization, the bladder does not fill rapidly and it is therefore not necessary to empty it often. Each time the operator stops to empty the bladder, valuable time is consumed. While the inward excursion of the loop is being made, visualization is not necessary, for the loop is buried in the tissue and the piece being resected is so close to the lens that visualization is not possible. Therefore, at this time no necessity for water inflow exists. The water is shut off the instant the loop is in place to begin removal of a piece, and is turned on again when visualization is necessary to place the loop for removal of the next piece. This control of water inflow can be made with the petcock manipulated by the index finger of the left hand, by kinking the inflow tube, by the use of a Timberlake valve or in some other way. It is coordinated with the other maneuvers previously described.

d) Logical sequence of procedure

More important than the specific method or technique used is the necessity for the operator to follow the method he uses in a logical sequence of steps, and

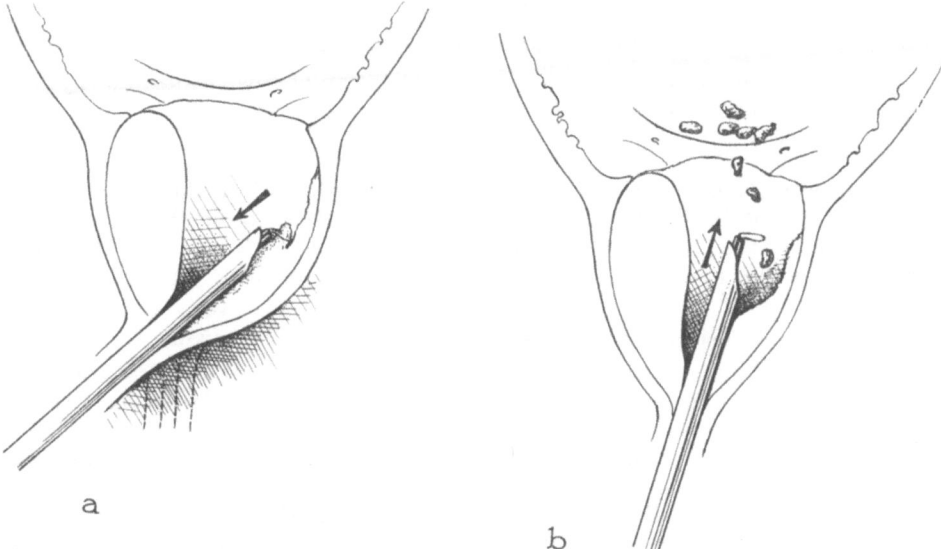

Fig. 147. Inner end of sheath and loop are swing a) against tissue while a piece is being resected, and b) away from tissue while loop is being extended. Arrows indicate direction of movement of loop

to develop a rhythm and coordination of movements which will enable him to perform difficult endoscopic surgical procedures rapidly and accurately. The surgeon should use the planned approach best suited to him. Unorganized removal of a piece of tissue here followed by another some place else is not conducive to rapid and adequate endoscopic prostatectomy.

e) Sequence for bladder tumors

A similar method is used for endoscopic removal of bladder tumors. A row of pieces is removed from the top of the tumor, then successively deeper rows until the bladder muscle fibers are visible through the resectoscope lens. When the tumor is large, bleeding is controlled throughout the area just resected before another section is engaged.

XI. Visualization

1. Importance of clear visualization

Clear visualization is essential to the proper performance of endoscopic surgery. Without it the surgeon becomes lost in a maze of cloudy fluid, indefinite landmarks and unrecognizable tissue. The beginner finds it very difficult to see clearly, but the experienced resectoscopist can usually correct the causes of poor visualization and recognize landmarks and identify tissue accurately.

2. Causes and correction of poor visualization

The causes of poor visualization during cystoscopy are given on p. 42. These apply also to endoscopic surgery. The most common causes and procedures for improving visualization are given here.

a) Water or debris on ocular lens

The operator should develop the habit of wiping the ocular lens with a piece of sterile gauze the first time a telescope is used during the operation, and occasionally during the procedure.

b) Poor illumination

The causes and correction of insufficient brilliance from the bulb are given on p. 29.

c) Debris and air bubbles covering the objective lens

These can be displaced by increasing the inflow of irrigating fluid and by tipping the inner end of the sheath upward to allow the bubbles to float out of the fenestra into the bladder.

d) Pieces of tissue

These sometimes obscure vision. They may *cling to the loop* and be held in a position covering the objective lens. This is especially prone to occur in association with resection of small pieces and with excessive coagulation during the cutting excursion of the loop. Incision of larger pieces, more rapid traction of the loop through tissue, and utilization of a more powerful cutting current aid in avoiding this deterrent to good visualization. When a section persists in sticking, it may be disregarded if it does not obscure vision in the extremely extended position of the loop, or if the operator is certain that the next cut — even though vision be obscured — will be into prostatic or bladder tumor tissue. Resection of the next piece will remove from the loop the section which has clung to it. Sometimes pieces of tissue are not cut cleanly but *hang to the unresected tissue* even though the loop does recede entirely within the sheath at the end of its excursion. It is then necessary to rotate the electrotome or to advance it slightly while the loop is still within the sheath; in this way the adhering fragment is pulled away from the unresected tissue.

A piece of tissue may be *held by the beak of the sheath* against the unresected tissue at the conclusion of the inward excursion of the loop. By swinging the inner end away from the side of the prostatic wall when the loop is advanced, the resected piece is washed back into the bladder and out of the field of vision (Fig. 147). When very soft tissue is being resected, which is frequent in bladder tumors, the resected piece may persist in clinging to the unresected tissue; in this case the clinging fragment can be pushed out of the visual field by the loop as it is advanced for the next bite. When the entire enlargement of the prostate is intraurethral, the *vesical orifice may remain small* while the prostatic urethra is being enlarged by removal of the adenoma. The small bladder neck may hinder the free passage of resected pieces out of the prostatic urethra into the bladder. To obviate this, the dorsal lip of the bladder neck is incised early instead of at the end of the procedure (p. 213). By so doing, the vesical orifice is enlarged enough to allow the resected pieces to be washed freely into the bladder.

e) Debris clinging to loop

When it is necessary to fulgurate a surface covered by a clot, the loop will become covered with coagulated blood and partially obscure vision. It can be cleaned by resecting another piece — drawing the loop through the tissue with the cutting current in operation removes the debris which has adhered to it.

f) Inadequate inflow of irrigating fluid

Increasing the rate of inflow of clear fluid over the operative field helps to keep the field of vision clear of blood. The danger of rupturing through the prostatic capsule must be kept in mind when a high pressure of inflowing fluid is maintained. When the bladder becomes distended the intravesical pressure may become as great as that of the distending fluid; this stops the inflow through the instrument and the field of vision becomes cloudy. When air bubbles cover the field, they may be cleared by depressing the ocular end of the sheath while fluid is running in and also by increasing the pressure of the inflowing fluid.

g) Objective lens too far from tissue

In order to obtain clear visualization during endoscopic surgery, the inner end of the resectoscope must be held very close to the tissue while fluid is flowing in through the instrument. Blood normally obscures the visual field when there is more than 2 millemeters' distance between the objective lens and the operating field, and/or when a flow of clear fluid is not washing blood away. Very close vision is more essential for clear visualization when an isotonic solution is being used for irrigation than when water is the distending medium; the hypotonic properties of water lyse the red blood cells, thus clearing the fluid in which they are suspended. Whole blood cells becloud the fluid much more than do lysed cells.

h) Objective lens against tissue

When the objective lens is against the tissue, the field of vision is blurred (p. 42). This occurs when the resectoscope is too far advanced in the bladder wall, or when it is too far outside it and the mucosa of the membranous urethra lies against the lens.

i) Excessive bleeding

Visualization improves as bleeding is controlled. The experienced endoscopic surgeon is usually able to obtain adequate visualization even though bleeding is profuse, he can rapidly complete the removal of one section of the prostate or of a bladder tumor before stopping to coagulate the bleeding points. However, the surgeon who is inexperienced in performing endoscopic operations must electrocoagulate frequently to keep bleeding at a minimum, otherwise visualization becomes hopelessly blurred. See p. 207 for control of bleeding.

j) Clots covering the field of vision

Visualization of landmarks and identification of tissue are not possible when clots cover the field of vision. These are scraped off by the loop electrode. When the endoscopic procedure is expeditiously carried out, clots usually do not form over the resected area or in the bladder during the operation.

XII. Identification of tissue

1. Importance

It is more important and also more difficult accurately to identify tissue during endoscopic surgery than it is during open operations (BARNES and MARTIN). The ability to differentiate tissue as seen through the resectoscope lens improves with practice. The individual who is learning endoscopic surgery at first can distinguish no difference between the various kinds of tissue which may appear during the course of the operation. After having concentrated during numerous resections upon the visual images he obtains, he gradually becomes able to identify them more accurately. The appearance of bladder neck fibers, of prostatic tissue, of true capsule and of bladder muscle is characteristic and each different from the other. The importance of visual recognition of near perforation and complete perforation of the prostatic capsule and of the bladder wall cannot be over-emphasized.

2. Objective lens close to tissue

In order to recognize these tissues, the objective lens of the resectoscope must be held very near to the tissue. Close proximity of the lens to the tissue not

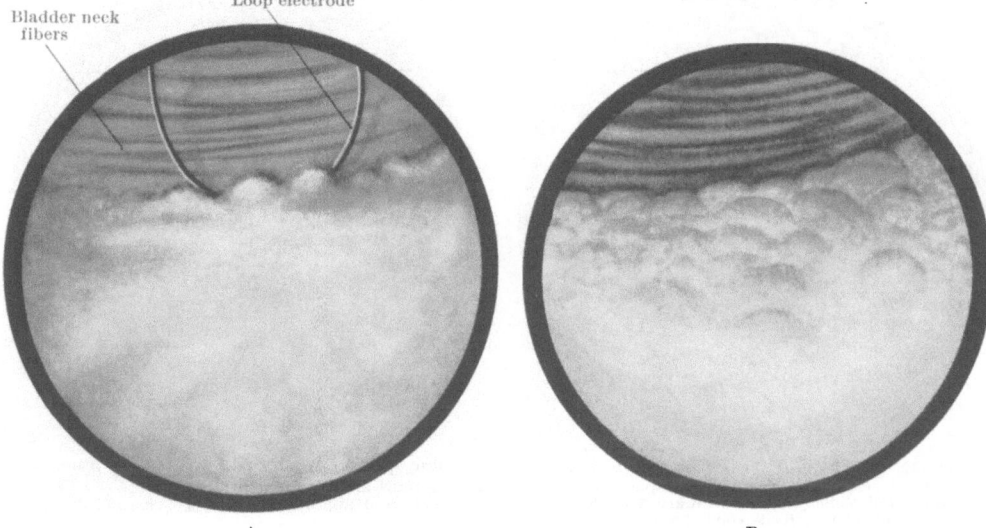

Fig. 148 A and B. Resected prostatic tissue as seen through the resectoscope lens. Very close view. Cut surface varies from A finely granular to B fuzzy or cotton-like. Bladder neck fibers in upper portion of field

only helps to keep the visual field clear, but also magnifies the tissue and shows variations which would remain unrecognized without magnification.

Following is a brief description of a very close view of resected tissue.

3. Prostatic tissue

Normal or adenomatous prostatic tissue varies from finely granular to fuzzy or cottonlike in appearance. The presence of infection increases the fuzziness. This is the first cut surface seen after starting the resection (Fig. 148) unless the bladder neck fibers are exposed with the first cut.

4. Highly malignant tissue

The cut surface of highly malignant tissue, from either the prostate or the bladder, is soft and coarsely granular. Very short fine fuzzy strands of tissue often project from the edges of the cut surface. Tissue of lower grade malignancy is more likely to be firm in appearance and its cut surface is sometimes even smoother than that of the adenomatous prostate.

5. Bladder neck fibers

Bladder neck tissue consists of coarse concentric parallel fibers which are slightly concave toward the bladder (Fig. 149). These fibers are seen after removal of the prostate covering them. They are the landmark which delimits the upper edge of the resected area; tissue should not be removed above them except in certain cases such as bladder neck resections for neurogenic bladder. Whenever the resection is continued above them, it is extremely important to identify and then to avoid the ureteral orifices.

6. False or surgical capsule

This is prostatic tissue compressed against the true capsule by the adenoma, and is difficult to differentiate from adenomatous tissue. In appearance it is

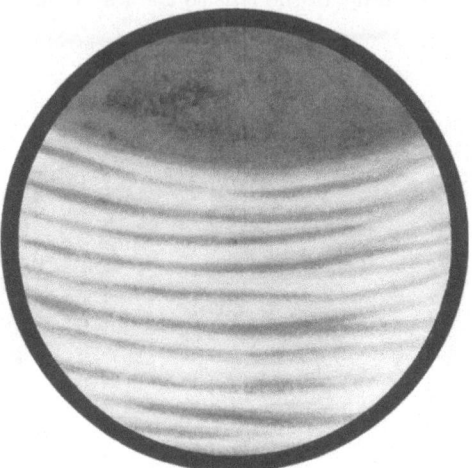 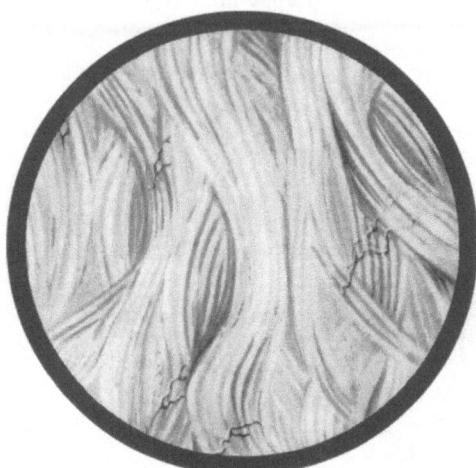

Fig. 149. Bladder neck fibers as seen through the resectoscope lens. Very close view. Coarse fibers are parallel and slightly concave toward bladder. Dark area is bladder cavity beyond rim of bladder neck

Fig. 150. True capsule of prostate. Fine strands of fibrous tissue are irregularly arranged. Surface reflects light and appears shiny. Very close view

slightly smoother than adenoma, not quite so granular, and unless there is marked infection it is not very fuzzy or cottonlike.

7. True capsule

The true capsule of the prostate is identified by the irregular arrangement of its fine strands of fibrous tissue (Fig. 150). When the resection extends into the capsule, the fibrous strands are coarser and more distinct. The surface of some of the tissue reflects light from the resectoscope bulb, making it appear more shiny than prostatic tissue.

8. Bladder muscle

This is exposed during the endoscopic removal of bladder tumors, and is seen as coarse fibers which are irregularly arranged (Fig. 151). The strands of muscle are thicker than the strands of fibrous tissue in the prostatic capsule. The muscle tissue is shiny, whereas the cut surface of the neoplastic tissue is dull and granular, more like the adenomatous prostate. When bladder muscle fibers are seen, care must be taken to avoid taking a deep bite of tissue which might result in a large perforation of the bladder wall.

9. Near perforation

When the resection is continued into the prostatic capsule, the fibers become coarser and begin to spread apart. There are very fine spiderweblike strands extending between the spreading coarser bundles; these are seen against the

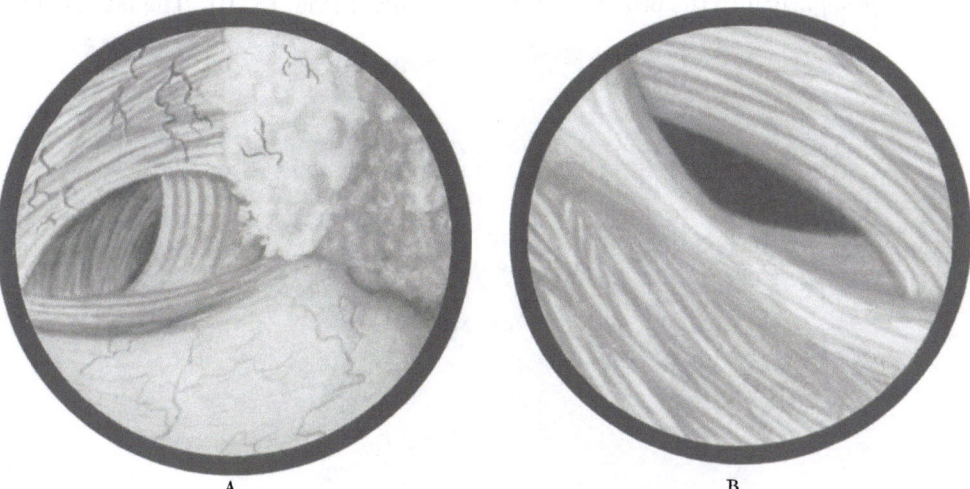

A   B

Fig. 151 A and B. Bladder muscle fibers seen during endoscopic removal of bladder tumors. A Small amount of remaining tumor tissue is seen on right. B Perforation through bladder wall. Very close view

Fig. 152. Near perforation of prostatic capsule. Coarser fibers deep in capsule spread apart, and very fine spider-web-like strands extend between them and are seen against the darker background. Very close view. Another view of near perforation of capsule

darker background between the coarse fibers (Fig. 152). During resection of bladder tumors, a similar appearance indicates that the bladder wall is thin and about to perforate.

10. Complete perforation

At first sight the appearance of complete perforation of the prostatic capsule or of the bladder wall is similar to that of near perforation. Then a black hole comes into view between the coarse fibers, and the spiderweblike strands do not cross the perforation (Fig. 153 A).

Pericapsular fat

This may be seen through the opening of a complete perforation or may even protrude slightly into the prostatic urethra through it (Fig. 153 B). The fat is light

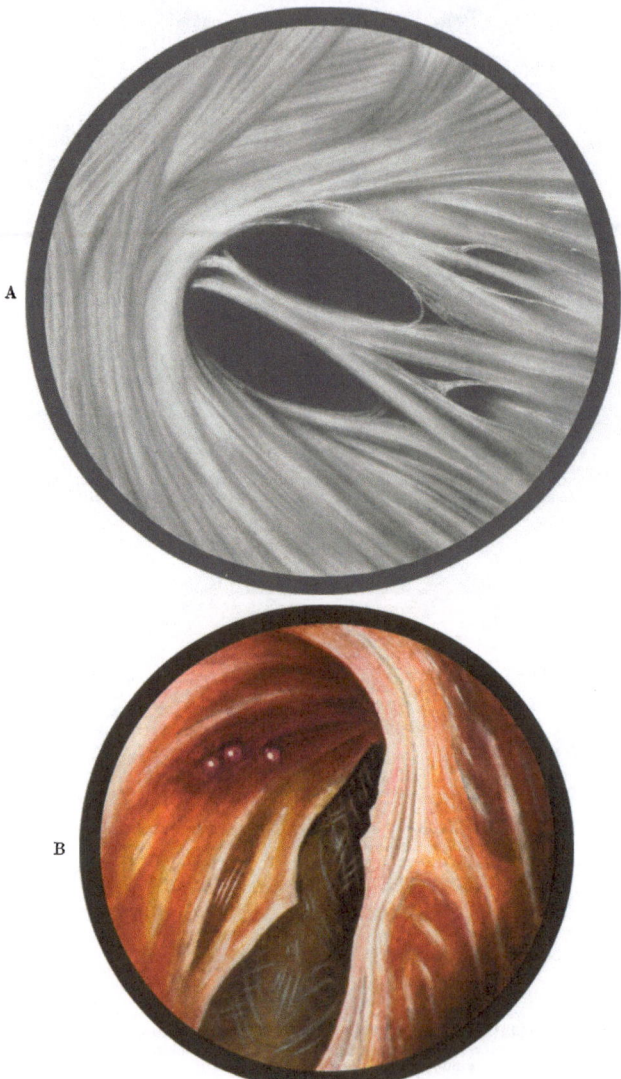

Fig. 153 A and B. Complete perforation of prostatic capsule. A Dark area is hole through the capsule. B Yellow glistening pericapsular fat can be seen through the perforation

yellow in color, the surface is irregularly globular and light is reflected from small smooth areas on its surface. (See p. 231 for procedure when perforation occurs.)

11. Openings which are not perforations

Openings in the prostatic tissue which are not perforations may be encountered.

a) Venous spaces

These are thin walled, shallow, oval or circular cavities the edges of which are well defined (Fig. 154). Blood flows from them when the intravesical pressure

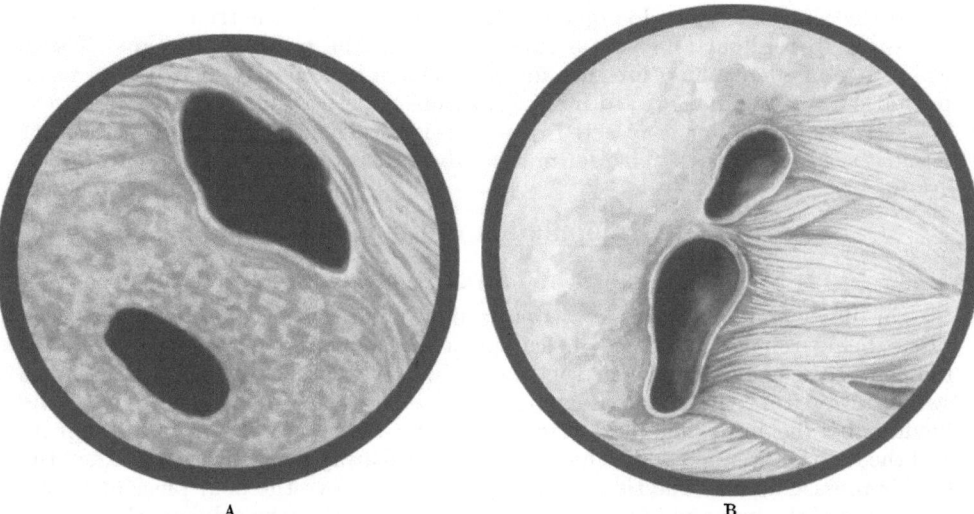

Fig. 154 A and B. Venous spaces uncovered during endoscopic prostatectomy. A Spaces filled with blood. B Irrigating fluid enters spaces when bladder overdistended

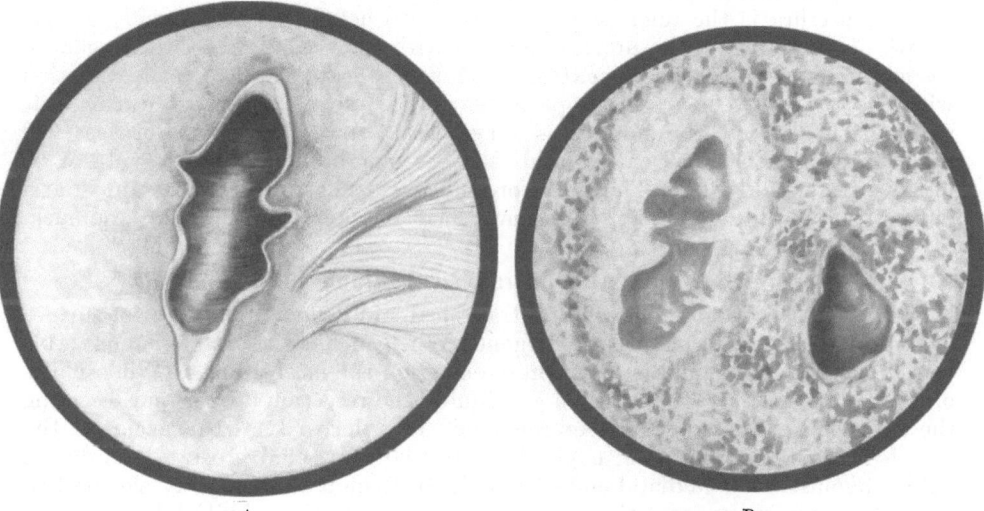

Fig. 155 A and B. Ejaculatory ducts which have been cut into. A One non infected duct. B Both ducts; severe infection of prostatic tissue and of duct on left side of field. One on left has been unroofed and that on the right is severed. Seminal fluid is flowing from one on left. Moderately close view.

is low and distending fluid is not flowing in. Sometimes a blood clot fills the opening.

b) Ejaculatory ducts

While resecting the dorsal portion of the prostate near the midline, these may be unroofed or cut across (Fig. 155). The resulting opening is circular if the duct is cut entirely through or oblong if it is only unroofed. The walls are thicker than those of venous spaces and are often fuzzy. Sometimes seminal fluid is seen to exude from them.

12. Survey at conclusion of operation

At the conclusion of endoscopic prostatectomy fibers of the true capsule should be visible throughout almost the entire extent of the prostatic urethra. Tissue which is seen encroaching into the urethra is examined closely, if it is seen to be prostatic tissue and not inward bulging capsule or pericapsular fat, it is removed. When a bladder tumor is resected, the area is scrutinized closely; if any tumor tissue is seen covering the bladder muscle fibers, it is removed deeply and widely.

XIII. Manipulation of the resectoscope

1. Swinging against and away from nonresected tissue

The inner end of the resectoscope is manipulated so that it is against the tissue being removed during the cutting excursion of the loop electrode. At the end of the excursion, when the loop is within the sheath, the inner end is swung away from the tissue allowing the resected piece to be washed into the bladder by the inflowing fluid (Fig. 147). The resectoscope is rotated slightly and the loop is extended while the inner end is still away from the tissue. Then the loop is pressed against the tissue again, ready to remove the next piece of tissue. The swing of the inner end of the resectoscope against the tissue during the cutting excursion of the loop, then away from it as the loop is extended, is accomplished by moving the ocular end of the instrument in the opposite direction; the membranous urethra is the fulcrum point. It is also helpful to press firmly on the sheath at a point near the urethral meatus with the second and third fingers of the left hand while pressing the extreme end of the sheath in the opposite direction with the thumb. The resectoscope must not be moved in or out during this manipulation, unless the surgeon is very certain that the ureteral orifices will not be cut off if he moves it in, and that the membranous urethra will not be invaded if he pulls it out. The safest procedure is to keep it in a constant in and out position until reorientation is accomplished by observing a familiar landmark.

2. Removal of intravesical middle lobe

Sometimes in the presence of an intravesical middle lobe, the posterior surface of this lobe lies near or against the trigone. When such is the case, it is preferable to remove the entire middle lobe before engaging the lateral lobes. When it lies on the trigone, the danger arises of continuing the resection deeper and resecting the trigone after the lobe has been removed. In order to avert this accident, the loop is hooked over the inner edge of the partially resected prostate and drawn lightly against it. The distal end of the electrotome is then slightly elevated to raise this edge away from the trigone (Fig. 156). If the middle lobe has been entirely removed and there is no tissue projecting up over the trigone, the loop will slip off the edge at this maneuver. If, however, there are tissue tags here which should be removed, the raised position of the loop will safeguard the trigone when the cutting current is applied. Each time a section is taken from this location, the same technique should be repeated. Another method of safeguarding the trigone while resecting an overhanging middle lobe is to resect this portion of the gland only when the bladder is distended; the trigone is then depressed away from the overhanging prostatic tissue. If at any time it is questionable that adequate tissue has been removed from the bladder neck and that the trigone is being encroached upon, bleeding should be checked sufficiently to permit clear visualization, the retrograde lens inserted and the vesical neck viewed from within the bladder.

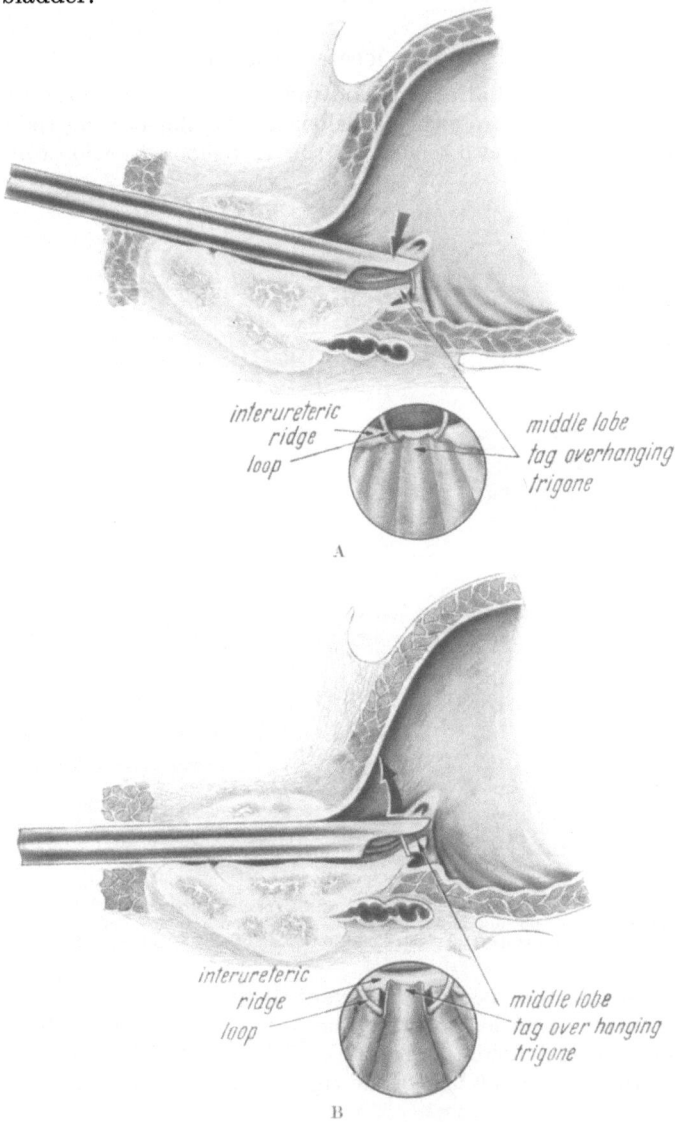

Fig. 156 A and B. A Method of removing middle lobe tag overhanging trigone without resecting trigone. Loop is placed over edge of prostatic border, vesical end of electrotome depressed as shown by arrow. Inset shows cystoscopic view. B. After loop has been hooked under overhanging edge of prostatic border, interior end of sheath is elevated, as shown by arrow, and overhanging tag removed with cutting current. This avoids resecting trigone

3. Tissue located ventrally

a) Ventral lobe

An intravesical ventral lobe may be difficult to reach with the cutting loop, especially after most of it has been removed and there is only a small piece left on a pedicle which recedes as the loop is placed against it. Subrapubic pressure exerted by an assistant will help to displace this tissue downward so that it can be more easily attained and removed with the loop. Care must be exercised, however, to avoid cutting into the ventral bladder wall.

b) Tags located ventrally

Tags of unresected prostatic tissue located on the ventral aspect of the bladder neck or prostatic urethra are easily overlooked. While rotating the resectoscope in either direction from the 6 o'clock to the 12 o'clock position during removal

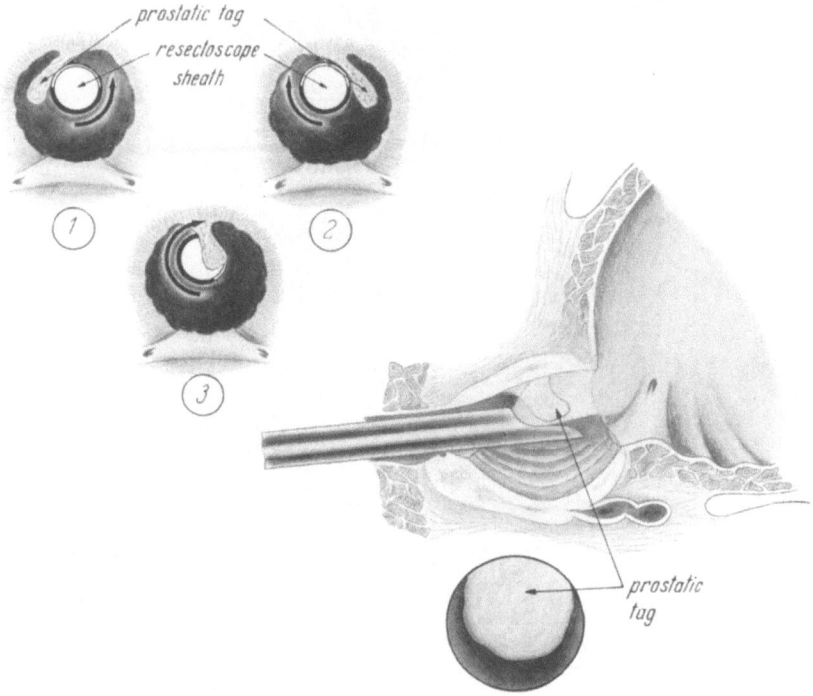

Fig. 157. Prostatic tag on pedicle hanging from roof. *1* Rotation of electrotome to right (from position of operator) pushes tag to right, away from opening in end of sheath. *2* Rotation in opposite direction pushes tag away to left. *3* Sheath rotated past 12 o'clock position which brings tag into opening of sheath

of a series of pieces, the tissue at the ventral aspect is rotated away from the fenestra (Fig. 157). In order to bring the unresected tag into the fenestra and field of vision, the instrument is rotated past the 12 o'clock position and then rotated back; this maneuver brings the tag into the opening of the sheath.

c) Tumors located ventrally

Technique for resecting these is described on p. 74.

4. Undermining the trigone

This is an accident which can be prevented by proper manipulation of the electrotome. It sometimes occurs when the sheath is pushed back into the bladder during resection of the prostatic urethra in the presence of a high dorsal lip of the bladder neck. It is due to the beak of the sheath's being kept too low (Fig. 158) while it is being pushed inward, and can be avoided by depressing the eyepiece of the electrotome, thus elevating the beak (Fig. 159) each time the inner end of the sheath is replaced into the bladder from the prostatic urethra. This accident

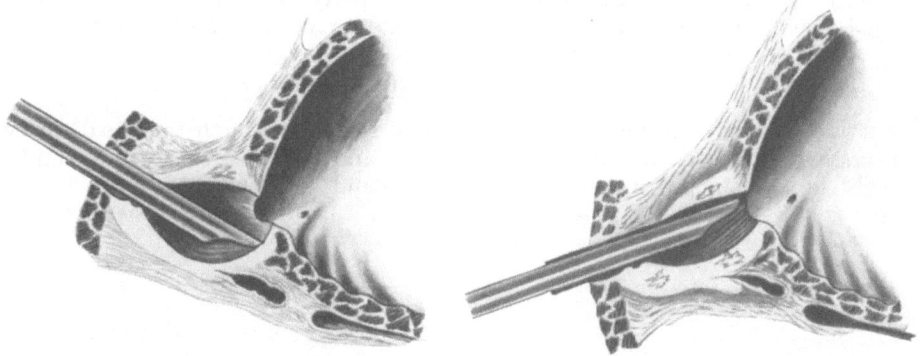

Fig. 158. Beak of sheath directed downward while being pushed back into bladder undermines trigone

Fig. 159. Beak of sheath elevated while being pushed back into bladder allows it to enter bladder above trigone

is more likely to occur in cases of median bar or small prostate, for here the trigone is stretched tighter and is thus more easily separated from the subtrigonal tissues than in the case of larger prostates.

5. Resecting tissue about the verumontanum

Some of the encroaching tissue near the verumontanum can be removed more accurately by first visualizing the verumontanum, then placing the loop just proximal to the veru and removing the section with a backward motion of the loop toward the bladder (Fig. 160). When this procedure is followed, great care must be exercised to elevate the inner end of the electrotome as the cut is being finished (Fig. 161). If this is not done, the loop will plunge deep into the floor of the prostatic urethra, with the danger of undermining the trigone or penetrating the prostatic capsule at this point.

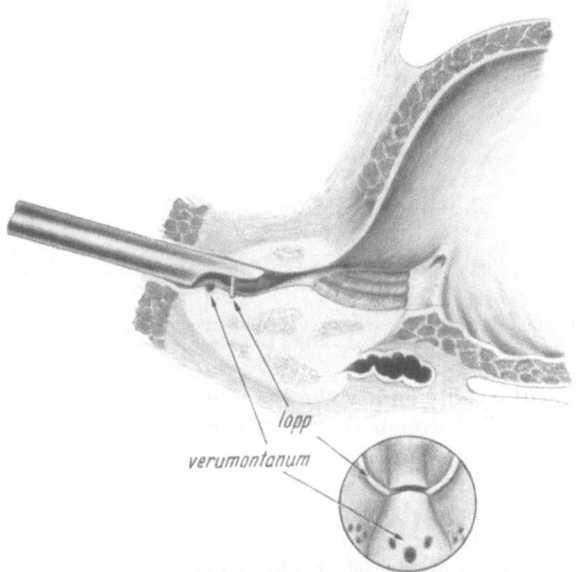

Fig. 160. Prostatic tissue adjacent to verumontanum is removed with a backward movement of loop after placing it against prostatic tissue at edge of verumontanum

Infringement just lateral to the verumontanum may also be accurately removed in this way, but the operator should never forget to bring the loop out of the tissue by elevating the distal end of the electrotome away from the tissue being cut.

6. Evacuation of tissue and clots

a) By manipulation of the sheath

The resected pieces which are washed into the bladder by the inflowing fluid during the resection, are washed out each time the bladder is emptied. This is done by alternately lowering and raising the fenestra against the bladder floor where the pieces collect; the outflowing fluid takes them with it through the sheath (Fig. 162). When this manipulation is done properly each time the bladder is emptied, usually no pieces will remain in the bladder at the end of the operation.

b) By suction, pressure or alligator forceps

Suction, pressure or the use of an alligator forceps through the sheath may be necessary; 1. when the *bladder is hypotonic* and its contents do not empty through the sheath, pressure of an assistant's hand over the suprapubic region

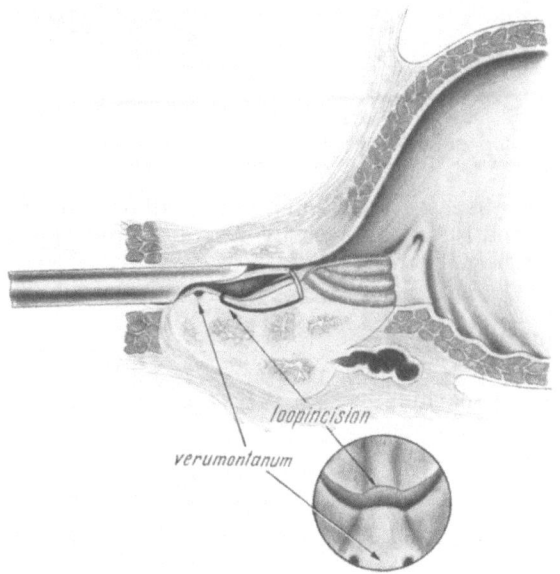

Fig. 161. Backward excursion of loop to trim prostatic tissue away from verumontanum. Note elevation of interior end of sheath to bring loop out of prostatic tissue and avoid cutting too deep

will usually empty the bladder contents; 2. when there are *pieces or clots too large* to pass through the sheath, forceful aspiration with a piston syringe or the use of alligator forceps (p. 24) is necessary; 3. when there is a large *vesical diverticulum* into which pieces and clots gravitate, the end of the resectoscope sheath can usually be passed through the diverticular orifice into its cavity and its contents aspirated.

Many urologists prefer to use *suction routinely* to remove all pieces and clots. However, removal with outflowing fluid by manipulation of the sheath is less traumatic to the bladder mucosa and is more easily accomplished if the manipulation is properly performed. Clots very seldom form in the bladder when the resection is expertly performed; there is insufficient time for blood to congeal if the procedure is rapidly executed.

c) Technique for use of suction

When an evacuating syringe is being used and the return flow into the evacuator is not free, the inner end of the sheath may be resting against the vesical wall, or the amount of water in the bladder may be inadequate. The evacuator should be held in such a position that air is not pumped into the viscus. By manipulating the inner end of the sheath to different intravesical areas, all resected pieces and clots can be removed during evacuation of remnants of tissue and clots from the bladder. It is beneficial for the operator to be able to visualize continuously in his mind's eye the distal end of the sheath in its relation to the

Fig. 162. Manipulation of sheath to remove resected pieces of tissue from bladder. Fenestra is alternately placed against bladder floor, then raised, about 2 cm away from it as fluid is flowing out through sheath

interior of the bladder, and to think of what is going on inside at least as much as outside where he can actually see the instruments and their manipulation. If the surgeon can train himself to do this throughout the entire operation, he will find it of great aid in successful accomplishment.

XIV. Locating and controlling bleeding

1. Pinpoint electrocoagulation

Bleeding is controlled by identifying the bleeding points in the section just resected and electrocoagulating each of them. Extensive electrocoagulation over a large area is damaging to tissues and is usually ineffective in controlling hemorrhage.

2. Lens close to tissue

While searching for bleeding points the objective lens of the resectoscope must be held very close to the tissue. The inflowing fluid clears the field of vision unless an artery is spurting directly at the lens.

3. Systematic search for bleeders

When a bleeding point is elusive, the resected area is searched systematically by alternately advancing, then withdrawing the resectoscope as it is rotated, thus covering all of the resected area. The planned approach of removing the prostate or a bladder tumor in sections and controlling bleeding in each section reduces the area which must be searched for a bleeder; it is probably located in the section which is being removed at the time the bleeding occurs.

4. Pressure of the sheath against a bleeder

Sometimes a general idea of the location of a bleeding point can be obtained by a distant view while the inflow of fluid is reduced; a cloud of blood may be seen advancing into the field of vision from a certain direction. After the general location of bleeding has been identified, the resectoscope is advanced into the bladder; then the inner end of the sheath is pressed firmly against the general area of bleeding and the resectoscope slowly withdrawn until the bleeding point is seen to emerge from under the edge of the sheath into the field of vision, when it can be identified and electrocoagulated (Fig. 163).

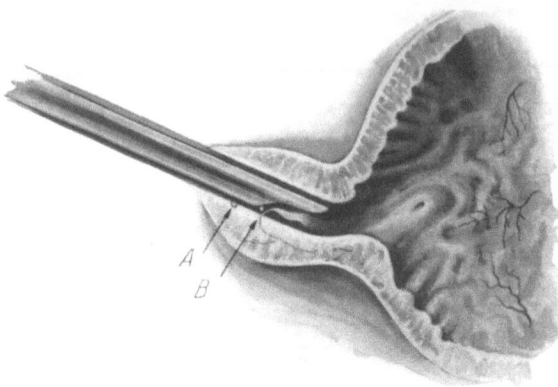

Fig. 163. Location of bleeding points. Sheath is pressed firmly against area of bleeding as resectoscope is slowly withdrawn. Bleeding points are identified and electrocoagulated as they emerge from under edge of sheath. At *A* there is a bleeder being compressed by sheath. At *B* a bleeder which has just emerged from under edge of sheath is being electrocoagulated

5. Rebound bleeding

Sometimes blood from an artery spurts across the prostatic urethra, striking the opposite side, and then rebounds, making it appear that the source of the bleeding is from the area of rebound. A close scrutiny of all possible areas of bleeding may be necessary before the point of bleeding is correctly located.

6. Bleeding under clots

When blood clots form over bleeding points and obscure them, it is necessary to scrape these clots off with the loop before the bleeders can be located and coagulated. Clots are more likely to occur when the resection is done slowly, for rapid operation does not allow them time to form.

7. Bleeding behind tags of tissue

Tags of tissue may project between a bleeding point and the lens of the instrument (Fig. 164); when these tags are resected the points are easily seen and coagulated (Fig. 165).

8. Vessel spurting into lens

A large bleeding vessel may spurt directly into the lens with such force that it will entirely obscure the field of vision even though the water inlet is turned on full force.

Temporary increase in water pressure created by elevating the irrigating jar may help in locating such a bleeder. Another aid is to *rotate the electrotome* first in one direction and then in the other. When it is rotated to the right past the bleeding point, the left part of the field of vision will be the more obscured by blood, and when it is rotated to the left the right side will be the more obscured.

In this way the approximate point of bleeding is located and then the loop with the coagulating current applied is drawn slowly over the surface of this area. The spurter may not be completely stopped the first time this is done, but may be controlled to such an extent that more definite visualization is possible, in which case the loop is then placed directly on the bleeding point and fulguration effected. Sometimes by *resecting* a little more tissue in the *general area of bleeding*, the direction of an arterial bleeder may be changed so that it does not spurt directly into the lens. This should not be done, however, if the capsule has already been exposed in that area. As stated above, it is not good technique to coagulate over a large area,

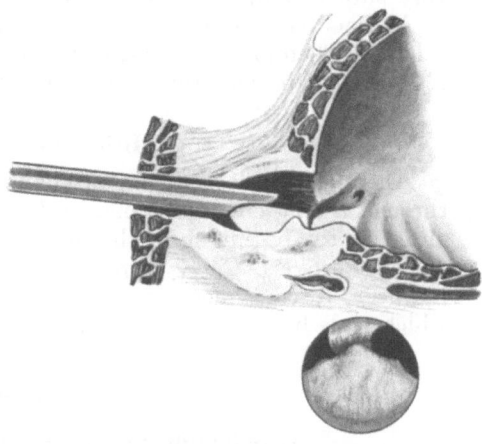

Fig. 164. Tag of prostatic tissue obstructing view of bleeding vessel. Insets show cystoscopic view

and whenever possible the exact bleeding point should be located and the loop applied directly to that spot before the coagulating current is turned on.

9. Bleeding behind the bladder neck

The retrogarde lens is useful in localizing bleeders about the bladder neck, especially on its anterior portion where they are sometimes entirely out of the field

of vision with the foroblique lens. After the position of the bleeding point has been noted with the retrograde lens, it can be found and coagulated more easily. When the retrograde lens is used, water is temporarily substituted for isotonic irrigating fluid because vision through this telescope is obscured by the cloudiness of the fluid mixed with blood.

10. Venous bleeding

This cannot be controlled by electrocoagulation; it persists regardless of heavy coagulation directly on the venous sinuses. The bleeding from venous sinuses is much less profuse while irrigating fluid is flow-

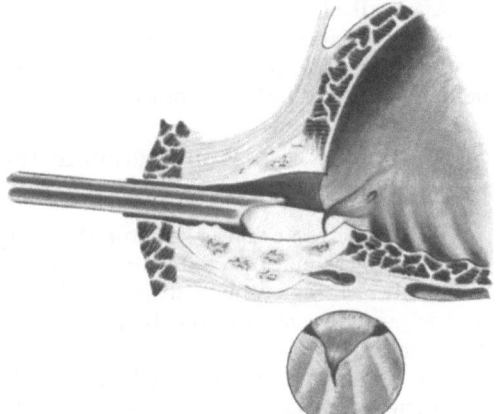

Fig. 165. Prostatic tag removed. Bleeding vessel visualized. Inset shows cystoscopic view

ing in over the bleeding area. As the bladder becomes distended and intravesical pressure is increased, bleeding from venous spaces stops entirely. The area from which the bleeding occurs can be located by shutting off the irrigating fluid before the bladder becomes distended; the blood is then seen to well up into the field of vision from the area of the venous sinuses. Bleeding from these spaces is easily controlled, however, by slight tension on a hemostatic bag catheter which is inserted after completion of the endoscopic prostatectomy. Pressure from a bag catheter, however, does not control arterial bleeding. The surgeon must be very sure to stop all bleeding from arteries before removing the resectoscope.

Venous bleeding does not often occur during endoscopic removal of bladder tumors. Therefore, it is very important to secure adequate hemostasis before terminating the resection of a vesical neoplasm.

11. Injection of vasoconstrictors

The injection of pitressin, adrenalin or other vasoconstrictors into the prostate through a long endoscopic needle helps to control bleeding. This may be done either at the beginning of the operation or during the procedure when bleeding becomes profuse. Creevy reported no ill effects from this procedure. The author, however, discontinued its use because of rapid variations in blood pressure which were believed to be due to absorption of the vasoconstrictor drug.

XV. Concluding the operation

1. Selection of pieces for microscopic examination

Just before completion of endoscopic prostatic resection, several pieces of tissue should be saved from the dorsal portion of the *prostate* near or including the capsule and close to the verumontanum. These pieces could show malignancy whereas pieces from other portions of the prostate might be benign. In approximately 20 per cent of cases of prostatic carcinoma, the microscopic diagnosis will be missed unless these special pieces are saved and given to the pathologist separately (BARNES 1947). When soft fuzzy tissue or small bright yellow nodules are encountered during a resection, pieces should also be saved for biopsy study; such tissue is often malignant.

After all of a *bladder tumor* has apparently been removed, it is well to take a few small pieces from the edges and from the bottom of the resected area and send them to the pathologist separate from the other mass of tissue, thus it can be determined whether or not all of the involved bladder wall has been removed.

2. Examination at the end of operation

a) Prostatic urethra

When the resection is supposedly completed and bleeding controlled, careful examination is made to determine whether a sufficient amount of tissue has been removed, or if tags remain which might cause subsequent trouble. The retrograde lens is used to examine the bladder neck, which should be smooth and regular. The ureteral orifices should be visible above the cut edge of the bladder neck (in the lower part of the visual field of the retrograde lens), and there should be no tissue projecting into the bladder. This lens may also reveal any prostatic encroachment into the posterior urethra. Inasmuch as this area should have concave posterior and lateral walls when the operation is completed, the verumontanum—at least the region of the verumontanum—should be visible through the retrograde telescope. If some projecting tissue still remains, its location and estimated amount is verified, the lens removed, the working element reinserted and the excess tissue resected. The electrotome is then withdrawn sufficiently so that the verumontanum is barely visible in the lower edge of the field, and the floor of the prostatic urethra is viewed from this point. The straight forward

(Vest) telescope is useful for observation at this point (Fig. 166). Prostatic tissue may be identified by its dull, granular appearance, as described above, and encroachment or irregularities are easily seen. When tissue is to be resected from the urethral floor, the electrotome is pushed slightly inward from its position at the verumontanum so that the veru lies just outside the lower margin of the visual field; otherwise this landmark might be removed and the danger of resecting the

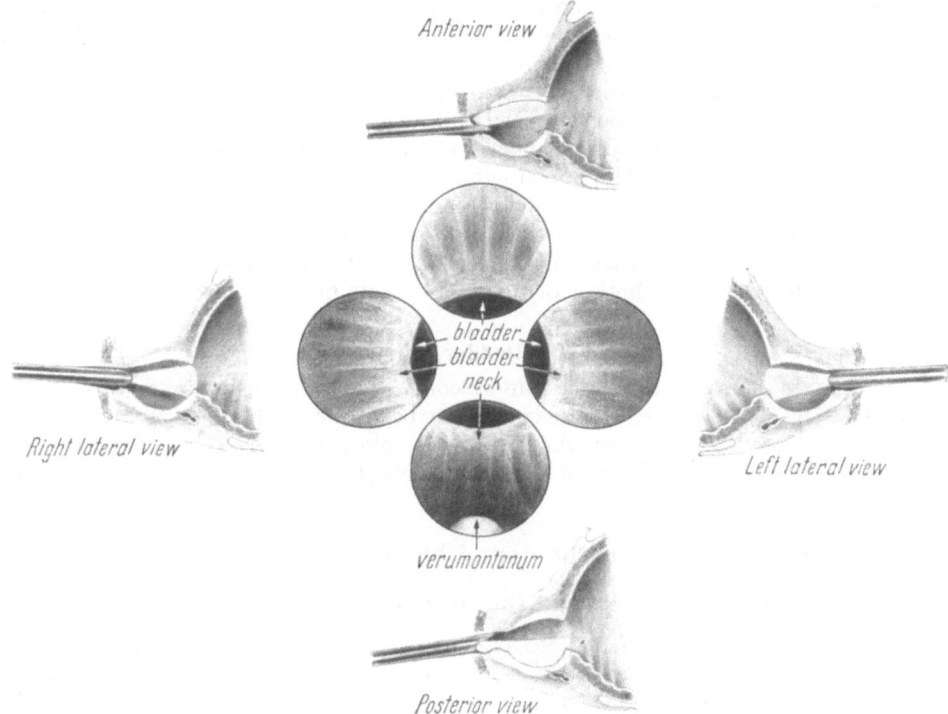

Fig. 166. View of prostatic fossa from the region of the verumontanum after resection is completed. Note deep concavity of posterior portion of prostatic urethra, and very little concavity of anterior portion as seen through foroblique or direct vision lens

external sphincter increased. Excavation of the prostatic urethra may safely be carried considerably deeper in the floor just above the verumontanum than on the sides or nearer the bladder neck. While resecting the apical portion of the gland, the operator must use great care to avoid slipping the electrotome either inward or outward the slightest degree. To rest his left elbow on a stationary object, as noted above, will aid in this regard. After the urethral floor has been well cleaned out, the electrotome is again withdrawn to its former position at the verumontanum, and rotated first one way entirely to or slightly past the 12 o'clock position, then the other way for the same distance. In this manner, tags or larger pieces of tissue are often discovered which had remained unnoticed because rotation of the electrotome during resection of this region had pushed them out of sight.

b) The inside of the bladder

This is examined for pieces of tissue and clots; if any are found which have not washed out by manipulation of the sheath during the outflow of irrigating fluid, aspiration may be necessary (p. 206).

14*

3. What constitutes adequate removal of tissue

a) Prostate

The prostatic tissue is removed down to the true capsule throughout most of the prostatic urethra. There are no tags or areas of thickened capsule left. Some of the surgical or false capsule, however, is allowed to remain. Healing is more rapid and postoperative fibrosis less extensive when thin patches of glandular tissue are left covering the true capsule (WEYRAUCH et al). The mucosa of the prostatic urethra grows from the prostatic duct orifices and covers the denuded area more rapidly than when the only source of musoca is that located at the two extremities of the prostatic urethra—the bladder neck and the membranous urethra. When the resection has been completed, a view through the foroblique telescope whose objective lens is placed at the verumontanum should encompass the interior of the bladder; no tissue should obscure the view of the bladder cavity. Sometimes the dorsal lip of the bladder neck is elevated enough to obscure this view. In this case it is incised as described below. Digital palpation of the prostatic urethra through the anterior rectal wall over the resectoscope may be helpful in determining whether or not more tissue needs to be removed.

b) Bladder tumors

Adequate removal of bladder tumors consists of extirpation down into bladder muscle. The area from which the tumor has been removed is thoroughly searched by a very close view through the foroblique telescope. If any tissue other than the characteristic coarse fibers of bladder muscle (Fig. 151, p. 199) is seen, it should be resected.

XVI. Incision of the dorsal bladder neck and trigone

1. Combined hypertrophy of the trigone and elevated bladder neck

Hypertrophy of the trigone is frequently seen in patients who have had obstruction at the bladder neck over a long period of time; it is a result of the oft repeated effort of the trigonal muscles to open the vesical orifice to initiate micturition. Elevation of the interureteric ridge is produced, and results in a culdesac above the trigone, which often causes urinary obstruction even after the prostate has been well removed. Sometimes the interureteric ridge is displaced toward the bladder neck, and the shortened trigone actually forms a part of the vesical neck. This may be the cause of a persistent suprapubic urinary fistula which will heal promptly after incision of the trigone and bladder neck with a COLLINGS' knife through the panendoscope. When the endoscopic route is chosen, elevation of the bladder neck or hypertrophy of the trigone can be easily detected with the retrograde lens, and is incised after the resection has been completed (Fig. 167). This is done with a resectoscope knife inserted in the electrotome in place of the loop (Fig. 168). The latter obviates the necessity of removing the resectoscope and then introducing the panendoscope for use with the COLLINGS' knife. Because of the fact that the blade of the resectoscope knife is not long, an incision several layers deep must be made with this instrument in markedly hypertrophied trigones. This type of knife is easy to construct by filing the lateral surfaces of the ball electrode supplied with the Stern-McCarthy electrotome to make a blade.

The trigone is not resected with the loop electrode, for frequently the ureteral orifices are so close together that they may be injured by so doing. This danger is eliminated by incising in the midline between the ureteral orifices with a knife electrode. The incision is carried deep enough so that the culdesac back of the trigone is on a level with the prostatic urethra. Inasmuch as the trigonal muscles are considerably thickened in these cases, there is no danger of perforating the bladder wall if the incision is not carried deeper than the floor of the culdesac above the trigone. After the trigone has been incised, examination with the foroblique lens will determine whether the proper depth has been attained, and also whether too much tissue remains between the edge of the incision and the ureteral orifices. If this is the case, the loop is replaced in the electrotome, and a small piece of tissue is removed from each cut end of the incised interureteric ridge.

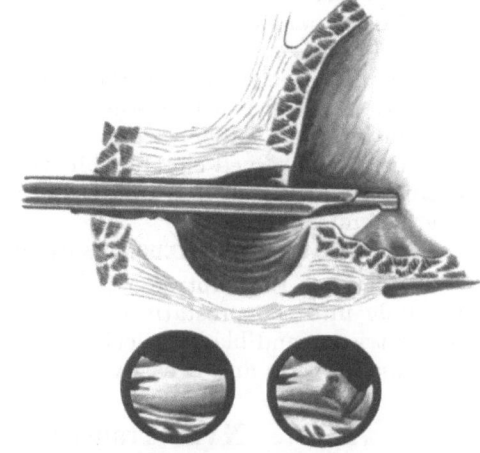

2. Elevation of bladder neck only

Incision of the bladder neck at the apex of the trigone may be indicated even though there is no elevation of the interureteric ridge. Combined intraurethral and intravesical prostatic hypertrophy as it enlarges dilates the bladder neck; but an entirely intraurethral hypertrophy does not have this effect and the diameter of the neck

Fig. 167. Hypertrophied trigone as seen through retrograde lens following prostatic resection. Insets show cystoscopic view of trigone before and after incision (retrograde lens)

remains normal. Therefore, after the intraurethral protrusion has been removed

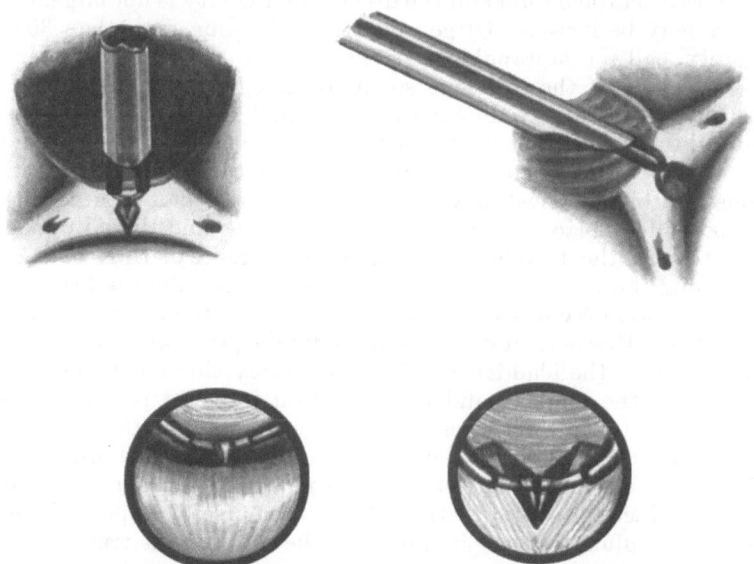

Fig. 168. Incision of trigone with resectoscope knife. Insets show view through foroblique lens before incision is made and after first excursion of knife. Incision must be carried deeper than shown here to allow adequate drainage of culdesac above trigone. Inset shows cystoscopic view

the bladder neck is comparatively small, the prostatic urethra large and the floor of the latter is considerably lower than the bladder neck at the apex of the trigone. In these cases, an incision is made with the resectoscope knife in the mid-dorsal line through the bladder neck to bring the floor of the prostatic urethra level with the trigone or fundus above, allowing free passage of a catheter into the bladder. If this procedure is not followed, there is obstruction to the passage of the catheter, and the prostatic urethra is left as a cavity which predisposes to infection and even calculus formation. The bladder neck is also likely to contract further, resulting in recurrence of urinary obstruction. When the bladder neck is incised, the edges of the incision retract and leave a groove between the fundus and the prostatic urethra. In making this incision, there is little danger of perforating the bladder wall to such an extent that extravasation will occur. If the tissue is observed closely while the cut is being made, perforation can be recognized immediately and will result in no harm if the bladder is not subsequently overdistended.

3. Exposure of subtrigonal adenoma

Sometimes the subtrigonal portion of an hypertrophied prostate does not protrude into the prostatic urethra sufficiently to be removed. After incision of the trigone and bladder neck, however, any subtrigonal adenoma which might be present everts and is easily recognized and resected.

XVII. Transurethral diverticulotomy

Diverticula of various sizes are frequently found in cases of bladder neck obstruction, but usually these are small and will empty and cause no symptoms after the obstruction has been removed. There are a few, however, which continue to retain fluid, and these must be treated. Preoperative cystograms aid in determining whether a diverticulum drains or not. If the contract cystogram shows retention of cystographic fluid and the diverticular cavity is not large, the diverticular orifice may be incised. Larger diverticula holding more than 35 cc., which do not empty, call for suprapubic diverticulectomy. If the capacity is less than 35 cc., — especially if the orifice is so small that it closes down entirely when contracted by the bladder wall—the opening may be enlarged with the resectoscope knife through the electrotome at the time of endoscopic resection (BARNES & BERGMAN 1938; SENGER et al). This is done at the end of the resection, after the trigone and bladder neck have been incised if this has been necessary. The resectoscope knife is placed in the diverticular orifice, and the lower edge incised downward toward the bladder neck. The cut edges will then be seen to pull apart, widening the orifice appreciably. If the diverticulum is located near the trigone, the incision may be carried to a depth of more than 1 cm. without danger of extravasation. However, if it is high in the fundus, the cut should not penetrate farther than 1 cm.; the bladder wall is considerably thinner in the fundus than in the region of the trigone, and extravasation into the perivesical tissues or even into the peritoneum may result from deep section in this area. If the diverticulum is near the bladder neck, the incision is carried downward into and sometimes through the neck so that the bottom of the diverticulum is on a level with the floor of the prostatic urethra. This, of course, is impossible in the case of a large diverticulum, because its floor may lie beneath the trigone and bladder neck.

XVIII. Insertion of the catheter

1. Hemostatic bag catheter

After all pieces of tissue and clots have been evacuated from the bladder and hemostasis has been secured so that the return irrigating fluid is only pink or light red in color, a few ounces of fluid are left in the bladder, the resectoscope sheath is removed, and a No. 24 thirty cubic centimeter hemostatic bag catheter is inserted. It is best to use a wire stilet or mandrin inside the catheter to guide it over the floor of the prostatic urethra and through the vesical orifice with a minimum of trauma. The bag is inflated while in the bladder cavity, and tension is applied to the catheter to pull the bag down against the edges of the vesical orifice (Fig. 169). The bag in this position maintains hemostasis better than when it is entirely within the prostatic cavity. Much of the bleeding emanates from the rim of the bladder neck, and if the bag is not against this edge, it is ineffective in controlling bleeding from it. Another disadvantage of the position of the bag in the prostatic cavity is the danger of injury to the external sphincter when the bag is distended and tension applied. In the presence of venous bleeding it is advisable to apply tension for one to two hours. This is simply and easily accomplished by tying a gauze sponge around the catheter just external to the meatus; as light tension is applied to the catheter, the knot is gently pressed against the glans. Tension should not be great enough to collapse the shaft of the penis or to cause excessive pressure of the gauze against the glans; only sufficient to hold the hemostatic bag gently against the

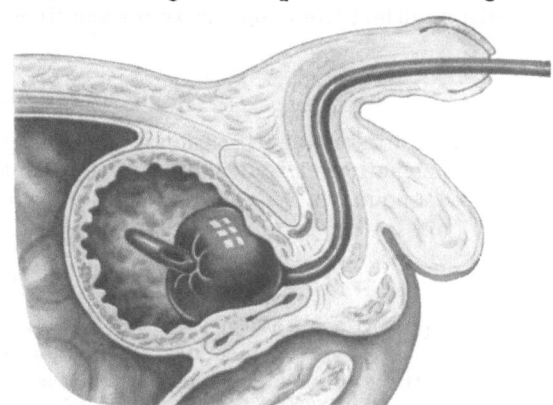

Fig. 169. Hemostatic bag catheter pulled down against rim of bladder neck following endoscopic removal of prostate

bladder neck; this is sufficient to control venous bleeding. No amount of tension will stop arterial bleeding; it must all be controlled before removal of the resectoscope. The gauze sponge tied around the catheter gradually loosens, and if it is not tied too tightly it slips down the catheter so that the tension is automatically released in a hour or two.

2. Catheter passed through the resectoscope sheath

If, at the beginning of the operation passage of the resectoscope has been difficult, it may be safer at the close to pass a No. 22 regular soft rubber catheter through the resectoscope sheath and then remove the sheath over it; the surgeon can thus be assured that the catheter is in the bladder. A coude tip catheter may pass more easily if difficulty is encountered in passing a catheter after removal of the sheath.

3. Immediate bladder irrigation

After the catheter has been passed the bladder is thoroughly irrigated before the patient is removed from the table, in order to be sure that all clots and pieces of tissue have been removed and that free drainage through the catheter will occur when the patient is returned to bed. A few ounces of irrigating fluid are left in the bladder so that when the catheter is connected to the drainage bottle at the bedside, it will drain immediately.

XIX. Rapid resection of large prostates and bladder tumors

1. Swift technique

When an operator attempts endoscopic removal of a Grade III or larger gland, his technique must be very swift in order not to prolong the procedure unduly.

In fact, he must be able to remove more than one gram per minute if he is dealing with glands of 75 to 100 grams. Some operators are able to work at this tempo. It is, of course, easier to remove more tissue in a shorter length of time from these larger glands due to their greater volume. A good share of the operating time is always consumed in trimming up around the edges and in careful resection close to the capsules. The situation in the case of large bladder tumors is similar. Endoscopic removal of them is likewise faster because of their known volume. The more meticulous and time consuming work comes after extirpation of the tumor tissue down into bladder muscle, when any remnants invading tissue must be distinguished from the coarse muscle fibers and resected.

2. Rapid identification of tissue

Rapid tissue identification is imperative; a quick glance through the telescope must be sufficient to decide whether or not the true capsule has been exposed at a given point, or the bladder wall perforated or nearly perforated.

3. Powerful electrosurgical unit

A powerful electrosurgical unit, such as the Birtcher model No. 2000 "Personally Certified" which has an output of 500 watts from the tube circuit and 275 watts from the spark gap circuit, is a necessity if rapid removal of tissue is to be accomplished; with it the cutting loop can be drawn through the tissues as swiftly as desired, and it does not slide over the tissue when tags or thickened capsule are being removed.

4. Control of bleeding

Bleeding must be controlled in each section (p. 192). When resection of the intravesical section of the enlarged prostate has been completed, bleeding should be controlled throughout that area with the exception of the distal edge. This is left uncontrolled because removal of the next or intraurethral section on the same side will immediately cut into the same vessels; it saves time to leave them until all tissue has been resected down to the capsule on that side. To electrocoagulate every bleeding point as it appears retards the procedure until the time limit is exceeded for removal of large glands by the endoscopic route. The formation of clots over the resected area impedes progress and clots form when there is delay from any cause; interruptions to the smooth continuance of the procedure must be avoided. Control of bleeding by sections during removal of large bladder tumors makes it possible to remove these endoscopically within the allotted time.

5. Large pieces of tissue

The pieces of tissue removed must be large; pressure of the inside of the resectoscope against the tissue helps to sink the loop into it. Moving the resectoscope sheath in and then out as a piece is being resected results in the removal of longer pieces; this should be done, however, only if the prostatic urethra is long and if the surgeon is sure he is not resecting in or out too far. Rectal counterpressure helps to remove thicker pieces of tissue. In resecting bladder tumors, the endoscopic technique for removing large pieces is applied until the vesical wall is approached or until the telescope reveals a bite into muscle fiber.

Chapter XX

Variations in technique of endoscopic prostatic resection

The basic principles of the technique of endoscopic surgery are the same regardless of the variations employed. The technique with the Stern-McCarthy electrotome as described in Chapter XIX is not necessarily the only correct technique to use with that instrument. When other instruments are used, even more variation in technique is necessary. Each technique has its particular merits, and the one which is successful in the hands of the individual operator experienced in the performance of endoscopic surgery is, in most instances, the best for him to use.

I. Rectal palpation and counterpressure

1. Purpose

The purpose of rectal palpation and counterpressure during endoscopic prostatectomy is; 1. to displace prostatic tissue toward the cutting instrument in the urethra, making it more available for resection; 2. to estimate the amount of prostatic tissue between the anterior rectal wall and the resectoscope; 3. to help determine when sufficient tissue has been removed, and 4. to prevent injury to the rectum.

2. Technique

The use of the one hand operated resectoscope such as the Nesbit and its modifications permits the surgeon to insert the index finger of his free hand into the rectum during the endoscopic procedure. When the Stern-McCarthy electrotome is used, both hands are needed to operate the instrument. The surgeon may insert his finger into the rectum, however, while he is not engaged in resecting tissue. The punch resectoscopes such as the Braasch-Bumpus and Thompson instruments usually require both hands to operate, but occasionally tissue is resected with one hand; leaving the other free for palpation. It is advantageous for the operator to use his own finger rather than to rely upon an assistant's; by this "three dimencional perception" he can estimate the amount and position of tissue which needs to be removed and can press it toward the cutting loop or blade.

Some endoscopic surgeons rely upon rectal palpation more than upon the appearance of the tissue in order to determine when adequate removal of the prostate has been accomplished. It is necessary for them to keep one finger in the rectum during the last stages of the operation; otherwise some tissue which should be excised is left in place, or some is removed which should have been left.

When the punch resectoscope, which has the fenestra in the side of the sheath, is being used, it frequently becomes necessary to press tissue into the fenestra by means of rectal counterpressure. In this way the dorsal and dorsolateral portions of the prostatic urethra can be excavated. Partially resected masses of tissue which move about in the prostatic urethra may need to be stabilized and displaced toward the cutting loop by rectal counterpressure.

II. Encirclement of prostatic tissue

1. Technique

Instead of beginning the resection at the dorsal midline margin of the prostate, as explained in Chapter XIX, some operators partially "enucleate" the lateral lobes by removing tissue at or near the 12 o'clock position first, exposing the capsule at this point, then continuing laterally and downward in the same direction as is followed when separating the lateral lobes from the capsule in the course of suprapubic enucleation. The capsule is identified after each piece is removed, and a deep, narrow groove is thus made between the lateral lobe and the surgical capsule from the 12 o'clock to the 4 or 5 o'clock position on one side first, and then to the 8 or 7 o'clock position on the other. This allows the lateral lobes to fall in over the middle lobe, after which they can be rapidly resected, for the preliminary formation of the lateral groove has dispensed with the immediate need for further hemostasis and for cautious dissection.

The intravesical section or zone of the prostate is first removed on one side then on the other by the encirclement technique, then the intraurethral or subvesical section in the same manner and lastly the apical zone. The resection is started at or near the 12 o'clock position in each of these zones and is continued downward in the arc of the internal sphincter, exposing the capsule and allowing the mass of prostatic tissue to fall away from the capsule. Fragmentation of the mass is accomplished in each zone before starting to encircle the tissue in the next zone. The tissue of the apical zone is not encircled but is removed entirely — first on one side then on the other, starting at the 12 o'clock position and removing the first series of pieces down to the 6 o'clock position.

2. Advantages

The boundaries of the operative field are delineated early. The bleeding vessels at the vesical neck are exposed and electrocoagulated before the mass of tissue is removed; therefore, fragmentation of the encircled mass of tissue results in almost no bleeding and can be accomplished rapidly. The bladder neck or "internal sphincter" (p. 198) fibers are exposed first at the ventral portion of the vesical orifice where they are close to the urethral mucosa; the prostatic tissue is not often thick at this point. These fibers constitute the landmark delineating the upper edge of the resection; they are exposed around the arc of the vesical orifice from the 12 o'clock to the 4 and 8 o'clock positions. After the lateral masses have been encircled they are fragmented; then the prostatic tissue which is located dorsally is removed. When the one hand operated resectoscope is being used, the operator's finger is kept in the rectum during removal of the dorsal portion of the gland; this helps: 1. to prevent extending the resection too deeply toward the rectum, and 2. to give the operator an estimate of the amount of tissue yet to be removed.

3. Disadvantages

a) **Landmarks on the floor.** Three of the four landmarks wich are used for orientation during endoscopic prostatectomy are on the floor of the operative field; the trigone and ureteral orifices help to identify the proximal limit of the resection and the verumontanum its distal boundary. The other landmark — bladder neck or internal sphincter fibers — are visible throughout the circumference of the vesical orifice after the overlying mucosa and prostatic tissue has been resected. Orientation is easier to obtain when the resection of each series of pieces starts on the floor of the operative field at the 6 o'clock position, because more landmarks are there to guide the surgeon.

b) **Obstructing masses of tissue.** The mass of partially encircled tissue lying in the prostatic urethra may be confusing; it sometimes covers all of the landmarks; even the bladder neck fibers may be lost if the mass falls between the resectoscope and the groove which has been made at the bladder neck. An orderly removal of tissue from within the urethra outward toward the capsule enlarges the prostatic urethra symmetrically; there are no tissue masses to obstruct the view.

c) Early perforation. When the capsule is exposed during the first few excursions of the loop, by the encirclement technique, there is more danger of partial or complete perforation early in the operation; extravasation through a perforation made near the end of operation is less extensive than when the opening occurs early.

III. Punch prostatectomy[1]

1. Technique

a) Manipulation of the instrument. The operator, having passed the punch instrument into the bladder without force, withdraws the obturator and replaces it with the tubular knife. He now turns on the irrigating fluid and, filling the cuplike glass window with water, places it like a cork in the proximal end of the tubular knife, guarding against admitting air bubbles. Should bubbles appear, lowering the proximal end of the punch allows the current of water to carry them into the bladder and prevents the necessity of removing the window in order to clear the instrument. Time should always be taken before beginning the operation to examine carefully the entire field, to ascertain which portion of the gland is causing the obstruction, to determine how much tissue it will be necessary to resect and from what location. Landmarks such as the ureteral orifices, the interureteric ridge and the verumontanum should be located and carefully examined to determine their relationship to contiguous structures, as the tremendous deformity to the prostatic urethra incident to enlargement of the gland greatly alters their normal relationship. Such a preliminary survey will save much time later in the procedure and insure the operator an accurate knowledge of the entire area before any tissue is resected.

Once the various landmarks have been observed and mentally recorded in their abnormal relationship, it is time to start the operation. The knife is withdrawn far enough to open the window in the side of the outer sheath; then the instrument is moved forward so that whatever tissue is chosen to excise protrudes into the window. This procedure is carried out under full vision, and as the tissue to be excised protrudes through the opening, the knife is thrust forward and a cylindrical portion of tissue excised. This may be repeated as often as is neccesary to remove the obstructing lobe. A punch resection should always begin as far toward the bladder as there is hypertrophic tissue to be removed in that area, and after each bite an edge should remain to be grasped in the fenestra for the next bite. In this way one works back toward the verumontanum, which in most cases is the external limit of the hypertrophied intraurethral lateral lobes. It may be necessary to remove large amounts of tissue in order to obtain a wide enough channel, for the mere tunneling of a passage into the bladder by excising the median lobe and resecting the lower borders of the lateral lobes is insufficient. The attempt at this halfway measure has been undertaken by most operators early in their work. It will almost invariably result in ultimate failure, although the obstruction may for a time be relieved. Failure results because such a tunneling process leaves the roof of the urethra encumbered with overlapping lateral lobes; whereas the urethra must be unobstructed throughout its circumference in order to function normally.

---

[1] Adapted from Chapter XI, *Punch prostatectomy.* written by H. C. Bumpus jr., in Barnes' Endoscopic prostatic surgery. Corrected to date by Ben Massey.

If the obstruction is chiefly median, being confined to an intravesical enlargement of the subcervical glands or to hypertrophy of the commissural portion of the prostate, resection is simple and good functional results are assured no matter how great the obstruction. This pertains, of course, only if care has been taken to leave no nodules in the region of the internal sphincter and if the vesical lip has been left clean and free of tags — both of which precautions insure a smooth, even healing which tends to enhance functional results.

b) Control of bleeding. Because tissue is removed in cylindrical pieces, the incised surface bleeds where small arteries or veins have been severed. When bleeding is judged to be excessive it is controlled by advancing the electrode into the field of vision and applying it to each bleeding point, using a minimum of coagulating current. Under no circumstances is the entire raw surface cauterized. It is a most important point in technique to prevent clots from forming, for they obscure bleeding points and make satisfactory control of hemorrhage exceedingly difficult. If the operator is so unfortunate as to have clots form and adhere to the operative area, he should attach a plunger syringe to the instrument at once and irrigate the bladder vigorously until the clots are washed off. After this, to trace bleeding to its origin will not be difficult. A rubber bulb syringe is not used for evacuation of clots because it does not provide adequate suction, and bladder overdistention and ultimate rupture may occur from repeated insertion of more fluid than is withdrawn. A bladder filled with clots can be easily emptied through a punch instrument in three to four minutes when a powerful plunger syringe is used or suction applied from the operating room equipment.

Clots usually form as a result of faulty irrigating during operation, or from the cutting of a considerable sized artery or venous sinus outside the prostatic capsule. As experience with resection is developed, however, the added speed which the operator acquires practically obliterates the complication of clot formation, and only the matter of controlling the larger spurters demands consideration. If the operator does not cut beyond the prostatic capsule, he will encounter very little difficulty with bleeding.

It is well to cut rapidly rather than to stop to control venous bleeding until the portion of the gland which is being resected has been removed. If the operator stops to coagulate any but the larger bleeders he is bound to delay the procedure, prolong the operating time, and in the aggregate cause greater blood loss than if he went right on until all the tissue from the particular area on which he is working has been removed. This is a point to be emphasized, because as the operation progresses the same vessels are cut through several times, and if no immediate attempt at hemostasis is made, it is possible to excise an entire lobe within a few minutes.

c) Method of resection. Usually some degree of hypertrophy of all the lobes is present, in which case the median obstruction should be resected first. Once this is removed the lateral lobes will more easily fall into the operative field. Early removal of the median obstruction also makes manipulation of the instrument simpler and has the added advantage that, if for any reason it becomes desirable to terminate the operation before completion, the possibility of symptomatic relief is greater if the median obstruction has been removed than if either lateral lobe has been excised.

Having removed one lobe and satisfactorily controlled bleeding, the operator locates another lobe and repeats the procedure until all obstructing tissue has been removed. No pedunculated tissue should be left and every effort should be made to leave the floor of the urethra smooth and free of tags. A smooth surface heals rapidly if not devitalized by too excessive coagulation. If it has been necessary for the sake of hemostasis to coagulate an area more than is desirable, it should be carefully examined and perhaps it will be possible to remove some or all of the devitalized tissue without producing a second hemorrhage. Tissue should be removed widely enough from all portions of the vesical orifice to form an adequate channel from the verumontanum to the trigone. Once this has been accomplished the operator will have an unobstructed view from the verumontanum into the bladder. When doubt exists concerning the advisability of excising more tissue once a satisfactory channel has been constructed, it is always advisable to be conservative. Cutting too deep may cause extravasation of urine later and is bound to produce annoying bleeding. Tissue once taken out cannot be replaced, but if insufficient prostate has been removed at the initial stage to give good functional results a repeat procedure may easily be done.

d) Adequacy of the resection. A very reliable test of the adequacy of operation is at its conclusion to fill the patients's bladder, remove the instrument and exert suprapubic pressure. The fluid should gush from the urethra in a copious stream. If it does not, sufficient tissue has not been removed.

The most difficult cases to resect are those in which the hypertrophy is entirely confined between the two sphincters. The difficulty encountered here is the tendency of the lateral lobes to fall away from the instrument, the procedure being not unlike bobbing for apples at Halloween. If an assistant places his finger in the rectum in such cases he can greatly facilitate matters by repeatedly forcing these lateral lobes into the fenestra of the instrument. This makes excision of the lateral lobes relatively easy, although to the beginner this part of the punch operation usually offers the greatest difficulty. Care must be exercised not to cut too deep and only on the bladder side of the verumontanum, for this structure marks the beginning of the external sphincter and a single incision beyond it may produce incontinence.

There need be no fear of removing as much prostatic tissue as possible between the veru and the bladder in order to make an adequate channel. The removal of large masses of tissue from the bladder improves but slightly the free passage of urine through the urethra if remnants of intraurethral lateral lobes have been left in place. Although these remnants may be small in proportion to the whole amount removed from the bladder, if their location is near the verumontanum they interfere with uninterrupted urinary flow to a greater extent than their size would signify were it not for their position here where the lumen of the urethra is already somewhat occluded by the verumontanum. An insignificant amount of tissue hanging from the roof of the urethra may cause the most persistent symptoms of obstruction even when an apparently adequate channel has been formed. Such neglected tissue may in occasional cases be found in the anterior lobe of the prostate which undergoes hypertrophy less frequently than the lateral or median lobes. The overlooking of an hypertrophy of the anterior portion of the gland frequently occurs because the operator fails to turn his instrument over for observation of that region. Fortunately, such anterior lobe hypertrophies are rare, and a careful preoperative survey will always reveal them and put the operator on his guard. If such a survey should reveal an anterior lobe projecting into the urethra, it will have to be punched out as the others were. Care, however, must be exerted when working in this location for, except in cases of adenomatous hypertrophy of the anterior lobe, the bladder wall here is very thin and is easily perforated.

2. Advantages

a) **Little trauma.** The punch resectoscope causes the least amount of tissue trauma because it excises with a knife without inflicting thermic destruction. Heat is the most potent of all traumatic agents, and heat which is generated by electricity is more corrosive in action than any other type. Surgeons who use the punch instrument are dependent upon this form of current for the control of bleeding, while those who prefer the loop electrode must use it for cutting as well as for hemostasis. Since the procedure which causes the least tissue destruction will leave the most rapidly healing scar, one may expect convalescence following punch prostatectomy to be considerably more rapid than after loop resection. The only thermic destruction produced by punch prostatectomy occurs in small areas about bleeding vessels when hemostasis is applied in the form of electro-coagulation with a single wire tip electrode — a process not unlike spot welding. In fact, the rapidity of convalescence following endoscopic resection is in direct ratio to the degree of operative trauma.

b) **Volume of water inflow.** Clear observation of the operative field is more easily acquired by means of the punch instrument, for the volume of water flowing from it is so great that blood is washed entirely out of view. Such a volume of irrigating fluid is not possible with lens resectoscopy, as the lumen of the instrument has of necessity to carry the system of lenses and the shaft of the loop electrode, leaving less free space than the unencumbered tube mechanism. Vision obtained through the direct cystoscope is further desirable in that what is seen is not magnified, and hence does not need immediate interpretation as to its size or position.

c) **Tactile evaluation.** The knife punch would seem to present another advantage over the wire loop electrode, because it permits tactile evaluation by the operator. This is almost as important as vision, and may be relied upon to some extent when visualization is poor. With an electric wire loop operated by a lever and ratchet, all tactile sense is lost and one cannot readily tell from the feel what type of tissue is being cut. When tissue is excised by electrocoagulation, the cut surface of all tissue through which the loop has passed may appear very similar to the novice. When the tubular knife is used, the operator's tactile sense immediately informs him what type of tissue the knife is entering, the cheeselike consistency of adenomatous hypertrophy being entirely different from the cicatricial tissue of prostatitis or the fibrous feel of bladder musculature. Vision is also of the utmost aid, for the characteristic appearance of adenomatous tissue makes it possible to excise it right up to, but not including, the so-called capsule.

d) **Direct vision.** The direct vision obtained through the punch resectoscope is advantageous to the urologist who has been trained in the use of direct vision cystoscopes; interpretation of the findings through a lens system is difficult when he is not accustomed to looking through an endoscopic telescope. This difficulty is compounded when he begins to perform endoscopic surgery with an instrument to which he is not accustomed.

3. Disadvantages

a) **Increased bleeding.** There is usually more bleeding when the punch resectoscope is used; the cold knife does not close as many cut ends of vessels as does the loop electrode which is activated by an electrical current.

b) **Difficult excavation.** When the prostate is very large it is more difficult to excavate the prostatic urethra and leave it concave because the fenestra is on the side of the punch resectoscope and tissue must be displaced up into it.

c) **No magnification.** There is no magnification of the tissue which is being examined; for operators accustomed to lens instruments it is therefore more difficult to recognize tissue through the direct vision instrument; Urologists who use the punch resectoscope, however, are accustomed to the appearance of tissue by direct vision and can identify tissue through it.

d) **Direct vision.** Urologists who have been trained to use a lens instrument find it difficult to change to a direct vision resectoscope; they are accustomed to interpret the view through an optical system.

e) **Bladder tumors.** The punch resectoscope is not suitable for removal of some bladder tumors. The position of the fenestra in the side of the instrument precludes its use for removal of tumors any place in the bladder except on the floor and at the vesical neck. It can not excavate tumors such as large squamous cell neoplasms which invade deep into the bladder wall.

Chapter XXI

Endoscopic resection of the bladder neck in the female

I. Indications

1. Urinary obstruction

Obstruction to urinary outflow due to collar contracture of the vesical orifice is the most definite indication for endoscopic resection of the bladder neck in women and girls (DAVIS). Symptoms and findings are the same as in small collar type prostatic hypertrophy or marked fibromuscular contracture of the vesical orifice. Residual urine, trabeculation and cellules in the fundus of the bladder and intravesical protrusion of the entire circumference of the rim of the bladder neck are present. Obstructive indications, however, may not be so clear cut. In North America one of the most common syndromes in the female which the urologist is called upon to treat is dysuria, frequency and pain in the region of the urethra. Often thorough urological examination fails to reveal the cause. If there is any lack of force or hesitancy in starting the urinary stream, collar contracture of the bladder neck should be suspected and re-examination performed with this in mind. When elevation of the dorsal lip of the bladder neck is present (p. 275) and no other cause for the symptoms can be found, and when symptoms persist after a course of urethral dilatation, bladder neck resection is indicated.

Although the finding of residual urine is usually a definite indication for endoscopic resection of the bladder neck, obstruction to urinary outflow due to collar contracture may exist in the absence of this finding. Residual urine is a measure of a decompensating bladder; early or compensated obstruction may be overlooked when this one finding is used as a measure of obstruction.

2. Chronic inflammation of the bladder neck

Persistent chronic inflammation, granulations and polypoid growths at the bladder neck (p. 13) can be removed more easily and throroughly by endoscopic resection than by electrocoagulation of the individual lesions. Often some collar contracture of the bladder neck accompanies these evidences of chronic infection; endoscopic resection excises all these lesions together.

3. Hyperplasia of the periurethral glands

These glands are located at or near the vesical neck and are sometimes referred to as the "female prostate" (FOLSOM); they produce true glandular obstruction. The appearance is similar to the more common fibromuscular hyperplasia, but the edges of the protruding vesical neck are usually more irregular. Endoscopic resection of this lesion is the treatment of choice.

4. Neurogenic vesical dysfunction

This is sometimes an indication for endoscopic bladder neck resection in women. The criteria are similar to those which indicate the operation in men (p. 177). If residual urine is present, the operation may help the patient to empty her bladder; when this finding is accompanied by a tight bladder neck or collar protrusion of the rim of the vesical orifice, the operation is almost certain to be of benefit. When the resection is properly performed, there is no danger of causing urinary incontinence provided none has existed before operation (NELSON, BARNES et al.).

5. Collar contracture of the bladder neck and elevation of the posterior lip

Collar contracture is often found in women who have no symptoms referable to it. There is usually an elevation of the dorsal lip of the vesical neck in the presence of cystocele, especially when depression of the floor of the bladder is marked. The presence of collar contracture or of elevation of the posterior lip is not an indication for endoscopic resection unless symptoms and/or residual urine accompany these findings.

II. Preliminary conservative care

Most female patients who present definite symptoms and findings (including residual urine) of vesical neck obstruction should be treated by endoscopic resection without preliminary conservative therapy. Most of those in whom obstructive indications are less marked should first be treated conservatively for several months. Urethral dilatations, application of medicaments to the bladder neck and urethra, urinary sedatives and relaxing agents, and antibacterial drugs in the presence of frank urinary tract infection may clear up the symptoms. A complete urological investigation, including upper tract study and urethroscopic examination, is indicated in all cases before endoscopic resection is carried out; there are many possible causes of bladder and urethral symptoms in women.

III. Cystoscopic appearance of collar contracture

The appearance of collar contracture in the female is not constant when viewed through different optical systems.

1. Right angle lens (Fig. 170 A)

The cystoscope is advanced into the bladder and then slowly withdrawn, providing visualization first of the trigone, then of the bladder neck in sequence. The dorsal lip of the vesical neck comes into the lower edge of the field of vision suddenly; there is a definite transverse line of demarcation between the apex of the trigone and the blur caused by the close proximity of the bladder neck to the objective lens. Normally the transition from the apex of the trigone to the bladder neck is gradual.

2. Foroblique lens (Fig. 170 B)

The appearance through this lens is similar to that through the right angle lens, except that the bladder neck appears to be higher. Inasmuch as the view is obliquely forward, the apex of the trigone is hidden by the elevated bladder neck; the middle of the trigone is in the lower edge of the field of vision. An erroneous diagnosis of elevation of the bladder neck is more likely to be made

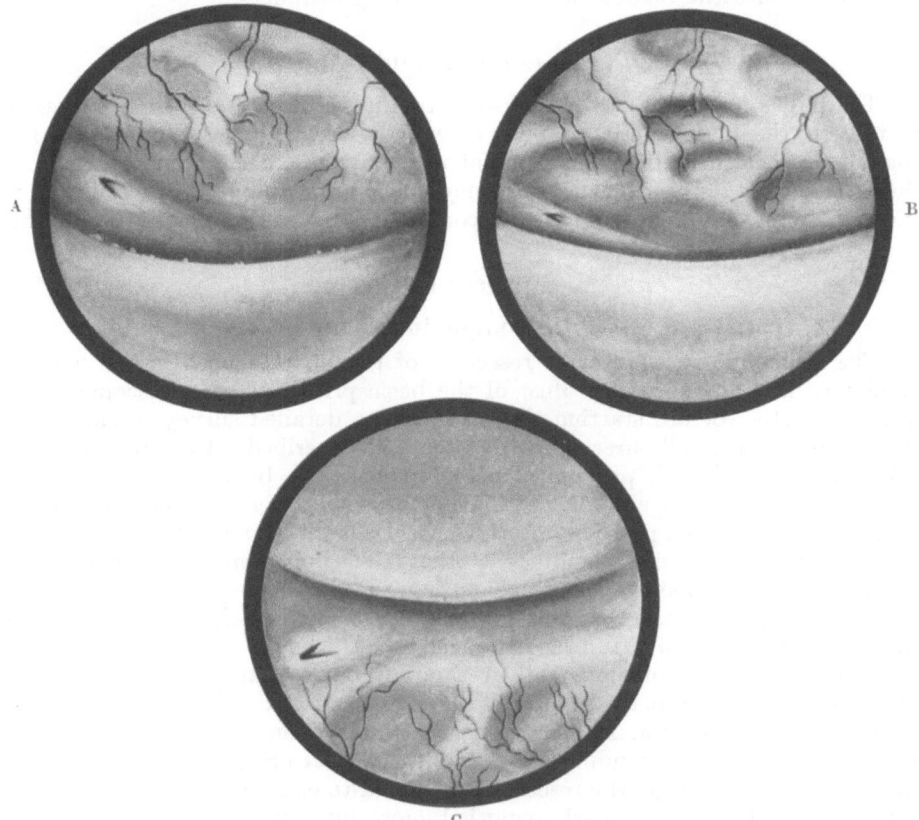

Fig. 170 A—C. Appearance of collar contracture of bladder neck in female as seen through different optical systems. A View through right angle lens. B View through foroblique lens. C View through retrograde lens

through the foroblique than through the right angle lens. Conversely, a slight elevation may be seen through the foroblique but missed by the right angle lens. Pseudopolypi, granulations and other lesions in the urethra near the vesical orifice can be seen more clearly through the foroblique than through the right angle optical system.

3. Retrograde lens (Fig. 170 C)

A more accurate diagnosis of collar contracture of the bladder neck can be made with the retrograde lens than with either of the others. The cystoscope is advanced into the bladder until the objective lens is 1 to 2 cm. away from the vesical outlet and the bladder is moderately distended. The elevation and intravesical protrusion of the dorsal lip of the bladder neck can be seen in the upper edge of the field of vision when the objective lens is directed dorsalward[1]. A narrow

---

[1] See p. 16 for orientation with the retrograde lens.

shadow is cast on the apex of the trigone by protrusion of the collar out into the bladder. As the cystoscope is rotated the entire rim of the vesical outlet is seen in the sequence of visual fields; the amount of intravesical protrusion of the collar can thus be determined. When the lesion is causing symptoms there is usually increased redness and sometimes mottled red patches on the protruding collar; these are due to subacute inflammation. Sometimes the inflammatory change extends up onto the bladder mucosa for varying distances.

IV. In children and infants

The indications for bladder neck resection in children and infants are similar to those in the adult female. In most cases the more definite symptoms and findings of difficult urination and residual urine pertain in children; the more indefinite complaints of frequency, urgency and pain in the urethra usually do not indicate endoscopic resection unless accompanied by the finding of residual urine.

V. Surgical technique

1. Technique in general

The technique of endoscopic resection of the bladder neck in the female is similar to that in the male. Most of the basic principles of endoscopic surgery apply to both. Before starting the operation, a detailed survey of the bladder neck is made with all three optical systems as described. The ureteral orifices are identified and their proximity to the collar at the bladder neck noted. Resection is started by placing the cutting loop electrode over the edge of the bladder neck at the six o'clock position; a series of pieces is then removed to the 12 o'clock position, first on one side, then on the other. The inner end of the resectoscope is pressed firmly against the tissue during the cutting excursion of the loop. The loop is not advanced far into the bladder; the total excursion is short, not more than one half of a full length excursion. Although it may be possible to resect the inner third of the female urethra without causing urinary incontinence (EMMETT et al 1950), it is usually not necessary to extend the resection outward that far into the urethra. A distance of $1^1/_2$ cm. between the inner and outer edges of the resected area extending all around the vesical orifice is usually sufficient. When the collar is large, the resection may need to extend inward a little farther into the bladder and outward somewhat more into the urethra. The pieces of tissue removed are smaller than those obtained during endoscopic resection of the prostate. It is usually necessary to remove tissue three to four layers deep. When the protruding bladder neck is thick and/or the resected pieces of tissue thin, it may be necessary to remove more than four depths. The bladder neck fibers are exposed throughout the circumference of the vesical orifice. When the resection is being performed chiefly for the purpose of removing granulations and pseudopolypi, the tissue is removed more superficially. The intraurethral protrusion can be brought into the fenestra of the resectoscope by withdrawing the instrument about one half centimeter into the urethra, then advancing it inward again; the redundant tissue at the outer edge of the resected area protrudes over the edge of the sheath and can be resected.

2. Adequate removal of tissue

After removing what appears through the foroblique lens to be a sufficient amount of tissue, examination is made with the retrograde lens. If a rim of tissue which casts a shadow is seen (Fig. 171 A), the resection should be continued a little deeper and a very short distance higher on the trigone. When the apex of the trigone extends smoothly into the proximal urethra (Fig. 171 B), it is not necessary to remove more tissue.

It is often difficult to determine the optimal limits of the resection. The landmarks are not as definite in the female urethra as they are in the male prostatic urethra. It is better, however, to remove an insufficient amount of tissue than to resect too much; the operation can easily be repeated to remove more, but it is difficult to cure incontinence or a vesicovaginal fistula which may be caused by resecting too extensively. The ureteral orifices are identified frequently during the operation to prevent injuring them by resection too high on the trigone.

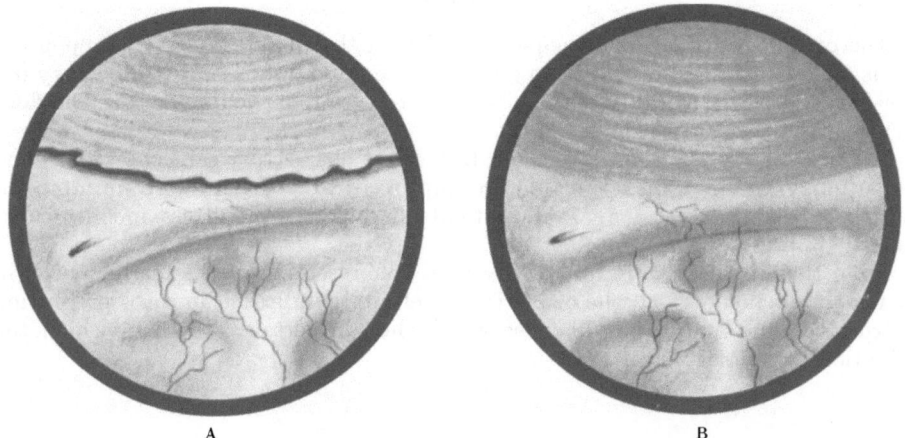

A   B

Fig. 171 A and B. Examination with retrograde lens to determine whether sufficient tissue has been removed. A Edge of unresected tissue whichprojects enough to cast a narrow shadow. B Adequate resection of tisssue; floor of urethra on a level with apex of trigone; no ridge of tissue between

3. Incision of the interureteric ridge

Hypertrophy of the interureteric ridge is incised with the resectoscope knife in the same manner as in the male (p. 213).

4. Bladder neck resection in children

The technique of bladder neck resection in children and infants, either boys or girls, is similar to that in women. The resectoscope must be of smaller calibre, and the short one is a little easier to handle than the one of standard length. The pieces of tissue which are removed are smaller. There is no retrograde lens to use; the appearance through the foroblique lens must be relied upon to determine when sufficient tissue has been removed.

5. Postoperative catheterization

At the conclusion of the operation a No. 24, 30 cc. hemostatic bag catheter is inserted. Only 10 to 15 cc of water are injected into the bag. Tension is not necessary because bleeding is usually considerably less than that encountered during endoscopic resection of the prostate.

VI. Postoperative care

This is similar to care following endoscopic prostatic surgery (p. 233). The catheter can usually be removed on the second postoperative day. There is danger

15*

of fibrous contracture of the bladder neck; therefore sounds or bougies are passed once a week for four to five times beginning two weeks postoperatively. The use of a Kollman dilator in place of the sound may be indicated if there is a tendency to stenosis or if symptoms of pain in the urethra persist more than four to five weeks. Six weeks are usually required for the resected area to heal, and the patient may continue to have dysuria during that period of time. When symptoms persist more than three months, cystoscopic and urethroscopic examination should be made, and if there is evidence of inadequate removal of tissue, the operation may need to be repeated.

VII. Results

The overall results of endoscopic resection of the bladder neck in women are not as good as those obtained from endoscopic removal of the prostate. They do, however, compare favorably with endoscopic resection of median bars and bladder neck contractures in the male.

In our series of 123 cases (NELSON, BARNES et al.), 51 per cent showed a good result (no symptoms), 35 per cent were improved, 11 per cent were unimproved and 2 per cent were worse. Four patients had persistant stress incontinence; one had it preceding surgery.

Even though the results following endoscopic resection of the bladder neck in the female are not perfect, the operation is definitely indicated in cases presenting obstructive symptoms; it is likewise worthwhile in certain others in which the indications are not as clear cut.

Chapter XXII

Immediate complications

I. Frequency

Because of the technical vicissitudes incident to endoscopic prostatic resection, accidents and immediate complications are very likely to occur when this operation is improperly performed. The difficulty of teaching the technique and the inability of the instructor to observe the operative field while the student is at work make it almost impossible to prevent every mishap. Even the experienced endoscopic surgeon occasionally encounters, or is the direct cause of, an immediate complication; but if this is identified and treated at once, serious consequences will probably be averted. Unrecognized surgical emergencies may become very grave, and often progress to a fatal termination. The technique used to avoid operative complications is detailed in Chapter XIX; however, prevention is so important that some of it will bear repeating here.

II. Injury to the urethra and bladder

1. Pendulous urethra

Injury to the pendulous urethra has its highest incidence in small caliber urethras which are divulsed by the overaggressive surgeon in his eagerness to pass the electrotome. Such damage always results in eventual stricture formation which is very difficult to keep dilated. If the urethra cannot be stretched preoperatively to permit easy passage of a No. 30 steel sound, there are several alternatives to forceful dilatation. For a small hypertrophy a No. 24 electrotome can be used, but this size instrument is unsatisfactory for the more extensive resections. In the larger glands, when intraurethral obstruction is due to fibrosis rather than to a congenitally restricted urethra, dilatation can be accomplished by gradually increasing the size of an indwelling urethral catheter for several days before surgery. This method of enlarging the urethral calibre is not indicated, however, when the urethral lumen is consistently narrow throughout its entire length. In the latter case, a perineal urethrotomy, as described by NESBIT, may be advantageous. This is done by first inserting a sound through the urethra to its perineal portion, and exerting pressure on the handle so that the curve of the sound pushes the perineum outward. An incision is then made through the perineum on to the sound, and each edge of the urethra is grasped with Allis forceps and transfixed with a suture. The electrotome can be passed into the bladder through this opening. When none of the above variations of technique is applicable to the case with a small urethral calibre, the endoscopic method of approach should be abandoned and a perineal or suprapubic exposure substituted.

2. Prostatic perforation

Perforation of the prostatic urethra while introducing the endoscope is not a common accident, but if the curve in this portion of the urethra is acute the electrotome — even though a Timberlake obturator is used — may puncture the prostate and its capsule, and either undermine the trigone or penetrate anteriorly into the prevesical space. Preparing for the passage of the electrotome by preliminary introduction of a No. 28 and a No. 30 F. sound into the bladder, and gentle handling of the electrotome with avoidance of all force, will help to prevent this accident. If the electrotome is allowed to enter the bladder by itself rather than aided by external force, it will follow the urethral curve more exactly and reduce the danger of perforation. Should such an untoward event occur, if the surgeon is sure that the puncture has invaded the prostate alone, and not its capsule, the operation may be continued if it is possible to pass the electrotome into the bladder. If the capsule has been perforated, it is best, in most cases, to discontinue the procedure; it may be completed four or five days later.

3. Bladder wall perforation

Very rarely, in cases presenting a small bladder and a short urethra, the electrotome is advanced too far into the bladder and perforates the bladder wall either through or above the trigone. To preclude this eventuality, great care should be exerted to guard against penetrating too deep. Removal of the obturator several times while the electrotome is being introduced, to determine when the bladder has been entered, will aid in preventing this accident.

4. Resection of the trigone

A frequent operative complication of endoscopic prostatic surgery is the accidental resection of part or all of the trigone, including one or both urethral orifices. This has its most common incidence in the hands of the inexperienced operator, and is a result of his zeal to remove every fraction of the bladder neck obstruction. Inasmuch as the dorsal portion of the prostate bulges toward the rectum, this convexity must be excavated; the floor of the prostatic urethra is thus made to lie deeper than the bladder neck. When this has been effected, a transverse ridge remains at the prostaticovesical junction which is higher than both the floor of the prostatic urethra and the trigone immediately above it, and may appear sufficiently elevated to cause obstruction. The unskilled operator, therefore, may place the resectoscope loop over the edge of this ridge and remove the upper portion of the trigone and a ureteral orifice (Fig. 7, Chap. XIX). Fibrosis of the orifice, frequently complicated by upper urinary tract infection and impairment of kidney function, is the unhappy result; if both orifices are injured uremia may ultimately develop.

5. Ventral bladder wall resection

Sometimes during endoscopic prostatic resection, while the intravesical ventral portion (11 to 1 o'clock) of the prostate is being removed and the bladder is empty, folds of vesical mucosa may encroach down on to the prostatic border and entail accidental resection through the bladder wall at this point. This may be avoided by definitely identifying tissue before it is cut, and if a tendency to mucosal redundancy is perceived, by partially distending the bladder before any sections are removed.

6. Perforation at the prostaticovesical junction

This is of frequent occurrence in the annals of the inexperienced operator, and also happens occasionally in the hands of the expert. In the majority of prostatic hypertrophies, the external circumference of the capsule measures considerably less at the prostaticovesical junction than in the midportion of the prostatic urethra. The prostate is, therefore, in most cases thinner at the bladder neck than elsewhere, and this is the point where perforation is most to be feared by carrying resection too deep. If careful scrutiny for bladder neck fibers and capsule (Fig. 12, Chap. XIX) is maintained, the danger of such an accident is minimized.

III. Recognition of perforation and extravasation

1. Importance of recognition

It is very important that perforation be recognized (p. 199) when it does occur, although a minute puncture offers little risk when an isotonic fluid is being used and when intravesical pressure is not sufficient to induce extensive extravasation into the perivesical tissues. When a small perforation is identified, the operation may be carried to completion under constant guard of distending the bladder to its full capacity. It is much safer to continue the operation regardless of the perforation when an isotonic solution is used than when water is the irrigating fluid.

2. Suprapubic or perineal pain and rigidity

Irrigating fluid often extravasates through perforations, especially when these are more than one-half centimeter in diameter or when the bladder is overdistended. The first knowledge the operator may have of an unrecognized perforation is the patient's complaint of pain in the suprapubic region, or in the perineum if the perforation is dorsal. This pain occurs even though spinal or other regional anesthesia is adequate. Abdominal rigidity develops at the time or soon after the pain is noted, and frequently shock ensues. These symptoms are characteristic enough to enable the surgeon to make a definite diagnosis of ruptured bladder, and when they occur cystograms are not necessary for confirmation. Indeed, sometimes a cystogram will fail to show extravasation through the point of perforation, and may be more confusing than helpful in establishing a correct diagnosis.

3. Cystourethrograms

Some urologists, however, depend upon cystourethrograms for diagnosis of extravasation (KENYON).

4. Appearance of the area of perforation

Extensive extravasation of fluid through a perforation can usually be determined by watching the opening of the perforation. When the dark area becomes larger, it is evidence of increasing extravasation. If the small dark area remains the same and the fibers of the capsule do not spread apart, it is safe to proceed to completion of the operation; drainage of the prevesical space is not necessary in this case if an isotonic irrigating fluid is being used.

IV. Treatment of perforation and extravasation

Whenever extensive extravasation through a perforation of the prostatic capsule or bladder wall is suspected, drainage of the extravasated area — whether suprapubic or perineal — should be instituted immediately. Most resectoscopists take recourse in prompt suprapubic cystostomy as well as drainage of the extravasated area; but the authors have found that patients who have been subjected to this accidental complication do as well if not better with simple drainage of the extravasated area rather than with the combined method. Inasmuch as they are already suffering from some degree of shock due to the ruptured bladder or prostatic urethra, it is not wise to add to their burden by performing cystostomy when simple drainage is adequate and does not provoke adverse reactions. If the site of extravasation cannot be verified resectoscopically, the location of the pain may aid in orienting it. If the patient complains of perineal and rectal discomfort and has the sensation that his bowels are about to move, it is probable that extravasation is dorsal. Treatment consists of incision of the perineum and, with one finger in the rectum to prevent injury to the rectal mucosa, careful blunt dissection carried into the region of the prostatic capsule, and insertion of a soft rubber tissue drain. If pain is predominantly suprapubic, or if there are other reasons to believe that extravasation is lateral or ventral to the bladder, a small incision is made in the midline immediately above the pubis. The recti are separated, the prevesical space down to the region of the prostate is opened by blunt dissection, and a soft rubber tissue drain is inserted. This may be done under local anesthetic while the patient is in bed if the complication is suspected following his return from surgery.

V. Undermining the trigone

1. Method of avoiding

Cases in which the trigone is elevated are susceptible to the constant danger of undermining it with the sheath of the electrotome while the resection is in progress. After the bladder neck has been resected, and especially after the floor of the prostatic urethra has been made concave by removal of most of the gland dorsal to the urethra, great care must be exercised to avoid inserting the point of the sheath into the subtrigonal tissue when the instrument is pushed back into the bladder. To avoid this contingency, the ocular end of the electrotome is depressed so that the distal end is raised and rides over the elevated trigone and into the bladder (Fig. 159, Chap. XIX).

2. Treatment

Undermining the trigone is bound to occur sometimes, and when it does it is treated by incising the trigone in the midline with the resectoscope knife; this allows the tissue on each side of the incision to flatten down on to the subtrigonal tissue. Thus, the blind pocket under the trigone resulting from the perforation is converted into a depression on the floor of the bladder, and facilitates subsequent passage of catheters and sounds into the bladder. If undermining of the trigone involves only the superficial subtrigonal tissues, the complication is not serious. However, if the injury is deep — particularly if it extends through the bladder wall — extravasation may occur and should be drained through the perineum as noted above.

VI. Injury to the external sphincter

Injury to the external sphincter or to the mucosa of the membranous urethra may result in urinary incontinence, which is usually temporary but may be permanent. To prevent this emergency, the landmark of the verumontanum is constantly kept in mind, and the resection is carried external to the veru only in the few cases presenting hypertrophied prostatic tissue distal to it. By removing very shallow sections in this area, and by careful observation of tissue, general contour of the urethral walls and verumontanum, the external sphincter is detected before damage can be done. One loopful of tissue removed from the membranous urethra will result in permanent urinary incontinence, but if only a small section of mucosa is injured the incontinence will be temporary.

VII. Excessive blood loss

1. Detection

Excessive blood loss during endoscopic prostatic surgery may be considered a complication. When the operation has lasted more than one hour, when a large amount of irrigating fluid has been used during the resection, and especially when the return irrigations are bloody and coagulate on standing, it is probable that hemorrhage has been excessive and a transfusion should be given. The quantity lost cannot be accurately estimated, however, unless it is measured in some way such as NESBIT & CONGER have recommended. If there is evidence that more than 300 cc have been lost during operation, transfusion is indicated, for the volume of postoperative loss approximately equals that of operative loss (PILCHER & SHEARD). It is unwise to wait until signs of shock and other evidence of hemorrhage appear before transfusion is given, for by that time considerable damage will have been done and transfusion then may not relieve the situation.

2. Treatment

When endoscopic surgery is expertly performed, blood loss is often less than when the same lesion is removed through an open approach (GOLDSTEIN et al.). A systematic approach, control of bleeding at the conclusion of resection of each section, and avoidance of delay during surgery are all factors in keeping hemorrhage to a minimum. The more expert the resectoscopist becomes, the less troublesome bleeding he will encounter.

3. Fibrinolysis

A rare cause of profuse hemorrhage during or immediately following endoscopic prostatic surgery is fibrinolysis due to the liberation of proleolytic enzymes from the prostate (LOMBARDO). The condition is suspected when a marked delay in coagulation time is observed and is verified by a lowered concentration of plasma fibrinogen. It is treated by administering whole blood, fibrinogen and toluidine blue.

VIII. Absorption of irrigating fluid

The different kinds of irrigating fluids and the results of their absorption are discussed in Chapter XV.

The transfer of large amounts of irrigating fluid to the blood stream or into the perivesical tissues during endoscopic surgery predisposes to shock. Water is a greater offender than an isotonic solution in this respect. The untoward resection elicited is probably due to a variety of causes, among which are intravascular hemolysis, increase in blood volume and dilution of blood plasma (HARRISON & BOREN). MURPHY et al., however, report that even though irrigating fluid enters the blood stream during endoscopic surgery, there is no immediate change in serum electrolytes, serum protein, serum volume or hematocrit.

Chapter XXIII

Postoperative care

I. Importance

The postoperative treatment of the patient who has undergone endoscopic removal of the prostate or of a bladder tumor is as important as the operation itself. An efficient nursing force in the hospital where a large amount of this type of work is done, or experienced special nurses in the smaller hospital, are essential to the successful postoperative course. During the first few hours after surgery, constant expert care is needed, and the average general duty nurse in the hospital where only an occasional resection is performed is not able to cope with the situation. Unless the house surgeon or the family physician is versed in the management of these cases, he is of little help, and grave complications may arise even though close watch is maintained. It is better to rely on a special duty nurse with previous experience and training than to entrust these patients to the care of a physician who is not familiar with their idiosyncrasies.

II. Catheter drainage

1. Aseptic closed system

Asepsis is rigorously adhered to in the postoperative routine (BUDDINGTON et al.). As soon as the patient is returned to bed, the catheter is connected with a previously sterilized closed drainage system (SCHNEIDER). If frequent irrigation of the bladder is indicated, the system includes a jar hung on a standard at the bedside from which a rubber tube leads to a "Y" glass connection. One branch of this is attached to the catheter, and the other to a rubber tube emptying into a drainage bottle on the floor. The system is kept closed by covering the irrigating jar, or preferably by using an intravenous bottle. The drainage bottle is stoppered, the air vent being turned to the side to prevent dropping of contaminated material into it.

Aseptic technique is maintained whenever the drainage tube is removed from the catheter for syringe irrigation or other purpose. If it becomes necessary to change the catheter, this is done aseptically.

2. Maintenance of free drainage

a) **Without irrigation.** Until a few years ago the bladder was frequently irrigated postoperatively to prevent stoppage of the catheter by clots. It was found by STEGEMAN, however, that the postoperative course was as good if not better when no irrigation through the catheter was done. A clot may form in the dependent portion of the bladder, but if it is not disturbed the urine often drains around it and out through the catheter. When the clot is broken up by irrigation, the pieces clog the eyes of the catheter and hinder drainage. Unless bleeding is excessive, no irrigation is necessary.

b) **With irrigation.** Sometimes clots stop the flow through the catheter. The nurse is instructed to watch the drip of urine into the drainage bottle. When urine fails to drain for one hour, or when patient complains of bladder distention and fluid is not dripping through the catheter, irrigation, preferably with a piston syringe, is indicated. The catheter is disconnected from the drainage tube and a piston syringe (p. 113) is used to irrigate and aspirate clots. A bulb syringe is not satisfactory for this purpose because it does not produce adequate suction. Warm sterile saline or a mild antiseptic solution is used for irrigating the bladder, and the procedure is continued until all clots have been evacuated. Alternately pushing the catheter inward one to two centimeters, then drawing it out again may help to break up the clots. If any clots are left in the bladder after they have once been broken up, they are almost sure to plug the catheter again in a short time. Bleeding is also controlled better after all clots have been thoroughy evacuated. Sometimes during aspiration with the piston syringe, there is sudden stoppage of the outflow of fluid, and at the same time the patient notices a twinge of pain in the bladder. When this occurs the operator knows that the bladder wall is being sucked against the eye of the catheter, and that the stoppage is due to this rather than to a clot or piece of resected tissue. He should, therefore, avoid forceful aspiration at that time.

c) **To prevent bladder overdistention.** The importance of preventing overdistention of the bladder by obstruction due to clot formation cannot be overestimated, for bladder spasm resulting from overdistention increases hemorrhage and frequently leads to shock, even though the actual blood loss may have been insufficient to account for shock. Rupture of the bladder or prostatic capsule with resulting extravasation may occur from postoperative bladder overdistention in cases which have sustained perforation or near perforation at the time of surgery.

d) **Change of catheter and use of evacuating tube.** When clots cannot be removed through the catheter, or when obstruction is due to pieces of prostatic tissue inadvertently left in the bladder at the time of resection, the catheter is removed. While it is being drawn out through the urethra, gentle suction is maintained with the piston syringe in order to keep hard clots or fragments of tissue in the eye of the catheter, thus removing them from the bladder in one operation. If it is believed that only a few small clots or pieces of tissue remain in the bladder, the catheter is reinserted and irrigation and aspiration repeated in an attempt to effect complete evacuation of all material which would occlude catheter drainage. Should this second attempt prove unsuccessful in establishing free drainage, or if it is believed that more than a few clots or pieces of tissue are in the bladder, more positive measures are used to evacuate the bladder. A lithotrite evacuation tube connected to a large nozzled piston syringe through a short piece of heavy rubber tubing serves this purpose. The evacuation tube is passed into the bladder, and the clots easily evacuated through it by suction with the piston syringe.

Before the evacuation tube is removed, the bladder is thoroughly irrigated, the inner end of the tube being moved about to different positions to make certain that all clots or sections of prostatic tissue have been discharged. In nearly all cases, hemorrhage is markedly decreased following evacuation of clots; then when a hemostatic bag is replaced, and tension applied in prostatic resection cases, bleeding is usually so reduced that irrigation every ten or fifteen minutes is sufficient to keep the catheter draining, and further blood loss will not be serious. The authors have not found that prevention of the formation of clots, as advocated by some urologists (McLELLAN), is of value.

III. Control of bleeding

1. Medication

Scores of different methods and materials have been recommended as being helpful in controlling hemorrhage following endoscopic surgery[1]. These range from preoperative administration of coagulants to postoperative rectal pressure bags, and the fact that they are so numerous is prima facie evidence that probably none of them is of great value. All resectoscopists agree, however, that the best way to prevent troublesome hemorrhage during the first and second postoperative days is to control all bleeding at the time of resection, and most of them concur that the hemostatic bag aids in controlling venous bleeding but is of little value in stopping hemorrhage from arteries.

2. Electrocoagulation

Sometimes excessive bleeding persists even after the clots have been evacuated; it is then necessary to return the patient to the operating room for either suprapubic cystotomy and packing of the bladder neck with gauze held in place by a hemostatic bag, or for fulguration of bleeding points through the suprapubic cystotomy or through the electrotome. The latter method, performed under light anesthesia such as pentothal sodium, is advocated if the operator is experienced and can quickly locate the bleeding points. The resected surface will be found covered with clots, which must be removed with the loop before the bleeding points can be located. It is only waste of precious time and effort to fulgerate indiscriminately through clot formation, hoping that the bleeding points will be reached. If hypertrophy has been extensive, greater difficulty will be encountered in freeing the entire resected area of clots, and more thorough search will be required to locate all bleeders. An inexperienced operator may find it hard to remove clots and locate all bleeding points, and may consume so much time in attempting to do so that cystotomy and packing would have been a more suitable method for him. However, for the experienced operator, it is quicker and less shocking to the patient to control bleeding by fulguration through the electrotome. A panendoscope or other type of cystoscope is not used here, because removal of clots from the resected surface is much more readily effected with the loop of the electrotome than with any kind of fulgurating tip which can be applied through a cystoscope.

3. Blood transfusion

Whenever blood loss is excessive as shown by measurement or as evidenced by clots in the bladder, low blood pressure, rapid pulse and other signs of shock, transfusion is essential to help stop the bleeding and to save life; two or more of 500 cc. each may be needed to prevent serious consequences.

---

[1] Milliken; Greenberger; Boyd; Olwin et al.

4. Delayed secondary hemorrhage

Secondary hemorrhage may occur at any time during the postoperative course, but is unusual — especially after the third week — unless resection has been inadequately performed. If it is treated immediately by evacuation of clots and insertion of a bag catheter, as described above, it is seldom serious and fulguration of bleeding points is not often necessary. It is essential that all clots be removed, for if even a few remain their presence will tend to prevent complete hemostasis.

IV. Postoperative extravasation

A ruptured prostatic capsule or bladder may not be recognized at the time of surgery, or it may occur from bladder overdistention during the first few postoperative days. The symptoms and findings which are suggestive of postoperative extravasation are: 1. suprapubic pain, tenderness and rigidity; 2. sudden pain in the suprapubic region or perineum on voiding the first few times after the catheter has been removed; 3. persistent fever and/or low blood pressure accompanied by suprapubic tenderness and abdominal distention. Treatment is the same as that applied to extravasation which occurs during surgery (p. 231).

V. Fluid intake

Three thousand cubic centimeters of fluid are given each 24 hours for two weeks following surgery. During the first two days it may be difficult for the patient to take this amount by mouth, so the deficiency is met by administering 5 per cent glucose in distilled water intravenously. Saline, unless given to make up for loss of salt from vomiting or perspiration, has been shown to produce fluid retention in the body with resulting edema. A record of fluid intake and urinary output is kept, and when the latter remains less than half the former for more than five consecutive days, fluid intake should be reduced to prevent edema and cardiorespiratory complications. During the first postoperative day, the intravenous administration of fluids also aids in sustaining blood pressure when it is low. Small doses of ephedrine (one-fourth of one $3/8$ gr. ampule every two hours), if necessary, may be used for this purpose and are preferable to larger doses given at longer intervals.

VI. Ambulation

Ambulation on the first or second postoperative day is indicated in most cases. The catheter may be clamped off while the patient is up. Sometimes, however, he is too weak to be on his feet that soon; in this case he is allowed to remain in bed until the third or fourth day, but is encouraged to move about and exercise his limbs in bed as much as possible. He is not allowed to be ambulatory if his temperature is above 100° F (37.8° C) or if bleeding is severe enough to make the urine dark red.

VII. Bowel care

Bowel movement is not encouraged, and enemas are never given during the first three postoperative days. Two drams of liquid petrolatum twice daily will aid in obtaining free bowel movements; if none occurs until after the second day two drams of cascara aromatica are given that evening. If there is no bleeding, enemas may be given after the third day, but before this time they tend to induce or increase hemorrhage. If bleeding continues on the fourth day, however, it is better to allow the bowels to remain closed than to risk accelerating the bladder hemorrhage by giving an enema at this time. Postoperative gaseous distention is uncommon, but when it does occur it can usually be controlled by the subcutaneous injection of a smooth muscle stimulant. A colon tube may be passed if great care is used to avoid excessive manipulation and if the tube is placed along the posterior rectal wall rather than anteriorly along the bed of the prostate. Occasionally, a gastric tube is indicated for distention of the upper abdomen, but this is rare.

VIII. Sedatives

Sedatives are seldom required in any great amount if there has been no extravasation and if the bladder is not overdistended. Whenever the patient complains of pain, it is important before giving sedatives to make very sure that the catheter is draining well and that the bladder is not distended. Only light sedatives should be given; oversedation causes inactivity and predisposes to passive congestion of the lungs and other complications.

IX. Hiccoughs

Hiccoughs commonly occur following any kind of prostatic surgery. The reason for this is not definitely understood, but poor renal function is probably a factor. Sometimes the treatment of hiccoughs becomes a problem — occasionally the the major problem in postoperative care. There are numerous remedies recommended for the control of hiccoughs, and it has been said that the best is the last thing which has been used. In the authors' experience, the inhalation of carbon dioxide is more efficacious than anything else in controlling this sometimes distressing complication. The gas is administered through a mask which fits the face snugly and allows no leakage, and the patient breathes the pure carbon dioxide deeply until respiration is stimulated and he becomes dyspneic. The inhalation of a small amount of the gas mixed with air or oxygen is not effective. Sometimes the hiccoughs are halted for only ten to fifteen minutes, and inhalation then has to be repeated; even this is better than allowing the patient to wear himself out by continuous hiccoughing.

X. Postoperative catheter management

1. Removal of the catheter

The catheter is removed on the third postoperative day unless: 1. temperature is above 100° F (37.8° C); 2. bleeding is severe enough to make the urine red (old blood is evidenced by a dark reddish brown color which does not contraindicate removal of the catheter); 3. complete perforation or several near perforations of the capsule have occurred. In these situations the catheter is left in place five days or until the temperature remains below 100° F (37.8° C) for 24 hours and urine has cleared to light pink.

Removal of the catheter from the patient who has had endoscopic resection of a large bladder tumor is the same as for prostatic resection cases. When the tumor is small and there has been no perforation or near perforation, the catheter is removed on the first or second postoperative day. When voiding is free and the interval between urinations is more than $1^1/_2$ hours, and when there is no severe dysuria, it is not necessary to reinsert a catheter after the first one has been removed.

2. Replacement of the catheter

When, after removal of the catheter, difficulty in voiding or marked frequency or urgency is experienced, a smaller sized catheter is inserted and the amount of residual urine determined. If this measures more than 100 cc. the bladder is lavaged with a warm (110⁰ F — 43.3⁰ C) mild antiseptic solution such as 1:4,000

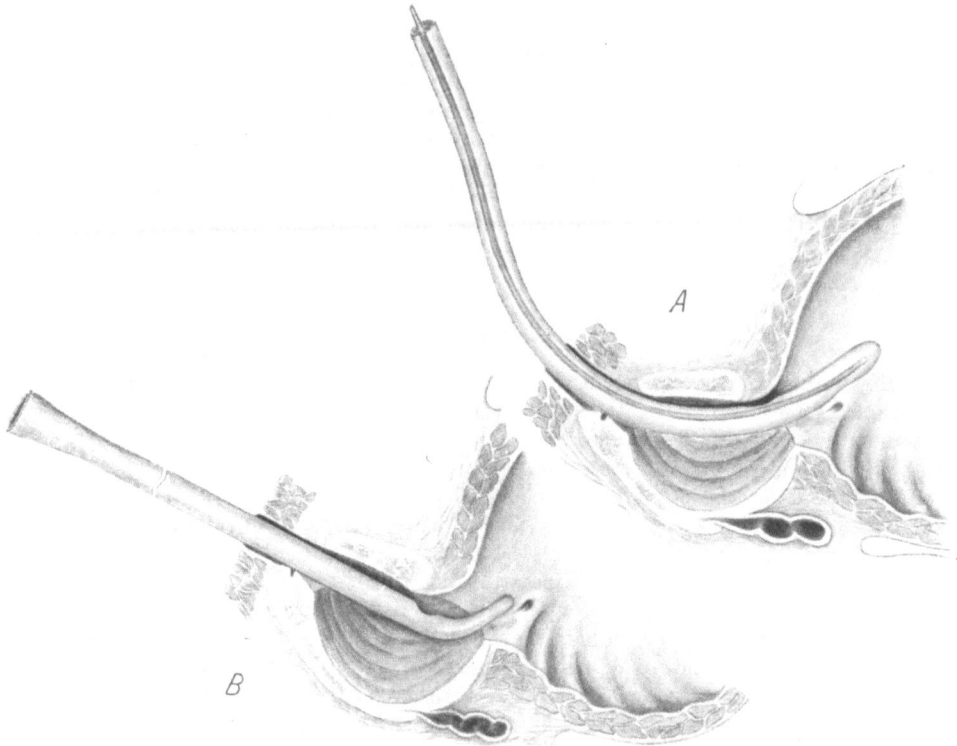

Fig. 172 A and B. A Stilet inserted through soft rubber "Robinson" catheter lifts it over projecting bladder neck
B Coudé tip soft rubber catheter also rides over bladder neck

potassium permanganate, and the catheter left in place until the next morning when it is removed again. If the residual urine is less than 100 cc. the catheter is removed after the bladder lavage.

3. Persistent residual urine

When the residual urine remains above 100 cc. and bladder tone is good, a repeat resection will probably be required; however, if bladder tone is poor, it is likely that the residual will gradually decrease in amount as bladder tone improves. Nevertheless, when hypotonia is marked, daily catheterization and irrigation may have to be continued for some time, and the patient or a member of his family may be taught to do this at home. Lavage of the bladder twice daily with one or more quarts of hot solution (110⁰ F — 43.3⁰ C) may aid in increasing bladder tone. The administration of a smooth muscle stimulant often helps to improve bladder tone, thus aiding complete evacuation of bladder contents (LEE).

4. Obstruction to passage of the catheter

A straight catheter, during passage into the bladder following a prostatic resection, may become obstructed in the prostatic urethra by the elevated bladder neck which is left after excavating the dorsal portion of the prostate. If a stilet is used to introduce the straight catheter, or a coudé tip catheter is passed, the tip will ride over the obstructing bladder neck and enter the bladder easily (Fig. 172). Even though there is no definite obstruction at this point, a soft rubber coudé tip catheter, with its tip pointed ventrally, will pass into the bladder with less trauma to the prostatic urethra than a straight catheter, and is better for routine postoperative catheterization.

XI. Infection and fever

Infection is always present to some extent during the immediate postoperative period, but unless it is severe enough to cause elevation of temperature above 100° F (37.8° C), it needs no special treatment. Since the advent of the antibacterial drugs, infection is easily controlled in most cases, and when the patient's temperature remains above 100° F (37.8° C), for more than three consecutive days, a wide spectrum antibacterial drug is administered. When the temperature remains elevated at more than 100° F (37.8° C) for one or two days after removal of the catheter, this is replaced and left until the temperature drops below this point.

XII. Hospitalization

The average endoscopic resection patient should be hospitalized for six postoperative days. If he is allowed to leave before that time, there is danger of fever from pyelonephritis due to overexertion, and the chances of secondary hemorrhage are increased. If he lives close to the hospital or can be transferred to a nearby hotel, it is safe to allow him to leave earlier; but he should be seen at least once daily by an experienced nurse or house physician. This is a decided reduction over the term of hospitalization required for either the suprapubic or the perineal operation, and the economic advantage is especially beneficial to the low income group as well as to the taxpayer who foots the bills of charity institutions. Patients who have had a small bladder tumor removed endoscopically can usually leave the hospital on the second or third postoperative day.

XIII. Dilatation of the urethra

1. Sounds

Inasmuch as urethral stricture is a sequela of endoscopic surgery, sounds are passed during the postoperative course in an effort toward prevention as soon as any slowing of the urinary stream is noted. A No. 24 F. may be passed on the sixth or seventh postoperative day; when it is inserted gently there is very little if any danger of exciting secondary hemorrhage. Urethral contraction may begin this soon, especially if trauma has been produced at the time of surgery; early dilatation will go far toward preventing later stricture formation. After this a No. 24 or No. 25 sound is passed once a week for three to six times, depending upon the degree of contraction.

2. Kollman dilator

When the endoscopic surgery has been performed for median bar or bladder neck contracture, especially if the contracture is secondary to previous surgery, stenosis of the vesical orifice tends to recur. An aid in preventing recurrence in these cases is the use of the Kollman dilator. The first dilatation is given two weeks postoperatively; the instrument is passed and the blades separated to a calibre of 35 to 40 Fr. This treatment is repeated once a week for 4 to 5 weeks. Dilating the vesical orifice in this way during the healing process will in most cases prevent recurrence of the stenosis.

XIV. Routine postoperative orders

A copy of the routine postoperative orders should be available to all persons responsible for supervision of the patient who has undergone an endoscopic prostatic operation. These should be complete and detailed, because the relief nurse or house surgeon may not have had previous experience in caring for such cases, and since their management differs so greatly from that of any other group of surgical cases minute instructions are necessary for proper acquittal. Following are the routine postoperative orders which the authors are using at present. Inasmuch as they have been modified several times in the past, it is probable that they will again be revised in the future as changing conditions indicate. The routine postoperative orders for bladder tumor cases is, for all practical purposes, the same as those for prostatic resection patients. If the tumor was small and was removed without penetrating deep into the bladder wall, the catheter may be removed the day after surgery. When the tumor was extensive, however, and penetration during surgery was deep into or through the bladder wall, the catheter should remain in place for 5 to 7 days.

Routine postoperative orders for prostatic resection cases

1. No. 24 or 26 hemostatic bag catheter inserted, and distended immediately after resection in operating room.
2. Force fluids — 3,000 cc. in 24 hours. Measure and record daily intake and output. Intravenous glucose (5 per cent) in water if necessary.
3. Connect catheter to sterile drainage tube leading to bottle on floor.
4. Do not irrigate, pinch tube to produce suction or stir up blood clots forming in the bladder.
5. If bleeding is excessive (so that return irrigation is dark red) place tension on bag catheter by tying gauze 4×4 around catheter and pulling it snugly against glans penis.
6. If clots block catheter, or if it does not drain, disconnect catheter from tube and irrigate and aspirate with piston syringe having a large bore nozzle. Continue irrigating with normal saline or boric acid solution until all clots are removed and catheter drains freely. Use aseptic technique.
7. Never inject more than one-half ounce of irrigating fluid at a time through catheter.
8. If clots obstruct drainage and cannot be aspirated through catheter, remove catheter and pass lithotrite evacuator carefully. Aspirate clots with Toomey syringe.
9. Liquid petrolatum, or plain petrolagar, drams 2 b.i.d. starting first postoperative day. If no bowel movement give cascara aromatic drams 2, or milk of magnesia, oz. 1 on second postoperative night. No enema until third postoperative day, as this may cause hemorrhage Mild catharsis preferable even then.
10. Dilaudid, gr. 1/32, or Demerol 100 mg if necessary for pain. *Be sure* bladder is not distended with clots before giving sedative.
11. On first postoperative day patient may be up if temperature is not above 100° F (37.8° C), and if no bleeding. Clamp catheter off while patient is up.
12. If there is no bleeding, and temperature is normal, remove catheter on third postoperative day. If there is any bleeding or temperature is above 100° F (37.8° C), leave catheter in until fourth postoperative day. Urine should be clear and temperature below 100° F (37.8° C) for 24 hours before catheter is removed. (Third exception to removal of catheter on 3rd postoperative day is complete or near perforation of capsule [p. 199].)
13. If patient voids freely and without pain, do not catheterize again.
14. If there is difficulty in voiding or painful urination, pass catheter on evening of the day catheter is removed. If more than 100 cc. residual urine, leave catheter in overnight and

remove in morning after irrigating with 500 cc. warm (110⁰ F, 43.3⁰ C) 1:4,000 KMnO₄ solution. Check for residual urine again that evening.

15. If residual urine is less than 100 cc., check once a week and irrigate with 1:4,000 KMnO₄ solution. Continue this for 3 weeks.

16. If secondary hemorrhage occurs, evacuate clots with evacuator, and proceed as outlined in No. 6 and No. 8 above.

17. If patient is unable to void, or purulent condition of urine persists, or temperature is above 100⁰ F (37.8⁰ C) replace retention catheter and leave in until temperature is normal.

18. Keep retention catheter in as long as temperature is above 100.4⁰ F (38.0⁰ C) or as long as there is much color from bleeding.

19. If temperature is above 100⁰ F (37.8⁰ C) for 3 or 4 days, have Gram stain made, institute antibiotic therapy—Achromycin, 250 mg. q.i.d. if not nauseated. When unable to retain anything by mouth give Achromycin, 100 mg. intramuscularly every 8 hrs. Erythromycin if coccal infection. Culture and sensitivity tests if fever persists more than three days after beginning antibiotics.

20. For nausea, give Dramamine, 100 mg. b.i.d. For headache, aspirin, gr. 10, repeat every 2 to 3 hrs. p.r.n. If blood loss marked or blood pressure below 100, type for transfusion.

Chapter XXIV

Results and sequelae

I. General discussion

The results obtained from any surgical operation, or from any type of treatment for that matter, are the most impelling factor in the scale of values determining whether or not that method should be continued. In fact, at the end of a sufficient adjustment period either contention will have automatically subsided or the force of consistently poor results will have doomed the procedure to ultimate abandon. It is undeniable that the wave of enthusiasm upon which a new operation rides into popularity may carry it farther than its merit would justify, but the tide soon recedes leaving the procedure to find its true level.

The burst of fervor which accompanied the popularization of endoscopic prostatic surgery in 1931 and 1932 resulted in a high mortality and considerable morbidity, because a multitude of inexperienced and clumsy operators used it promiscuously. At that time the number of skilled technicians could be counted on the fingers of one hand, and it seemed necessary for many to learn the routine by trial and error with what help could be secured from published reports and visits to clinics where the operation was demonstrated. It is probable that if its adoption had been slower in spreading throughout the country, and if more men had obtained training before attempting to execute it, the initial mortality and morbidity would have been considerably lower. McCARTHY was among the first to sound the warning against indiscriminate application and overoptimism, but was obligated to remind the profession of his admonition in a later communication.

II. Statistical reports

During the past twenty five years, scores of reports giving the results of endoscopic prostatic surgery have appeared in the literature. Dozens of series of cases ranging in number from less than one hundred to more than two thousand have been studied, analyzed and statistically recorded. The results of cases observed or treated by the different authors vary from the disastrous effects in the collected cases of LEWIS and CARROLL to reports of less than one per cent mortality (THOMPSON). The authors' mortality rate for 1,000 consecutive cases was

1.9 per cent (BERGMAN et al.). Vest reports a 2.2 per cent mortality for transurethral resection, and it is to be noted that he avoids preliminary cystotomy
whenever possible. This policy is corroborated by DORMAN, who attributes a large
percentage of the reduced death rate to elimination of this preliminary step.
ROLNICK and RISKIND published a 28.6 per cent mortality for all causes, although
among the survivors there were no deaths from subsequent transurethral resection in this series.

It is difficult to judge the efficacy of any procedure by statistics alone, for
there are so many actuating factors that a true comparison is practically impossible. On the other hand, about the only way we have of making comparisons
is by figures, and if it were possible to obtain standardized criteria upon which
statistics could be based they would be much more valuable. The variable elements which influence the reports of results obtained by endoscopic prostatic
surgery are numerous. For instance, the preoperative condition of the average
charity patient is worse than that of the private patient, which precludes any
parallelism in regard to outcome. Some surgeons deliberately operate upon the
patient who is in poor condition and who will not improve with further preoperative treatment, because they would rather risk a surgical mortality for the chance
of restoring normal bladder function than leave the sufferer in a condition requiring the constant attention of frequent catheterization or indwelling urethral
catheter. Another group of surgeons prefers to subject the prostatic to suprapubic cystotomy, even though the odds of his succumbing from this type intervention are as great as from endoscopic surgery, for they are jealous of their
resection statistics. The experience and dexterity of the operator, the type of
preoperative and postoperative care administered, the interpretation of "operative mortality", the length of postoperative survival and many other factors
affect statistical reports of the final outcome of endoscopic prostatic surgery.
Although a perusal of these reports per se is interesting and enlightening, a comparison of one with the other is futile.

If a worldwide survey could be conducted—representing results from the
inferior clinics as well as from those of high standing, and from unskilled as well
as expert surgeons—it would probably be discovered that during the first few
years after this operation came into vogue its effect on mortality and morbidity
was not one of improvement over the other methods of prostatectomy. If we can
judge by the fact that the series of cases which have been reported recently shows
a decided increase in favorable outcome, it is likely that the present status of
endoscopic prostatic surgery has been greatly improved, and that its results
when the operation is performed by expert resectoscopists are superior to those
obtained by the other types of surgery.

When it is considered that according to the American Experience Table of
Mortality the average death rate at 75 years of age is approximately 10 per cent
per year, or nearly 1 per cent per month, it is somewhat surprising as well as gratifying to note an operative mortality following endoscopic prostatic surgery of
around one per cent in reports from some of our best clinics. Results of this kind
are not possible, however, in institutions where a large proportion of the patients
are free or part pay and many are very poor risks, where most of the operations
are performed by house surgeons under supervision and where the postoperative
nursing care is inferior due to political or financial reasons. In some of these
institutions the operative mortality attending this form of surgery approximates
10 per cent, and it is probable that the death rate for the entire country lies between these two figures — somewhere around 5 to 6 per cent. This is undoubtedly
an improvement over the results of suprapubic prostatectomy, though possibly

somewhat inferior to the perineal rating, for the latter procedure is less frequently attempted by untrained surgeons than either of the other two.

III. Functional results

Although the majority of patients enjoy good functional results from endoscopic prostatic surgery, nevertheless there are sequelae which may occur. Inasmuch as some of these are of a disabling nature, it is felt that a detailed description may prepare the urologic surgeon for their incidence and management. Functional results depend almost entirely upon the precision and thoroughness with which the operation has been performed. The nearer the procedure approaches "endoscopic prostatectomy", the better will be both the immediate and late functional results. Because of the fact that complete removal of the hypertrophied gland down to the surgical or false capsule results theoretically in perfect function as far as emptying the bladder is concerned, endoscopic resection cannot give better functional results than enucleation when each is done correctly and completely, but at the present time a comparison of results is favorable (DAVIS and NESBIT).

IV. Incomplete removal of tissue

1. Symptoms and findings

Every urologic surgeon has witnessed persistence or return of symptoms resulting from inadequate removal of hypertrophied tissue at the time of suprapubic or perineal enucleation or from recurrence of the growth following these open surgical procedures, although up to the present time at least, the incidence of these sequelae has been higher after endoscopic removal. The symptom complex of urinary frequency, dysuria, bladder infection, incontinence and bleeding subsequent to endoscopic prostatic surgery is nearly always due to destruction of the blood supply to the unresected portion of the prostate. Cystoscopic examination will reveal encroaching prostatic tissue and sometimes tags at the bladder neck which may be necrosed and partially covered by calcareous plaques (Fig. 173).

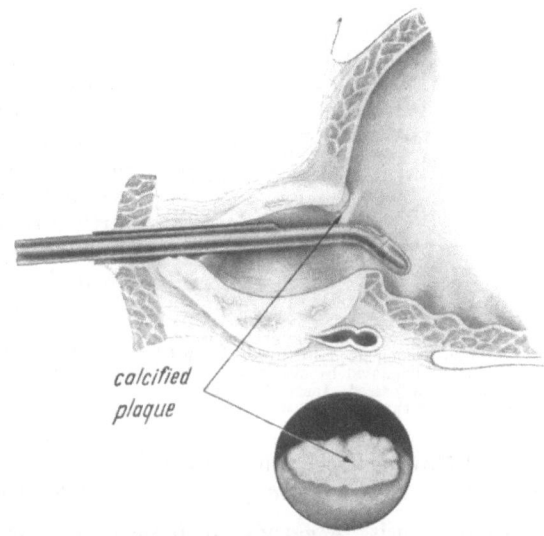

calcified plaque

Fig. 173. Calcareous plaques may form on incompletely resected prostatic tissue and result in persistent infection and irritation. Inset shows cystoscopic view

2. Repeat resection

a) **Early obstruction.** A few patients are unable to void satisfactorily during the first few days following postresection removal of the catheter, and require reoperation. If the subject's general condition is good, this may safely be done within a week of the first intervention, and complete removal of all pathologic prostatic tissue at this time will nearly always clear up the obstructive symptoms.

16*

In general, bleeding and other complications are less common and convalescence more rapid following the second stage.

b) Recurrence of the growth. Recurrence of urinary obstruction from six months to several years after endoscopic surgery sometimes takes place because of continued growth of an incompletely resected gland or from hypertrophy of a portion of the prostate which was normal at the time of operation. Such recurrences, however, cause intermittent bleeding more frequently than they do urinary obstruction, although both symptoms may appear in combination. Hemorrhage may occur several years after the operation, and is due to the formation of excessive granulation tissue in the prostatic bed or to rupture of dilated vessels on recurrent hypertrophied tissue. It may be so persistent and severe that a repeat resection becomes necessary, even though no other symptoms and no residual urine are present. Similar effects are sometimes observed following suprapubic or perineal enucleation of the prostate, but are rarer than following a poorly performed endoscopic removal.

V. Urethral stenosis

Urethral stricture occurs following endoscopic prostatic surgery, and is most frequently located at the external meatus, in the pendulous urethra or at the bladder neck (bladder neck contracture). Urethral dilatation with sounds, as described in Chapter XXIII, is the therapy of choice and may have to be continued indefinitely.

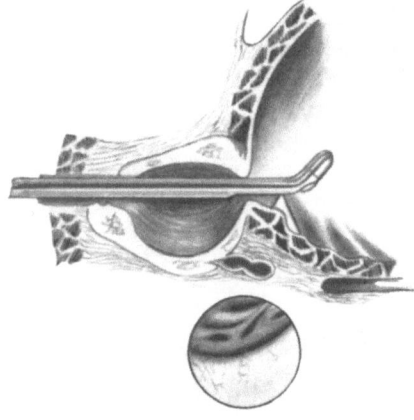

1. Meatal stenosis

Stricture at the meatus is dilated with less pain following injection of a local anesthetic through a fine needle into the ventral angle of the meatus (LANDES). Urethral meatotomy is sometimes indicated. If the stenosis extends farther than 2 or 3 cm. into the urethra a Johanson procedure (Vol. XIII) may be indicated.

2. Bladder neck stenosis

a) Causes. Bladder neck contracture (Fig. 174) makes its appearance from a few months to one year or more after endoscopic prostatic surgery, and is more prone to occur

Fig. 174. Postoperative bladder neck contracture is often accompanied by hypertrophied trigone, and is diagnosed by cystoscopy if instrument can be passed. Inset shows cystoscopic view through right angle lens

in the following cases: those in which the original pathology was *1. bladder neck contracture*; *2. median bar*; *3. small prostate* and *4. intraurethral hypertrophy.*

b) Diagnosis. A catheter is passed to calibrate the urethra and to determine whether residual urine is present. The catheter, however, will not advance entirely into the bladder. A sound inserted passes easily to the prostatic urethra, and since it encounters obstruction at this point the handle cannot be depressed between the thighs as freely as it should. A panendoscope is then passed into the dilated posterior urethra, and the diagnosis established with the aid of a foroblique lens system. A very small opening into the bladder can be seen from this position (Fig. 175)—an opening which could not have been visualized with a right angle lens system, for it would have been straight ahead in the "blind spot" of this lens.

c) Treatment. Such postoperative bladder neck contractures are treated with a Collings knife through the panendoscope, by inserting the blade through the aperture into the bladder and *incising* dorsalward (Fig. 176). The cut edges will

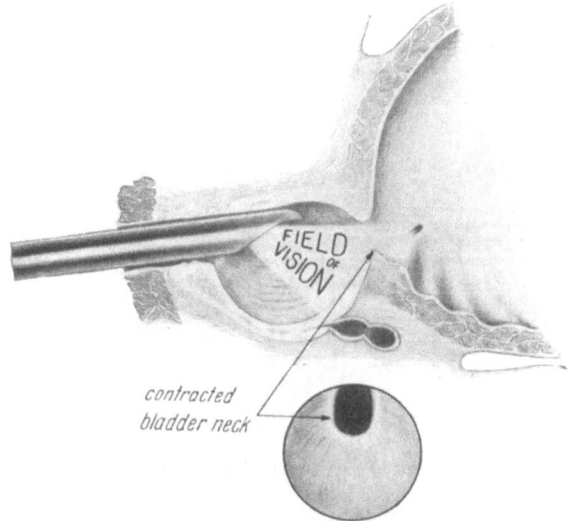

Fig. 175. Postoperative contracture of bladder neck and dilation of prostatic urethra. If cystoscope cannot be passed through contracted bladder neck, opening into bladder can be seen through foroblique lens. Inset shows cystoscopic view

retract, making it possible to advance the panendoscope into the bladder (Fig. 177). If at this time the contracture is found to involve the entire bladder neck, a total

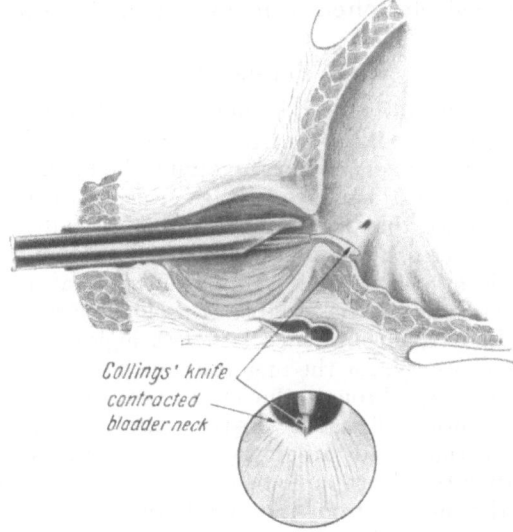

Fig. 176. Collings knife inserted into bladder through contracted vesical orifice, and incision made dorsally

resection is done; but if only the dorsal portion is elevated, the incision already made is carried deeper until the fundus above the trigone is level with the floor of the prostatic urethra. After incising the postoperatively contracted bladder neck, and in cases which do not contract down sufficiently to require incision, it is necessary to *dilate* the vesical orifice with sounds or preferably the Kollman dilator. Occasionally a *plastic procedure* performed through a suprapubic approach is necessary in cases which cannot be dilated adequately (Vol. XIII).

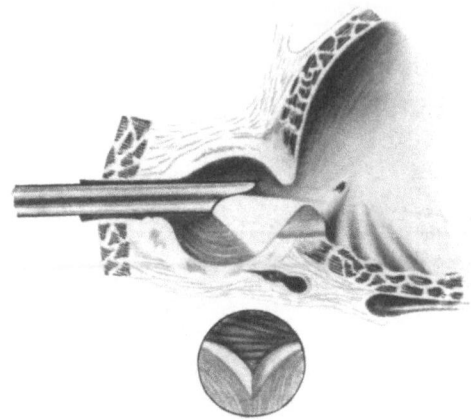

Fig. 177. Edges of incised bladder neck and trigone separate after incision is made

VI. Urinary incontinence

1. Temporary

Temporary urinary incontinence following endoscopic prostatic surgery is frequently seen, and may last from one day to several months. Occasionally, the leakage may persist for six months before it gradually begins to clear up. Exercise of the urethral sphincter often helps to correct the incontinence (VERGES).

2. Permanent

Permanent urinary incontinence does not occur following endoscopic prostatic resection unless there has been injury to the membranous urethra during operation [1]. Probably even a preexisting nerve lesion involving the bladder or external sphincter will not predispose to postoperative leakage of urine if it has not already given warning. It is undeniable that the advent of this most distressing sequela was far too common during the early days of endoscopic prostatic surgery, as evidenced by the increased sale of incontinence clamps during that period; but with improved skill on the part of a greater number of urologic surgeons its incidence has notably declined. Indeed, incontinence is a less logical outcome of the closed method than of open surgery, for the membranous urethra can be visualized and identified more accurately through the electrotome telescope than it can by vision or palpation during digital or instrumental enucleation. Incontinence occasionally follows perineal prostatectomy, even when performed by an expert, but rarely occurs after the suprapubic operation. When it does, it is the direct result of stripping the mucosa from the membranous urethra while the apex of the gland is being removed. There is no satisfactory treatment for this condition, and palliation consists of the patient's wearing a rubber urinal or a Cunnigham penic clamp.

---

[1] Carcinoma of the prostate may invade into the membranous urethra and cause incontinence.

VII. Sexual changes

1. Libido

Endoscopic removal of the prostate does not have a deleterious effect on libido. Every urologist has seen men who have undergone this operation and who still have the ability for sexual contact even though they are aged. It is probable that an impaired libido is due more to general debility or psychic influence than to the fact that the prostate has been removed.

2. Ejaculation

In most cases following prostatectomy by any route, when ejaculation occurs the seminal fluid flows backward into the bladder rather than outward through the urethral canal. When this occurs the patient passes seminal fluid with the urine the next time he voids.

VIII. Vesical hypotonia

1. Cause

Hypotonia of the bladder musculature sometimes causes residual urine to persist regardless of complete removal of obstruction at the vesical orifice. Hypotonia may be myogenic in nature, resulting from long continued preoperative bladder overdistention or general debility of the patient; or it may be neurogenic, arising from a central nerve lesion which coexisted with the prostatic obstruction. If inability to empty the bladder is due to the former, gradual improvement occurs in most patients as bladder tone returns and general health is restored. If it is due to the latter, some patients resume the ability to empty the bladder, but most of them continue to carry residual urine.

2. Management

a) **Catheterization and irrigation.** Treatment consisting of catheterization and bladder irrigation once or twice daily is necessary as long as residual urine of 100 cc. or more is present. As soon as improvement occurs, the interval between treatments is lengthened, but they should be continued once or twice a week until retention is less than 30 cc. If such a regime is not carried out, bladder infection becomes severe and soon the upper urinary tract is involved. Patients whose residual urine remains above 100 cc. for several weeks may be taught self catheterization and irrigation, or a member of the family may learn to do it for them; for sometimes this procedure must be prolonged indefinitely. The use of one quart of hot (110^0 F, 43.3^0 C) solution as irrigation sometimes aids in hastening the return of bladder tone, and in an occasional case cold solution (60^0F, 15.5^0 C) stimulates the bladder musculature to contract. Smooth muscle stimulants such as mecholyl bromide 10 mg. twice daily have been used. Sinusoidal current applied through the bladder may aid at least temporarily in increasing tonicity. Infection persists in the hypotonic bladder; chemotherapy in small doses continued over a long period of time, and alternated, keeps this at a minimum.

b) **Plastic procedure.** A plastic procedure performed through the suprapubic approach of overlapping the bladder walls of the dome of the bladder, thus strengthening the walls and reducing the bladder capacity, often helps these patients to void with more force (BARNES et al. 1953).

c) **Diagnosis of possible causes.** Some individuals with hypotonic bladders present a problem in management, for they cannot understand why they are

unable to pass all their urine when obstruction has been completely removed
from the bladder outlet. It is very important for the urologist to make certain
by cystoscopy and cystography that there is no other pathology such as a slight
bladder neck obstruction or vesical diverticulum which might be a causal factor,
before he tells his patient that the bladder walls are weak and that nothing further
can be done to help him to empty his bladder.

IX. Fibrosis of the ureteral orifices

Fibrosis of a ureteral orifice (DAVIS) resulting from injury at the time of
resection is treated by ureteral dilatation if a catheter can be passed into the
orifice. This may be impossible, however, and if only one orifice is involved and
if there is no persisting kidney infection on that side, no treatment is necessary.
If both orifices are fibrosed, upper urinary tract dilation with resulting uremia
and probably persistent infection will ensue, necessitating open ureteral meato-
plasty (O'CONOR).

X. Persistent infection

Urinary tract infection which persists more than three months postoperatively
is usually due to one of the conditions listed above; incomplete removal of tissue,
urethral stenosis, vesical hypotonia, fibrosis of the ureteral orifices, or to an
associated disease such as diverticulum, tuberculosis or preexisting upper urinary
tract disease. Sometimes numerous small cellules retain sufficient urine to cause
persistence of the infection, and occasionally the focus is in the seminal vesicles.

The treatment of postendoscopic surgical infection is the same as that of
any urinary tract infection — removal of the cause if possible, then application of
appropiate antibacterial agents.

Sometimes urinary frequency will persist because of fibrous contraction of
the bladder resulting from long continued infection. This is treated by bladder
overdistention and appropriate chemotherapy (MATHE).

XI. Recurrence of malignancies

1. Prostate

The results following endoscopic removal of bladder neck obstruction caused
by carcinoma of the prostate have been superior to those succeeding any other
treatment of this heretofore hopeless condition. Even since the advent of the
miraculous effects of endocrine therapy, it has sometimes been necessary to
establish more adequate bladder drainage, as well as to obtain tissue for biopsy
study by resecting the obstructing portion of the gland. Published reports concur,
and agreement among urologic surgeons is almost unanimous, in approving
this operation over any other in the treatment of obstruction due to the malignant
gland—even though it may have to be repeated for the relief of recurrent obstruction due to sustained growth of the malignancy. This persistent increment
often makes sounding necessary throughout the lifetime of the patient.

2. Bladder tumors

Most bladder tumors recur following removal regardless of the method of
their extirpation. The success of any form of treatment is measured more accurately by the survival rate than by the rate of recurrence. The overall five year
survival rate following endoscopic removal of bladder tumors is higher than that
following the open approach (FLOCKS). It is probable, however, that the smaller

and less invasive tumors are more often removed by endoscopic surgery and that the more extensive ones are more frequently operated upon through the open approach. This fact would weigh the statistics in favor of the endoscopic method. An overall five year survival of less than 20 per cent following cystectomy compared with a somewhat more than 50 per cent following endoscopic removal (BARNES et al. 1955) would, however, make it appear that the latter approach is indicated in all cases in which the tumor is not excessively large nor deeply invasive (p. 174). Even though some urologists decry the use of the endoscopic approach to bladder tumors (DEAN), others believe that the results following this method indicate its use in the majority of cases (HIGBEE).

The fact that most bladder tumors do recur following extirpation necessitates reexamination of the patient at regular intervals indefinitely. Cystoscopic examination of the bladder is performed at three month intervals. If there is no recurrence for six to nine months, especially when the tumor was not invasive and microscopic examination showed it to be of low grade malignancy, the interval between examination can be lengthened to six months, then to one year if no recurrences are found.

References

Chapter I

BINKLEY, J. S.: The use of dry cell batteries, rheostat and voltmeter for economy in endoscopic examination. J. Urol. (Baltimore) 41, 930—933 (1939). — BLASUCCI, M. D. F.: A new cystoscope holder for demonstration purposes. J. Urol. (Baltimore) 34, 337—339 (1935). — BODNER, H. and A. H. HOWARD, J. H. KAPLAN: A new development for light bearing surgical instruments. J. Urol. (Baltimore) 73, 879—882 (1955). — BUTTERFIELD, P. M.: New children's cysto-urethroscope. J. Urol. (Baltimore) 22, 347—350 (1929). — Experiences with new baby cystourethroscope. Amer. J. Surg. 8, 102 (1930). — CAMPBELL, M. F.: A leg holder for infants and children during cystoscopic examination. J. Urol. (Baltimore) 29, 627—629 (1933). — Three miniature cystourethroscopes. J. Urol. (Baltimore) 36, 183—188 (1936). — COUNCILL, W. A.: A new cystoscopic holder. J. Urol. (Baltimore) 24, 699—701 (1930). — The teleprobe: an instrument to detect calculi in the ureter and the kidney. J. Urol. (Baltimore) 47, 395 (1942). — DAVIS, D. M., and J. B. WHITEHEAD: On the importance of grounding in cystoscope lighting devices. J. Urol. (Baltimore) 25, 111—116 (1931). — ELLIK, M.: Stone in ureter: their extraction by loop catheter. J. Urol. (Baltimore) 57, 473—478 (1947). — FOLEY, F. E. B.: A completely rotatable resectoscope. J. Urol. (Baltimore) 62, 381 (1949). — FOURESTIER, M., A. GLADN, et J. VULMIERE: Étude technique de l'endoscope medical universel. Sem. méd. (Paris) 46, 13 (1956). — GIBSON, T. E.: Postol grip modification of Stern-McCarthy resectoscope. J. Urol. (Baltimore) 69 602—603 (1953). — GODDARD, D. W.: A new female air costo-urethroscope. J. Urol. (Baltimore) 67, 526—528 (1952). — GARSKE, G. L.: A sterile disposable plastic waterproof drape for transurethral surgery and cystoscopic procedures. J. Urol. (Baltimore) 69, 135 (1953). — HARKNESS, A. H.: Non-gonococcal urethritis, p. 365. Edinburgh: E. & S. Livingstone 1950. — HUNNER, G.: The etiology of ureteral calculus. Surg. Gynec. Obstet. 27, 252—270 (1918). — HUTCHINS, S. P. R.: A new resectoscope. J. Urol. (Baltimore) 69, 456—457 (1953). — JAUPITRE M.: Etat actuel de la photocinématographie endoscopique en urologie. Rev. Film médico-chirrurgical 1957, 2392. ~ La photocinématographie endoscopique en urologie. Sem. méd. (Paris) 46, 29 (1956). — KAHN, I. W.: New demonstration eyepiece for teaching cystoscopy. Amer. J. Obstet. Gynec. 22, 926—927 (1931). — KEYES jr., E. L.: The wax-bulb ureter catheter for routine use. J. Urol. (Baltimore) 4, 563—566 (1920). — KIEFER, J. H.: Alteration of the McCarthy cystoscope to permit use of the Mead telescope. J. Urol. (Baltimore) 75, 1025—1026 (1956). — KIRWIN, T. J.: A new cystoscope and its clinical uses. J. Urol. (Baltimore) 60, 537—541 (1948). — LATTIMER, J. K., M. KENNEY, G. ROSENBLATT and M. GOLDMAN: Inadequate sterilization of cystoscopes. J. Urol. (Baltimore) 76, 197—199 (1956). — LIVERMORE, G. R.: Presentation of new indwelling ureteral catheter. J. Urol. (Baltimore) 61, 753 (1949). — LOWSLEY, O. S.: A modification of Lowsley cystoscopic rongeur. Urol. cutan. Rev. 44, 366—367 (1940). — LOWSLEY, O. S., T. J. KIRWIN and A. DAVALOS: Urolocide, a new skin antiseptic and disinfectant for instruments. J. Urol. (Baltimore) 59, 966—969 (1948). — McCARTHY, J. F.: A new visual system for observation and operation in the urinary bladder. J. Urol. (Baltimore) **57**, 575—577 (1947). — Present concepts and future prospects of precision urology. J. Urol. (Baltimore) **65**, 1050—1055 (1951). — McCarthy, J. F. and J. F. Ritter: A new endourethral resectoscope. J. Urol. (Baltimore) **74**, 559—562 (1955). — McCrea, L. E.: Experimental studies in intravesical photography. J. Urol. (Baltimore) **47**, 148—155 (1942). — Value of cystoscopic photography in medicine. Amer. J. Surg. **56**, 622—625 (1942). — Natural color intravesical photography. J. Urol. (Baltimore) **50**, 673—679 (1943). — A new infant cystoscope. J. Urol. (Baltimore) **58**, 217 to 219 (1947). — McDonald, J. H.: A modification of the Bunge ureteral meatome. J. Urol. (Baltimore) **73**, 1098—1099 (1955). — Nation, E. F.: Facilitation of passage and extraction of ureteral calculi. J. Urol. (Baltimore) **70**, 373—375 (1953). — Nesbit, R. M.: An improved resectoscope in sizes 24, 26, and 28 French. J. Urol. (Baltimore) **63**, 191—193 (1950). — Nugent, J. J.: A simple rotatable electrotome. J. Urol. (Baltimore) **64**, 534—536 (1950). — Ohanneson, F., and W. K. Murphy: Device for controlling flow of water during cystoscopic procedures. Urol. cutan. Rev. **50**, 637 (1946). — Otis, W. K.: Concerning the new electrocystoscope. N. Y. med. J. **81**, 625—628 (1905). — Ravich, A.: Critical review of lithotriptoscopy. Urol. cutan. Rev. **50**, 625—627 (1946). — Riaboff, P. J.: Cystoscope and test tube holders. Amer. J. Surg. **29**, 487—488 (1935). — Riba, L. W., and J. B. Reynolds: A "baby" ureteral meatotome. J. Urol. (Baltimore) **66**, 578 (1951). — Ridley, J. H.: Indirect air cystoscopy. Sth. med. J. (Bgham. Ala.) **44**, 114—121 (1951). — Rinker, J. R.: Implanter for radon seeds for use with the McCarthy foroblique pan-endoscope. J. Urol. (Baltimore) **62**, 91 (1949). — Roome, N. W.: Instrument for endoscopic extraction of ureteral stones. J. Urol. (Baltimore) **66**, 575—577 (1951). — Rosenthal, A. A.: A new cystoscope holder. J. Urol. (Baltimore) **40**, 675—676 (1938). — Rowe, K. T.: Meatotomy electrode. J. Urol. (Baltimore) **70**, 125—126 (1953). — Segal, A. D.: A new cystoscopic holder. J. Urol. (Baltimore) **44**, 244—245 (1940). — Seid, B., and B. E. Cohler: A simple cystoscopic illuminator. J. Urol. (Baltimore) **75**, 1027 (1956). — Simons, I.: A self locking cystoscopic stone or foreign body forceps with detachable handle. J. Urol. (Baltimore) **34**, 190—192 (1935). — Sinclair, D. A.: Cystoscopic room deportment. Urol. cutan. Rev. **50**, 621—622 (1946). — Spence, H. M.: Cystoscopic clinic at advance naval base hospital. U.S. nav. med. Bull. **43**, 763—765 (1944). — Tatum, W. B.: Illumination of bladder diverticula with a new instrument — the Diverticulite. J. Urol. (Baltimore) **18**, 173—175 (1927). — Trattner, H. R.: Balanced suspension for facilitation of endoscopic instrumentation. J. Urol. (Baltimore) **68**, 862—864 (1952). — Tyvand, R. E.: Air valve for direct vision cystoscope. J. Urol. (Baltimore) **28**, 731 (1932). — Vest, S. A.: A new cystoscope for dilatation purposes. J. Urol. (Baltimore) **62**, 378—380 (1949). — A new inspection lens-sheath as an aid to transurethral resection. J. Urol. (Baltimore) **53**, 347—353 (1945). — Vlietstra, H. P.: Modification of loop resection principle adapted to an instrument for endoscopic bladder treatment. Arch. chir. neerl. **6**, 89—94 (1954). — Webster jr., H. N.: A flexible obturator for Stern-McCarthy resectoscope sheath. J. Urol. (Baltimore) **77**, 120—121 (1957). — Young, H. H.: Presentation of new double catheterizing cystoscopes for adults and for infants. J. Urol. (Baltimore) **27**, 249—264 (1932).

Chapter II

Alexander, H. L.: Reactions with drug therapy. Philadelphia: W. B. Saunders Company 1955. — Barnes, R. W.: The influence of scopolamine-morphine narcosis on renal function. Urol. cutan. Rev. **29**, 459—462 (1925). — Barringer, B. B.: Cystoscopic examination of female dogs. J. Urol. (Baltimore) **57**, 185—189 (1947). — Baurys, W.: Cystoscopic technique and instruments general consideration. Urol. cutan. Rev. **50**, 610—613 (1946). — Belt, E., and L. Bavetta: Intracain as a local anesthetic in urology. J. Urol. (Baltimore) **49**, 745—746 (1943). — Blasucci, P.: New radiologic cystoscopic table for urologic diagnosis. Policlinico, Sez. prat. **43**, 1692—1697 (1936). — Blusger, I. N., and J. H. Dixon: Pentothal sodium anesthesia for cystoscopy. Lancet **1943**, 111. — Boeminghaus, H.: Trocar for vesical fistula and suprapubic cystoscopy. Z. Urol. **34**, 335—338 (1940). — Briggs, W. T.: Choice of anesthetic in endoscopic procedures. Urol. cutan. Rev. **50**, 614—615 (1946). — Bumpus jr., H. C.: Indications for visual examinations of lower urinary tract. J. Amer. med. Ass. **123**, 615 to 617 (1943). — Coffett, R. W., and D. A. Charnock: Relation of ureteral length to trunk length and body height in children. J. Urol. (Baltimore) **46**, 667—670 (1941). — Corbus, B. C.: Intracaine lubricant a as urethral anesthetic. J. Urol. (Baltimore) **62**, 89—90 (1949). — Crabtree, E. G.: A one piece drapery for the cystoscopy table. J. Urol. (Baltimore) **22**, 449—452 (1929). — Crawford, O. B.: Comparative qualities of three new local anesthetic drugs: xylocaine, cyclaine and provocaine. Anesthesiology **14**, 278—290 (1953). — Dees, J. E., and S. S. Ambrose jr.: SKF 538-A, a new topical anesthetic for use in the urethra: A preliminary report. J. Urol. (Baltimore) **65**, 346—349 (1951). — De la Pena, A., and E.: Chromocystoscopy with neoprontosil. J. Urol. (Baltimore) **47**, 202 (1942). — Dietz, P.: Indications for chromocystoscopy. Z. Urol. **30**, 754—758 (1936). — Douglas, G. F.: Value of cystoscopic in female urology; preliminary report on use of intracaine as a topical analgesic. J. M. Ass. Alamba **12**, 137—144 (1942). — DUNPHY, J. E., and J. H. HARRISON: Intravenous evipal-sodium anesthesia in urologic surgery. New Engl. J. Med. **218**, 117—118 (1938). — FELBER, E.: Does cystoscopy necessarily need to be painful? Urol. cutan. Rev. **50**, 608—610 (1946). — FERRIN, J. W.: Use of diothane hydrochloride in urological cases. J. Urol. (Baltimore) **40**, 666—671 (1938). — FITZPATRICK, R. J., L. M. ORR and F. J. STUBBART: Antihistamines as local anesthetic agents for urethral manipulation. J. Amer. med. Ass. **150**, 1092—1094 (1952). — FRIEDRICH, H.: Value of chromocystoscopy as function test of engorged kidneys. Z. Urol. **31**, 145—152 (1937). — GRANT, O., G. LICH and G. ABAJIAN: Rectal evipal soluble anesthesia in urological procedures. J. Urol. (Baltimore) **42**, 204—210 (1939). — GRAVES, H. A.: Analgesic effect of nisentil in urological procedures. J. Urol. (Baltimore) **74**, 567 — 569 (1955). — GREENBERG, B. E.: Oil as cystoscopic medium. New Engl. J. Med. **204**, 1366—1367 (1931). — GUATHMEY, J. T., and J. ABAJION jr.: Safe, comfortable anesthesia for uroligic surgera. Anesth. et Analg. **22**, 228—232 (1943). — HARDING, V. J., and R. W. T. URQUHART: Urea clearance test in urology. J. Urol. (Baltimore) **29**, 1—13 (1933). — HAYWARD, J. C., L. M. ORR and H. F. LAGUETTE: Local anesthesia in the male urethra. J. Urol. (Baltimore) **70**, 624—625 (1953). — HEIDER, C. F.: Cystoscopy-neglected diagnostic procedure. Neb. St. med. J. **15**, 84—85 (1930). — HERMAN, L.: A modified shadowgraph ureteral catheter. J. Urol. (Baltimore) **15**, 478—482 (1926). — HORNE, F. G.: Trichlorethylene in office and hospital: with relation to urology. Sth. med. J. (Bgham, Ala.) **47**, 515—516 (1954). — JAHR, R.: Complete cystoscopy. Z. Urol. **21**, 641—646 (1927). — JASIENSKI, J.: Une nouvelle épreuve colorante combinée pour 1 éxploration du fonctionnement rénal. J. d'Urol. **38**, 518—529 (1934). — KAUFMAN, L. R.: Use of cystoscope in tuberculous of bladder. Urol. cutan. Rev. **35**, 294—302 (1931). — KEERBERGEN, G. VAN: Critical and comparative study of the various renal function tests. J. belge Urol. **15**, 621—749 (1947). — KEYES jr., E. L., and R. S. FERGUSON: Urology, p. 16. New York: D. Appleton-Century Co. 1936. — LAIBE, J. E. J.: Simplified analgesia in urology. Illinois med. J. **73**, 224—226 (1938). — LANDES, R. R., and C. L. RANSOM: Cystoscopy of the supine patient: a bedside cystoscope. J. Urol. (Baltimore) **74**, 556—558 (1955). — LEE, L. W., E. DAVIS jr. and J. BARMORE: Clinical use of cyclaine (hexylcaine) as an anesthetic agent in urological surgery — a report of observations of 1000 patients. J. Urol. (Baltimore) **72**, 261—266 (1954). — LEWIS, B.: Local anesthesia for urethra. J. Urol. (Baltimore) **12**, 527—533 (1924). — LOW, H. T.: Major indications for cystoscopy. Col. Med. **27**, 334—336 (1930). — MARION, G.: Cystoscopy with large perforation of the bladder. J. d'Urol. **36** (1933). — MARSHALL, B. M.: Surfacaine as urologic anesthetic. Urol. cutan. Rev. **53**, 86—87 (1949). — MERRITT, E. L.: Value of evipal anesthesia in urology. Urol. cutan. Rev. **39**, 709—710 (1935). — MORRISSEY, J. H., and A. N. SPINELLI: An experimental study of anesthetic and analgesic properties of pyridium. J. Urol. (Baltimore) **44**, 381—385 (1940). — MUSCHAT, M.: Urethral anesthetic lubricant. J. Urol. (Baltimore) **45**, 238—239 (1941). — NESBIT, R. M., and R. K. RATTIFF: A new anesthetic lubricant for urethra. J. Urol. (Baltimore) **34**, 394—395 (1935). — OCKERBLAD, N. F., and H. E. CARLSON: Nupercaine; it's place in urologic practice. J. Missouri med. Ass. **36**, 120—122 (1939). — PARKER, G.: Elimination of pain from urological investigation. Brit. J. Urol. **7**, 257—263 (1935). — PELOUZE, P. S.: Value of cystoscopy in general medical diagnosis. Pennsylvania med. J. **35**, 846—849 (1932). — Office urology, p. 208. Philadelphia: W. B. Saunders Company 1940. — PERSKY, L., and H. S. DAVIS: Xylocaine as a topical anesthetic in urology. J. Urol. (Baltimore) **70**, 552—554 (1953). — PONSOLD, A.: Fatal boric acid poisoning following perforation of urethra in attempted cystoscopy: case. Dtsch. Z. ges. gericht. Med. **34**, 321—326 (1940). — REISSER, M. P., and C. D. CREEVY: Nisentil in urologic procedures. J. Urol. (Baltimore) **77**, 880—883 (1957). — RIDLEY, J. H.: Indirect air cystoscopy. Sth. med. J. (Bgham, Ala.) **44**, 114—121 (1951). — RITTMANSBERGER, K.: Diagnostic inadequacy of elimination of blue dye. Z. Urol. **33**, 265—268 (1939). — SADOVE, M. S., G. M. WYANT, L. A. GITTELSON and H. E. KRETCHMER: Classification and management of reactions to local anesthetic agents. J. Amer. med. Ass. **148**, 17—22 (1952). — SECRETAN, M.: La valeur de la chromo-cystoscopie dons le diagnostic de localisation de la tuberculose rénale. J. d'Urol. **46**, 201—223 (1938). — STEINHAUS, J. E.: A comparative study of the experimental toxicity of local anesthetic agents. Anesthesiology **13**, 577—586 (1952). — THOMAS, B. A.: A new urethral anesthetic tablet depositor. J. Urol. (Baltimore) **23**, 491—492 (1930). — TOVEL, R. M., and G. J. THOMPSON: Pentothal sodium anesthesia in urologic practice. J. Urol. (Baltimore) **36**, 81—87 (1936). — YOW, R. M., J. E. MATHIAS and R. C. BUNTS: Use of nisentil in urology — a clinical evaluation. J. Urol. (Baltimore) **74**, 570—572 (1955).

Chapter III

D'ABREU, A. L., and A. C. LYSAGHT: Uraemic ulcerative colitis following cystoscopy. Brit. J. Urol. **7**, 336—344 (1935). — DIBLE, J. H.: Air embolism in urethroscopy and tubal insufflation. Lancet **1938**, 313—315. — DONAHUE, C. D.: Mortality in cystoscopy. Northw. Med. (Seattle) **31**, 561—563 (1932). — HENLINE, R. B.: Traumatic injuries of upper urinary tract following instrumentation (pyelography). J. Amer. med. Ass. **102**, 182—188 (1934). — KAHN, I. W.: Role of cystoscope in genecology. Urol. cutan. Rev. **52**, 152—156 (1948). — LEWIS, R. M.: Cystoscopic reactions. J. Urol. (Baltimore) **57**, 365—374 (1947). — PAGE, B. H., and C. WILSON: Acute mercurial poisoning after cystoscopy. Lancet **1940**, 640 to 764. — STOLZ, C. E.: Cystoscopic experience. Urol. cutan. Rev. **34**, 825 (1930). — WEHRBEIN, H. L.: Cystoscopy reactions. Ann. Surg. 87, 435—441 (1928). — WILHELM, S. F.: Perforation of bladder during cystoscopic examination. J. Urol. (Baltimore) **22**, 555—556 (1929). — WOLF jr., G. A.: Mechanism of reflex anuria. Ann. intern. Med. **23**, 99—102 (1945).

Chapter IV

BARNES, R. W.: Endoscopic prostatic surgery. St. Louis: C. V. Mosby Comp. 1943. — HOYT, H. S., and P. MESSIER: The mucosal blood bessels of the urinary bladder. Stanf. med. Bull. **9**, 34 (1951). — MACLEAD, D.: The trigonal muscle: its composition and urinary function. Brit. J. Urol. **13**, 135 (1941).

Chapter V

BAILEY, M. K., C. A. FORT and H. C. HARLIN: Papillary carcinoma of ureter and bladder arising after nephrectomy for tumor of kidney pelvis. J. Urol. (Baltimore) **62**, 44—47 (1949). — BENJAMIN, J. A., and C. E. TOBIN: Abnormalities of kidneys, ureters and perinephric fascia: anatomic and clinical study. J. Urol. (Baltimore) **65**, 715—733 (1951). — CAMPBELL, M. F.: Clinical pediatric urology, p. 217. Philadelphia: W. B. Saunders Comp. 1951. — GIBSON, T. E.: Rare reno-ureteral anomaly. J. Urol. (Baltimore) **46**, 517—519 (1941). — GRIM, K. B.: Bilateral renal and ureteral agenesis. J. Urol. (Baltimore) **44**, 397—400 (1940). — GUMMESS, G. H., D. A. CHARNOCK, H. I. RIDDELL and C. M. STEWART: Ureterocele in children. J. Urol. (Baltimore) **74**, 331—335 (1955). — KREUTZMANN, H. A. R.: Unusual ureteral anomaly, with discussion of embryology. J. Urol. (Baltimore) **38**, 67—73 (1937). — ORR, L. M., and J. B. GLANTON: Prolapsing ureterocele. J. Urol. (Baltimore) **70**, 180—186 (1953). — SORRENTINO, M.: Tumor of ureter prolaped into the bladder. J. d'Urol. **38**, 338—346 (1934). — WOODRUFF, S. R.: Complete unilateral triplication of the ureter and renal pelvis. J. Urol. (Baltimore) **46**, 376—379 (1941).

Chapter VI

HOYT, H. S., and P. MESSIER: Mucosal blood vessels of urinary bladder; cytoscopic study. Stanf. med. Bull. **9**, 34—38 (1951). — VASSALLO, S. M.: Urethrotrigonitis in relation to abacterial pyuria. Brit. J. Urol. 18, 189—195 (1946).

Chapter VII

BORS, E.: The anatomical and physiological basis of neurogenic bladder disturbances. J. Indian med. Prof. **2**, 954—964 (1956). — COMARR, A. E.: The problems of cord bladder rehabilitation. J. Urol. (Baltimore) **77**, 232—237 (1957). — COMARR, A. E., and A. A. KAUFMAN: A survey of the neurological results of 858 spinal cord injuries. J. Neurosurg. **13**, 95—106 (1956). — CROSS, W. W.: Sarcoma of bladder. Urol. cutan. Rev. **43**, 180—182 (1939). — DUZEN, R. E. VAN: Use of cystoscope as control in diagnosis and treatment of cystoceles. Urol. cutan. Rev. **36**, 187—194 (1932). — EDWARDS, C.: Congenital multilocular bladder. M. J. Aust. **2**, 443—447 (1933). — FARBER, S.: Neurofibromata. Amer. J. Dis. Child. **60**, 749—751 (1940). — FERRANDIZ, S. V.: Interureteric muscle forming a free band. Senate Rev. Cir. Barcelona **6**, 69—80 (1933). — HAGER, B. H., and V. C. HUNT: A report of a case of leiomyosarcoma of bladder. J. Urol. (Baltimore) **21**, 129—133 (1929). — HINMAN, F., and M. B. WESSON: Trigone of bladder as a factor in urinary obstruction, and operative treatment. Surg. Gynec. Obst. **43**, 1—9 (1926). — KEARNS, W. M., and S. M. TURKELTAUB: Hourglass deformity of urinary bladder. J. Urol. (Baltimore) **29**, 729—738 (1933). — KHOURY, E. N., and F. D. SPEER: Rhabdomyosarcoma of urinary bladder, a clinico-pathological case report with a review of literature, including a tabulation of rhabdomyosarcoma of prostate. J. Urol. (Baltimore) **51**, 505—516 (1944). — KOHLER, H. H.: Septal bladder, with multiple genitourinary anomalies and uremia. J. Urol. (Baltimore) **44**, 63—66 (1940). — KOLL, I. S.: Fibroids of urinary bladder. J. Urol. (Baltimore) **9**, 453—466 (1923). — KRETCHMER, H. L.: Leiomyoma of bladder. J. Urol. (Baltimore) **26**, 575—589 (1931). — Leiomyoma of bladder with report of a case and review of literature. Trans. Amer. Ass. gen.-urin. Surg. **24**, 363 (1931). — Leiomyosarcoma of bladder. Arch. Surg. (Chicago) **38**, 274—286 (1939). —

KUTZMAN, A. A.: Leiomyoma of urinary bladder. J. Urol. (Baltimore) 37, 117—132 (1937). — LAUGHLIN, C. V., G. H. DERIAN and P. F. BOYD: Incomplete frontal septum of bladder complicated by congenital urethral valves and complete reduplication of upper left urinary tract. J. Urol. (Baltimore) 68, 289—296 (1952). — LAZARUS, J. A., and A. A. ROSENTHAL: Myxosarcoma of bladder. J. Urol. (Baltimore) 27, 695—711 (1932). — LEVINGS, A. H.: Case of myofibroma of bladder simulating uterine myofibroma. Amer. J. Derm. 13, 477 (1909). — LICH jr., R., and O. GRANT: Vesical abnormalities incident to diabetes mellitus. J. Urol. (Baltimore) 59, 863—874 (1948). — LLANOS, M. A.: Diverticula of the bladder. J. Urol. (Baltimore 49, 628—638 (1943). — MACKENZIE, D. W., and W. H. CHASE: Rhabdomyosarcoma of urinary bladder with metastases: report of a case with brief review of clinical and pathological literature. Trans. Amer. Ass. gen.-urin. Surg. 20, 15 (1927). — MCCARTHY, J. F., C. T. STEPITA and S. J. HALPERIN: Sarcoma of bladder. Amer. J. Surg. 7, 229—233 (1929). — MINTZ, E. R.: Pedunculated neurofibroma of the bladder. J. Urol. (Baltimore) 43, 268—274 (1940). — MUNGER, A. D.: Primary mesothelial tumor (leiomyoma malignum) of bladder with secondary involvement of abdominal cavity. J. Urol. (Baltimore) 42, 229—235 (1939). — OCKERBLAD, N. F., and H. E. CARLSON: Congenital hourglass bladder. Surgery 8, 665—671 (1940). — O'CROWLEY, C. R., and H. S. MARTLAND: Is myxoma of bladder a pathological entity? J. Urol. (Baltimore) 11, 349—363 (1924). — SCARDINO, P. L., and T. E. UPSON: Inguinal hernia and bladder diverticulum. J. Urol. (Baltimore) 69, 282—283 (1953). SENGER, F. L., and V. J. SANTARE: Congenital multilocular bladder: a case report. J. Urol. (Baltimore) 68, 283—288 (1952). — SMITH, E. O., and D. H. ALLEN: Congenital double bladder of tremendous size. J. Med. 12, 89—90 (1931). — SMITH, G. G.: Experiences with submucous fibrosis of bladder. J. Urol. (Baltimore) 26, 455—464 (1931). — STEINHARDT, B.: A case of true giant bladder in fetus. Urol. cutan. Rev. 48, 261—266 (1944). — TREMBLAY, R. G., A. R. CRANE and A. HARRIS: Primary osteogenic sarcoma of bladder. J. Urol. (Baltimore) 51, 143—148 (1944). — VERMOOTEN, V.: Rhabdomyomyosarcoma of bladder: a case report. J. Urol. (Baltimore) 42, 126—130 (1939). — YOUNG, H. H.: Changes in trigone due to tuberculosis of kidney, ureter, and bladder. Bridge formation and floating trigone. Surg. Gynec. Obstet. 26, 608—615 (1918). — ZELLERMAYER, J., and H. E. CARLSON: Congenital hourglass bladder. J. Urol. (Baltimore) 51, 24—30 (1944).

Chapter VIII

ATCHESON, D. W.: Patent urachus: with a report of 2 additional cases. J. Urol. (Baltimore) 51, 424—430 (1944). — BEACH, E. W.: Urologic complications of cancer of the uterine cervix. J. Urol. (Baltimore) 68, 178—189 (1952). — CECIL, H. L., and J. H. HILL: Actinomycosis of the urinary organs. J. Amer. med. Ass. 78, 575—578 (1922). — CHERRY, J. W.: Patent urachus: review and report of a case. J. Urol. (Baltimore) 63, 693—697 (1950). — DEAN, A. L., and D. P. SLAUGHTER: Bladder injury subsequent to irradiation of uterus. J. Urol. (Baltimore) 46, 917—924 (1941). — FERRIS, H. W.: Tuberculosis of the bladder. J. Urol. (Baltimore) 25, 497—506 (1931). — HAGER, B. H.: Clinical data on alkaline incrusted cystitis. J. Urol. (Baltimore) 16, 447—457 (1926). — HATCH, W. E., and A. H. WELLS: Actinomycosis of the urinary bladder. J. Urol. (Baltimore) 52, 149—152 (1944). — HERGER, C. C.: Actinomycosis of the bladder. J. Urol. (Baltimore) 29, 739—743 (1933). — HIGGINS, C. C.: HUNNER's ulcer of the bladder. Urol. cutan. Rev. 34, 665—668 (1930). — KAUFMAN, L. R.: Use of the cystoscope in tuberculosis of the bladder. Urol. cutan. Rev. 35, 294—303 (1931). — LEBOWICH, J.: HODGKIN's disease of bladder. Amer. J. Cancer 30, 758—764 (1937). — MARQUARDT, C. R.: Blastomycosis of the genitourinary tract. J. Urol. (Baltimore) 35, 531—533 (1936). — MCCREA, L. E., and E. H. SPAULDING: Aerobic actinomycoses in urine. J. Urol. (Baltimore) 55, 428—434 (1946). — MCDONALD, H. P., W. E. UPCHURCH and C. E. STURDEVANT: Interstitial cystitis in children. J. Urol. (Baltimore) 70, 890—893 (1953). — MOULDER, M. K.: Thrush of urinary bladder. J. Urol. (Baltimore) 56, 420—426 (1946). — NITSCHKE, P. H.: Trichomonas infestation of bladder. West. J. Surg. 45, 278—281 (1937). — ORMOND, J. K.: Interstitial cystitis. J. Urol. (Baltimore) 33, 576—582 (1935). — POLLAK, W.: Simple ulcer of bladder. Z. urol. Chir. 39, 362—368 (1934). — PUGH, W. S.: Gonorrheal infections of the bladder and kidneys. Am. J. Surg. 9, 492—493 (1930). — RANDALL, A., and E. W. CAMPBELL: Alkaline incrusted cystitis. J. Urol. (Baltimore) 37, 284—299 (1937). — RHAMY, B. W.: Blastomycosis of bladder. J. Amer. med. Ass. 87, 405—406 (1926). — SAUER, H. R., and W. R. T. METZGER: Trush infection of the bladder: case report. J. Urol. (Baltimore) 59, 38—41 (1948). — SCHOLL, A. J.: Squamous cell changes and infection in the urinary tract. J. Urol. (Baltimore) 44, 759—767 (1940). — SMITH, G. G.: Bladder reactions following the application of radium to the uterus. New Engl. J. Med. 207, 822—823 (1932). — THOMPSON, G. J., and J. J. STEIN: Leukoplakia of the urinary bladder: a report of 34 clinical cases. J. Urol. (Baltimore) 44, 639—649 (1940). — THOMPSON, L. O.:

Syphilis, 2. edit. Philadelphia: Lea a Febiger 1920. — Syphilis of the bladder. Amer. J. Syph. 4, 50—90 (1920). — Young jr., E. L.: Leukoplakia of the bladder. J. Urol. (Baltimore) 18, 407—412 (1927).

Chapter IX

Abeshouse, B. S., and A. E. Goldstein: Primary carcinoma in a diverticulum of the bladder: A report of 4 cases and a review of the literature. J. Urol. (Baltimore) 49, 534—557 (1943). — Ballenger, E. G., O. F. Elder and H. P. McDonald: A case of cavernous hemangioma of bladder. Amer. J. Surg. 17, 409—412 (1932). — Barnes, R. W., and M. R. Hill: Intestino-vesical fistula. Calif. west. med. 56, 350—354 (1942). — Birnbaum, J.: Cystoscopic observations in cured gonorrhea. Urol. cutan. Rev. 46, 233—234 (1942). — Bowen, J. A., and G. A. Bennett: Solitary tuberculoma of bladder. Surg. Gynec. Obstet. 50, 1015—1017 (1930). — Boylan, R. N., L. F. Greene and J. R. McDonald: Epithelial neoplasms arising in diverticula of the urinary bladder. J. Urol. (Baltimore) 65, 1041—1049 (1951). — Burns, R. A.: Cystitis emphysematosa: a case report. J. Urol. (Baltimore) 49, 808—814 (1943). — Burrell, N. L.: Cystitis emphysematosa. J. Urol. (Baltimore) 36, 690—693 (1936). — Campbell, M. F.: Rupture of bladder. Surg. Gynec. Obstet. 49, 540—546 (1929). — Cauffield, E. W.: Dermoid cyst of bladder. J. Urol. (Baltimore) 75, 801—804 (1956). — Chrisholm, A. E. I., and A. R. Tudhope: Malakoplakia of urinary bladder. Edinb. med. J. 41, 626—629 (1934). — Christopherson, J. B., and R. O. Ward: Bilharzia disease in England. Brit. J. Surg. 21, 632—636 (1933). — Colby, F. H.: Embryonic rests of urinary bladder. Surg. Gynec. Obstet. 40, 528—530 (1925). — Collings, C. W., and F. Welebir: Osteoma of bladder. J. Urol. (Baltimore) 46, 494 —498 (1941). — Connery, D. B.: Leukoplakia of the urinary bladder and its association with carcinoma. J. Urol. (Baltimore) 69, 121—127 (1953). — Coppridge, W. M., L. C. Roberts and D. A. Culp: Glandular tumors of the bladder. J. Urol. (Baltimore) 65, 540—549 (1951). — Corbitt, R. W., A. C. Broders and T. L. Pool: Amyloidosis of the urinary bladder. J. Urol. (Baltimore) 52, 153—157 (1944). — Craig, L. G.: Cystitis cystica glandularis. J. Urol. (Baltimore) 42, 1197—1203 (1939). — Crane, A. R., and R. G. Tremblay: Primary osteogenic sarcoma of bladder. Ann. Surg. 118, 887—908 (1943). — Cristol, D. S., and A. C. Broders: Malacoplakia of bladder: report of 2 interesting cases. J. Urol. (Baltimore) 55, 260—266 (1946). — Cristol, D. S., and L. F. Greene: Gangrenous cystitis. Surgery 18, 343—346 (1945). — Culver, H., and W. F. Hoeppner: Vesical bilharziassi. J. Urol. (Baltimore) 27, 189—200 (1932). — Darget, R.: Deux cas d'herpes zoster de la vessie. J. d'Urol. 27, 229—231 (1929). — Davis, T. A.: Hamartoma of urinary bladder. Northw. Med. (Seattle) 48, 182—185 (1949). — Deming, C. L.: Complete urinary obstruction due to hydatid cyst. J. Urol. (Baltimore) 10, 1—43 (1923). — Ewell, G. H.: Nonspecific infected granulomata of bladder. J. Urol. (Baltimore) 41, 627—637 (1939). — Fagerstrom. D. P.: Perforation of urinary bladder by pelvic abscess. J. Urol. (Baltimore) 30, 207—220 (1933). — Fairlay, N. H.: Vesical schistosomiasis complicated by carcinoma. Brit. med. J. 2, 983—986 (1931). — Finestone, E. O.: Syphilis of bladder. Surg. Gynec. Obstet. 62, 93—113 (1936). — Fister, G. M., and A. J. Lund: Myosarcoma of the bladder. J. Urol. (Baltimore) 65, 401—407 (1951). — Fitzgerald, W. L., and M. A. R. Kuhn: Endometriosis of the bladder. J. Urol. (Baltimore) 62, 467—469 (1949). — French, A. J., and J. T. Mason: Malakoplakia of urinary bladder and sarcoidosis. J. Urol. (Baltimore) 66, 229—233 (1951). — Gemmell, H. D.: Cystoscopy in carcinoma of cervix. J. Obstet. Gynaec. Brit. Emp. 35, 465—497 (1928). — Grauer, T. P.: Leiomyoma of the bladder. J. Urol. (Baltimore) 40, 594—597 (1938). — Hayward, W. G.: Schistosomiasis japonicum with vesical involvement 37 years after infection. J. Urol. (Baltimore) 39, 722—726 (1938). — Heckel, N. J.: A study of the pathologic alterations in the female bladder and urethra resulting from infection with trichomonas vaginalis. J. Urol. (Baltimore) 35, 520—523 (1936). — Herger, C. C.: Actinomycosis of bladder. J. Urol. (Baltimore) 29, 739—743 (1938). — Higgins. C. C.: Primary lymphosarcoma of the urinary bladder. J. Urol. (Baltimore) 62, 317—321 (1949). — Hinman, F., and J. Cordonnier: Cystitis follicularis. J. Urol. (Baltimore) 34, 302 (1935). — Hoyt, H. S.: Cystitis cystica as a single tumor. J. Urol. (Baltimore) 59, 424—430 (1948). — Hurwitz, S. P., E. B. Jacobson and H. H. Ottenstein: Mucoid adenocarcinoma of the urachus invading bladder. J. Urol. (Baltimore) 65, 87—92 (1951). — Hyams, J. A., and I. Botvinick: Pemphigus vulgaris of urinary bladder. Amer. J. Surg. 56, 594—597 (1942). — Katzen, P.: Leiomyosarcoma of bladder: Report of case. J. Urol. (Baltimore) 67, 518—522 (1952). — Kirkaldy Willis, W. H.: Cystoscopy in diagnosis and treatment of bilharzia haematobium infection. Brit. J. Surg. 34, 189—194 (1946). — Kleitsch. W. P.: Parasitic fibromyoma of the bladder. J. Urol. (Baltimore) 65, 60—65 (1951). — Kretschmer. H. L.: The pathology and cystoscopy of cystitis cystica. Surg. Gynec. Obstet. 7, 274—279 (1908). — Endometriosis of the bladder. J. Urol. (Baltimore) 53, 459—465 (1945). — Lane, J. W., and P. Franke: Cystitis emphysematosa: case reports. J. Urol. (Baltimore) 75, 256—206 (1956). — Lower, W. E., and J. I. Farrell:

Appendicovesical fistula. Ann. Surg. **93**, 628—630 (1931). — LUND, H. G., F. G. ZINGALE and J. A. O'DOWD: Cystitis emphysematosa. J. Urol. (Baltimore) **42**, 684 (1939). —MAKAR, N.: Cystoscopic appearances of bilharziosis of the bladder. Brit. J. Urol. **4**, 209—216 (1932). — MARQUARDT, C. R.: Cystitis emphysematosa. Urol. cutan. Rev. **44**, 295—296 (1940). — MOORE, T. D., A. L. HERRING and D. A. McCANNEL: Some urologic aspects of endometriosis. J. Urol. (Baltimore) **49**, 171—177 (1943). — MORROW jr., R. P., L. B. WOOLNER and J. L. EMMETT: Metastatic melano-epithelioma of the urinary bladder: Report of a case. J. Urol. (Baltimore) **67**, 92—94 (1952). — MOSTOFI, F. K., and W. H. MORSE: Polypoid rhabdomyo-sarcoma (sarcoma botryoides) of bladder in children. J. Urol. (Baltimore) **67**, 681—687 (1952). — NEMSER, M. M., and H. A. WEINBERGER: Spontaneous rupture of the urinary bladder in a male. J. Urol. (Baltimore) **68**, 603—607 (1952). — NESBIT, R. M.: Is cystitis cystica an innocent or a malefic lesion? J. Urol. (Baltimore) **75**, 443—447 (1956). — NEWMAN, H. R.: transurethral surgery in relation to bilharziosis of the bladder. J. Urol. (Baltimore) **50**, 440 to 445 (1943). — NOURSE, M. H.: Primary osteogenic sarcoma of bladder. J. Urol. (Baltimore) **77**, 634—638 (1957). — OCKULY, E. A.: Bilharziasis of bladder. J. Urol. (Baltimore) **54**, 39—45 (1945). — ORMOND, J. K., and J. G. HEMMING jr.: Gumma of bladder: report of case. J. Urol. (Baltimore) **52**, 23—26 (1944). — ORTMAYER, MARIE: Cystitis emphysematosa: with a report of the 12th human case diagnosed as cystoscopy. J. Urol. (Baltimore) **60**, 757—762 (1948). — PESQUEIRA, M., and R. L. ENGELKING: Urachus and bladder actinomy-cosis: Presentation of one clinical case. J. Urol. (Baltimore) **62**, 163—167 (1949). — RATH-BUN, N. P.: Primary bladder tumore in infants a young children with a report of a case of hemangioma in a male child 21 month of age. Surg. Gynec. Obstet. **64**, 914—918 (1937). — REDEWILL, F. H.: Cystitis cystica emphysematosa. Urol. cutan. Rev. **38**, 537—543 (1934). — Malakoplakia of the urinary bladder and generalized sarcoidosis striking similarity of their pathology, etiology, gross appearance and methods of treatment. J. Urol. (Baltimore) **49**, 401—407 (1943). — ROEN, P. R., and J. WIENER: Primary amyloid tumor of the bladder: report of a case. J. Urol. (Baltimore) **66**, 119—121 (1951). — ROLNICK, D. F., and A. B. RAGINS: Malakoplakia of bladder. J. Urol. (Baltimore) **42**, 108—117 (1939). — SAUER, H. R., and M. S. BLICK: Cystitis glandularis: a consideration of symptomes diagnosis a clinical course of the disease. J. Urol. (Baltimore) **60**, 446—458 (1948). — SENGER, F. L., M. W. THOMLEY and R. G. McMANUS: Primary amyloidosis of the bladder. J. Urol. (Baltimore) **63**, 790—793 (1950). — SHIH, H. E., and G. Y. CHAR: Dermoid cyst ruptured into urinary bladder. J. Urol. (Baltimore) **38**, 165—172 (1937). — SMITH jr., B. A.: Palypoid rhabdo-myosarcoma of bladder: Sarcoma botryoides. Minn. Med. **38**, 215—219 (1955). — SMYR-NIOTIS, P. C.: Vessie bilharzienne calcifiee dilatation des ureteres et tumeur greffee sur la vessie. J. Radiol. Electrol. **21**, 489—493 (1937). — STIRLING, W. C.: Cystitis fallicularis. J. Amer. med. Ass. **112**, 1326—1331 (1939). — STIRLING. W. C., and J. E. ASH: Chronic proliferative lesions of urinary tract. J. Urol. (Baltimore) **45**, 342—360 (1941). — STERLING. W. C., and G. A. HOPKINS: Gangrene of bladder; review of 207 cases; report of two personal cases. J. Urol. (Baltimore) **31**, 517—525 (1934). — THOMPSON, L. O.: Syphilis, Secondaries like acute hemorrhagic cystitis. Philadelphia: Lea a. Fiebiger 1920. — Syphilis of the genital organs of the male and the urinary organs. Pamphlet. Collection of all papers by LOYD THOMPSON. — TRUC, GUILLAUME and BAUMEL: Disease of the neck of the bladder in neurological disease; necessity of urethrocystoscopy. J. d'Urol. **61**, 424—425 (1955). — WEAR, J. B.: End results of tuberculous cystitis. Arch. Surg. **37**, 821—826 (1938). — WEINER, I., L. BURKE and J. SHERMAN: Vesical calculi originating from penetrating calcified fibromyoma of the uterus. J. Urol. (Baltimore) **63**, 625—629 (1950). — WELLS, H. S.: Report of seven cases of cystitis emphysematosa. J. Urol. (Baltimore) **39**, 391—397 (1938). — WISHARD jr., W. M. N.: Surgical injuries of the ureter and bladder. J. Urol. (Baltimore) **73**, 1009—1014 (1955). — WYMAN, H. E., B. S. CHAPPELL and W. R. JONES jr.: Ganglioneuroma of bladder: report of a case. J. Urol. (Baltimore) **63**, 526—532 (1950). — YOUNGBLOOD, V. H., R. BANKS jr. and E. E. DENNEY: Rhabdomyosarcoma of the bladder. J. Urol. (Baltimore) **67**, 957—961 (1952). — ZIMMERMAN, E. L., and C. S. LEVY: Routine examination of bladder in secondary syphilis. J. Urol. (Baltimore) **3**, 407—410 (1919).

Chapter X

COUNCILL, W. A., and W. A. COUNCILL jr.: An aid to bladder clot evacuation. J. Urol. (Baltimore) **60**, 792—793 (1948). — FOWELL, A. H., and E. B. McLEAN: Effect of several urological irrigating fluids upon the clotting time of plasma and tensile strength of fibrin clots. J. Urol. (Baltimore) **73**, 888—890 (1955). — GARSHIVILES, W. P.: Foreign bodies in bladder: 5 cases. Amer. J. Surg. **22**, 199 (1933). — JOLY, J. S.: Stone and calculous disease of the urinary organs. St. Louis: C. V. Mosby Comp. 1940. — LETT, H.: On urinary calculus, with special reference to stone in the bladder. Brit. J. Urol. **8**, 205—232 (1936). — PETREN, G.: Sutures migrating into the bladder. Z. Urol. **24**, 748 (1930). — ROBERTS, G. M.: Prevention of blood clots in the bladder. J. Urol. (Baltimore) **52**, 563 (1944). — SMITH jr., E. P., W. H. TOULSOA and W. B. TENER: Simple method of evacuating blood clots from urinary bladder with streptokinase-streptodornase. Surg.-Gynec. Obstet. **96**, 171 (1953). — WINSBURY-WHITE, H. P.: Stone in urinary tract, 2. edit. St. Louis: C. V. Mosby Comp. 1954.

Chapter XI

BALDRIDGE, R. R.: A case of congenital hypertrophy of verumontanum. New Engl. J. Med. **213**, 46—49 (1935). — COLBY, F. H.: Varix of urethra with hematuria. New Engl. J. Med. **203**, 1011 (1930). — DAVIES, J. A.: Echinococcus cyst arising from prostate. Canad. med. Ass. J. **54**, 268—271 (1946). — EMMETT, J. L.: Obstruction of the vesical neck of a male infant produced by hypertrophy of the verumontanum: report of case. Proc. Mayo Clin. **15**, 364—366 (1940). — HARKNESS, A. J., and A. KING: Staphylococial infections of genital tract in male. Brit. J. Urol. **10**, 379—391 (1938). — HERTOGHE, C.: Peculiarities of vesical neck in MARION's disease. J. belge Urol. **15**, 774—775 (1947). — KEARNS, W. M., and E. B. JACOBSON: A pediatric-urologic problem: congenital valves of posterior urethra with case report. Wis. med. J. **39**, 603—606 (1940). — LATTIMER, J. K.: Tuberculous prostatic urethritis: A suggestive diagnostic sign. J. Urol. (Baltimore) **59**, 326—327 (1948). — LAZARUS, J. A.: Importance of routine cystoscopy in prostatism. J. Urol. (Baltimore) **55**, 79—83 (1946). — LICH jr., R., and J. E. MAURER: Vesical evidence of posterolateral spinal cord disease. J. Urol. (Baltimore) **66**, 213—217 (1951). — NATANSON, H.: Cystoscopy in nervous disease with special reference to SCHRAMM's spincter phenomenon. Z. Urol. **21**, 821—833 (1927). — PENA, A. DE LA , and E. DE LA PENA: Diverticular or cavitary chronic prostatitis. J. Urol. (Baltimore) **55**, 273—277 (1946). — SMITH, L. O.: Lymphosarcoma of prostate. J. Urol. (Baltimore) **38**, 375—382 (1937). — STEGEMAN, W.: Misinterpretation of enlarged prostates gained through cystoscopy. J. Urol. (Baltimore) **57**, 479—483 (1946).

Chapter XII

BRANNAN, D.: Stricture of the female urethra. J. Urol. (Baltimore) **66**, 242—253 (1951). — COOK, E. N.: Diverticulum of female urethra; Problems in diagnosis and treatment. Surg. Gynec. Obstet. **99**, 273—276 (1954). — FAGAN, G. E., and A. T. HERTIG: Carcinoma of female urethra: Review of literature; report of eight cases. Obstet. and Gynec. **6**, 1—11 (1955). — HAHN, G. A.: Primary carcinoma of the female urethra. J. Urol. (Baltimore) **67**, 319—325 (1952). — MOORE, T. D.: Diverticulum of female urethra: An improved technique of surgical excision. J. Urol. (Baltimore) **68**, 611—616 (1952). — NELSON, N. M., R. W. BARNES, H. L. HADLEY and R. T. BERGMAN: Transurethral resection of bladder neck in female. J. Urol. (Baltimore) **77**, 198—213 (1957). — RINGLEB, O.: Über die Möglichkeit, Tiefenerstreckungen in der Menschenblase sicher und richtig wahrzunehmen. Z. Urol. **33**, 65—73 (1939). — WISHARD jr., W. N., and M. H. NOURSE: Carcinoma in diverticulum of female urethra. J. Urol. (Baltimore) **68**, 320—323 (1952).

Chapter XIII

COUTTS, W. E.: Genito-urinary lesions in lymphogronuloma venerum. J. Urol. (Baltimore) **49**, 595—599 (1942). — GAILEY, H. A., and J. W. BEST: Primary carcinoma of the male urethra: Report of two cases. J. Urol. (Baltimore) **62**, 507—512 (1949). — HARA, H. J., and ROSENVOLD, L. K.: Cicatricial atresia of esophagus. Arch. Otolaryng. (Chicago) **34**, 574—582 (1941). — HARKNESS, A. H.: Non-gonococcal urethritis. London: E. & S. Livingstone 1950. — KLEIMAN, A. H.: A new sigmo-ureterostomoscope for treatment of implanted ureter. J. Urol. (Baltimore) **67**, 164—167 (1952). — KHOURY, E. N.: Diverticulum of the male urethra. J. Urol. (Baltimore) **69**, 291—298 (1953). — LAZARUS, J. A., and A. A. ROSENTHAL: Primary tuberculosis of the penis. J. Urol. (Baltimore) **35**, 361—377 (1936). — LEADBETTER, W. F.: Instrumental visualization of the renal pelvis at operation as an aid to diagnosis. Presentation of a new instrument. J. Urol. (Baltimore) **63**, 1006—1012 (1950). — SEELIG, A.: Ein Fall von Urethritis tuberculosa. Mber. d. Krankh. d. Harn. Sexualappar. **2**, 217—219 (1897). — TRATTNER, H. R.: Instrumental visualization of the renal pelvis and its communications: Proposal of a new method; preliminary report. J. Urol. (Baltimore) **60**, 817—837 (1948).

Chapter XIV

ADAMS, P. S.: Ureterocele, treatment by transurethral resection. Amer. J. Surg. **50**, 249 (1940). — ALCORN, K. A.: Cystoscopic management of ureteral calculi. Urol. cutan. Rev. **51**, 373—375 (1947). — ALYEA, E. P.: Cystoscopic removal of large ureteral calculi. J. Urol. (Baltimore) **40**, 83—100 (1938). — ANGLE, E. E., and L. F. PFEIFER: The bold excision of strictures of the urethra with the small resectoscope. J. Urol. (Baltimore) **67**, 695—697 (1952). — BALKUS, V. A.: Looped catheter in treatment of ureteral calculi. J. Urol. (Baltimore) **50**, 667—672 (1943). — BARRINGER, B. S.: Radium therapy of bladder cancer, retrospect and prospect. J. Urol. (Baltimore) **68**, 280—282 (1950). — Twenty-five years of radon treatment of cancer of bladder. J. Amer. med. Ass. **135**, 616—618 (1947). — BEATTY, R. P.: Cystoscope as an aid in office urology. Urol. cutan. Rev. **50**, 737—739 (1946). — BENEVENTI, F. A., and F. S. CREIGHTON: A complication following use of a ureteral stone extractor. Amer. J. Surg. **89**, 1086—1087 (1955). — BERRY, N. E., and R. C. BURR: Intracavity irradiation of carcinoma of bladder. Canad. med. Ass. J. **75**, 15 (1956). — BERRY, N. E., and J. D. HAMILTON: Diagnosis of bladder tumors. J. Urol. (Baltimore) **64**, 464—468 (1950). — BODNER, H.: Migratory foreign bodies in bladder. Urol. cutan. Rev. **52**, 457—460 (1948). — BROWNE, H. S.: Improved basket bougie for extraction of ureteral calculi. J. Urol. (Baltimore) **75**, 55—56 (1956). — CHAPMAN, T. L., and J. W. SUTHERLAND: The clinical significance of biopsy examination of bladder tumors. Brit. J. Urol. **26**, 369—374 (1954). — COLLINGS, C. W.: Treatment of impotence and premature ejaculation. Trans. west. Sect. Amer. urol. Ass. **13**, 97—100 (1946). — COPPRIDGE, W. M., L. C. ROBERTS and R. G. ROSSER jr.: Operative procedures within bladder conducted directly through the urethra outside the cystoscope. J. Urol. (Baltimore) **63**, 630—637 (1950). — COUNCILL, W. A.: Treatment of ureteral calculi: a report of 504 cases in which Councill stone extractor and dilator was used. J. Urol. (Baltimore) **53**, 534—538 (1945). — DARGET, and DE CASTELMUR: Value of loop catheter in the extraction of calculi from lower third of ureter. J. Urol. méd. chir. **61**, 446—447 (1955). — DAVIS, E., L. W. LEE and E. DAVIS jr.: Transurethral endovesical ureterolithotomy with resectoscope. J. Urol. (Baltimore) **67**, 634—636 (1952). — DAVIS, T. A.: Removal of ureteral calculus by a new catheter type extractor. J. Urol. (Baltimore) **72**, 346—349 (1954). — DEAN jr., A. T., and J. E. ASH: The disadvantages of the electroresectoscope for bladder tumor biopsy. J. Urol. (Baltimore) **72**, 652—655 (1954). — DOURMASHKIN, R. L.: Cystoscopic treatment of stones in the ureter with special reference to large calculi; based on study of 1550 cases. J. Urol. (Baltimore) **54**, 245—283 (1945). — ELLIK, M.: Cystoscopic transureteral extraction of a stone located in renal pelvis. J. Urol. (Baltimore) **56**, 46—48 (1946). — Stones in ureter: then extracted by looped catheter. J. Urol. (Baltimore) **57**, 473—478 (1947). — ELLIK, M., and J. GETZ: Ureteral stone in a child extraction by looped catheter. J. Urol. (Baltimore) **70**, 716—719 (1953). — FOLEY, F. C. B.: A new method for cystoscopic surgical treatment of ureterocele. Urol. cutan. Rev. **35**, 49 (1931). — GARVEY, F. K., and H. J. BRADLEY: Cystoscopic management of stones in lower third of ureter. Urol. cutan. Rev. **52**, 697—700 (1948). — HERBST, R. H., and J. W. MERRICKS: Transurethral drainage of the seminal vesicles in seminal vesiculitis. Illinois med. J. **86**, 190—195 (1944). — HOBERG, J. E.: Removal of large medicine dropper from the bladder by means of cystoscope. Urol. cutan. Rev. **50**, 620—621 (1946). — HUNNER, G. L., and L. R. WHARTON: Pathologic findings in cases clinically diagnosed as ureteral stricture. J. Urol. (Baltimore) **15**, 57—91 (1926). — IMBERT, M.: Corkscrew catheter (Chevassu) for extraction of ureteral calculi. J. d'Urol. **54**, 223—228 (1948). — IWANO, J. H., and R. C. BUNTS: Complications arising from transurethral manipulation of ureteral calculi. J. Urol. (Baltimore) **70**, 708—715 (1953). — JOELSON, J. J.: Instrumental removal of wax foreign bodies from bladder. J. Urol. (Baltimore) **64**, 572 (1950). — JOHNSON, F. P.: New method of removing ureteral calculi. J. Urol. (Baltimore) **37**, 84—89 (1937). — KIEFER, J. H.: Manipulation of ureteral calculi; meatotomy and immediate delivery. J. Urol. (Baltimore) **72**, 644—645 (1954). — KRUGMAN, P. I., and C. RIESER: Thermometer in the urinary bladder. Fertil. and Steril. **3**, 263—265 (1952). — McCARTHY, J. F.: Recent advances-instrumental urology. J. Urol. (Baltimore) **33**, 303—309 (1935). — McKAY, R. W.: A ureteral stone dislodger. J. Amer. med. Ass. **95**, 794 (1930). — MILLIN, T.: Cystoscopic surgery of ureter. Irish J. med. Sci. **1932**, 36—38. — MILNER, W. A.: Transurethral biopsy, accurate method of determining true malignancy of ca. — MOORE, T. D.: Carcinoma of the bladder; an improved technique for the cystoscopic implantation of radium element. J. Urol. (Baltimore) **51**, 496—504 (1944). — NATION, E. F.: Facilitation of passage and ectraction of ureteral calculi. J. Urol. (Baltimore) **70**, 373—375 (1953). — NESBIT, R. M.: Litholapaxy. J. Urol. (Baltimore) **70**, 594—599 (1953). — O'CONOR, V. J., and J. K. SOKOL: Vesicovaginal fistula from the standpoint of urologist. J. Urol. (Baltimore) **66**, 579—585 (1951). — ORMOND, J. K.: Endoscopic treatment of uretrotrigonitis in women. Urol. cutan. Rev. **50**, 645—647 (1946). — PETERSON, A.: Vesicovaginal fistulas treated by electrocoagulation. Amer. J. Surg. **17**, 247—248 (1932). — ROOME, N. W., and R. H. FLETT: Biopsy of intra-ureteral tumors by endoscopic means. Brit. J. Urol. **23**, 23—28 (1951). — RUSCHE, C. F., and S. K. BACON: Injury of the ureter due to cystoscopic intraureteral instrumentation: report of 16 cases. J. Urol. (Baltimore) **44**, 777—793 (1940). — SCHLOSS, W. A., and M. SOLOMKIN: Foreign body in the bladder; removal of thermometer with Stern McCarthy resectoscope. J. Amer. med. Ass. **143**, 804—805 (1950). — SCHULTE, W. G.: A novel way of ridding the bladder of paraffine. J. Urol. (Baltimore) **34**, 313 (1935). —

SEMPLE, J. E.: Papillomata of bladder treated with podophyllin. Brit. med. J. 1, 1235—1237 (1948). — TORRES, P. P.: Foreign body in the bladder. Rev. argent. Urol. 24, 380—382 (1955). — TWINEM, F. P.: Surgical removal of vesical calculus. J. Urol. (Baltimore) 41, 360 (1939). — VLIETSTRA, H. P.: A method for treatment of higher ureteric stenosis by means of catheter-led plastic dilating sounds. Urol. int. (Basel) 1, 47—62 (1955). — WINSBURY-WHITE, H. P.: Removal of ureteric stone by transcystoscopic manipulation. Lancet 1931, 793—794. — Slitting of ureteric orifice cystoscopically for impacted calculus. Brit. med. J. 2, 136 (1954).

Chapter XV

BARNES, R. W.: Instruments and technique for transurethral prostatic resection. Urol. cutan. Rev. 37, 637—639 (1933). ~ Evacuation with transurethral prostatic resection. Urol. cutan. Rev. 37, 106—107 (1933). — The necessity of meticulous technique in transurethral prostatic resection. Med. J. Australia 1, 207—209 (1955). — BAUMRUCHER, G. O.: Prostatic resection in vitro and in vivo. J. Urol. (Baltimore) 49, 660—664 (1943). — CREEVY, C. D.: Observations of hemolysis during transurethral resection: effects of urea. J. Urol. (Baltimore) 68, 324—328 (1952). — CUMMINGS, E. F.: Knee-action cord holder. J. Urol. (Baltimore) 71, 239 (1954). — FERGUSON, C., and C. D. MILLER: Heat factor as a cause of hemoglobinemia in transurethral resections. J. Urol. (Baltimore) 69, 128—134 (1953). — GARSKE, G. L.: A sterile, disposable, plastic, waterproof drape for transurethral surgeryand cystoscopic procedures. J. Urol. (Baltimore) 69, 135—141 (1953). — GRIFFIN, M., L. DOBSON, and J. C. WEAVER: Volume of irrigating fluid transfer during transurethral prostatectomy studied with radioisotopes. J. Urol. (Baltimore) 74, 646—651 (1955). — HAGSTROM, R. S.: Studies on fluid absorption during transurethral prostatic resection. J. Urol. (Baltimore) 73, 852—859 (1955). — KRAMER, S. E.: A new interchangeable tumor resecting instrument. J. Urol. (Baltimore) 72, 267—268 (1954). — MALUF, N. S. R.: Absorption of water, urea, glucosa, and electrolytes through human bladder. J. Urol. (Baltimore) 69, 396—404 (1953). — NESBIT, R. M.: Advantages of perineal urethrotomy in prostatic resection. Sth. Surg. 7, 501—504 (1938). — NICOLAI, C. H., and J. J. CORDONNIER: Serum electrolyte changes during transurethral prostatic resections. J. Urol. (Baltimore) 74, 118—122 (1955). — PITTS, H. H., and F. HINMAN jr.: Safety of water irrigation in transurethral prostatectomy. J. Urol. (Baltimore) 72, 925—927 (1954). — SCHULTE, T. L., H. J. HAMMER and L. R. REYNOLDS: Clinical use of cytal in urology. J. Urol. (Baltimore) 71, 656—659 (1954).

Chapter XVI

BELT, A. E., D. A. CHARNOCK, A. W. FOLKENBERG and R. A. FALCONER: The current activator in electrosurgical resection of the prostate. Urol. cutan. Rev. 37, 687—692 (1933). — BLECH, G. M.: Clinical electrosurgery. New York, Toronto and London: Oxford University Press 1938. — CLARK, W. L., E. J. ASNIS and J. D. MORGAN: Clinical and histological observations in the treatment of neoplastic diseases by combined methods. Atlantic med. J. 27, 541—549 (1923). — CLARK, W. L., J. D. MORGAN and E. J. ASNIS: Endothermic methods in treatment of neoplasms and other lesions with clinical and histological observations. Radiology 2, 233—246 (1924). — KELLY, H. A.: Advances in electrosurgery. I. Arch. phys. Ther. 12, 461—463 (1931). — KELLY, H. A., and G. E. WARD: Electrosurgery. Philadelphia: W. B. Saunders Comp. 1932. — WARD, G. E.: Advances in electrosurgery. II. Arch. phys. Ther. 12, 463—476 (1931).

Chapter XVII

ADAMS, P.: Prostatic obstruction complicated by diverticula of bladder. J. Urol. (Baltimore) 49, 558—574 (1943). — ALBUQUERQUE, P. F. DE, and E. TORRES: Occult carcinoma of prostate: clinical and pathologic considerations. Rev. bras. Med. 12, 520—523 (1955). — ANTONIO jr., D.: The operative management of hypertrophy of prostate with complicating coronary heart disease. J. Urol. (Baltimore) 50, 344—354 (1943). — BAKER, R.: Correlation of circumferential lymphatic spread of vesical cancer with depth of infiltration; relation to present methods of treatment. J. Urol. (Baltimore) 73, 681—690 (1955). — BARNES, R. W., and R. T. BERGMAN: Transurethral diverticulotomy: preliminary report. Urol. cutan. Rev. 42, 15—17 (1938). — BRUNKOW, C. D.: Evaluation of size of bladder neoplasms. J. Urol. (Baltimore) 70, 234—236 (1953). — EMMETT, J. L., and J. B. BEARE: Transurethral resection for vesical dysfunction in cases of tabes dorsalis. J. Amer. med. Ass. 136, 1093—1096 (1948). — GRAHAM, J. B., and G. J. BULKLEY: Angioma of bladder. J. Urol. (Baltimore) 74, 777—779 (1955). — GREENBERGER, M. E., and J. H. WINER: Transurethral resection for fibroadematous hypertrophy of prostate in tuberculous patients. Quart Bull. Seas View Hosp. 5, 161—166 (1940). — HECKENBACH, W.: Die Behandlung der Blasengeschwülste. Z. Urol.

33, 268—287 (1939). — KIEFER, J. H.: Bladder tumor recurrence in urethra; a warning. J. Urol. (Baltimore) **69**, 652—656 (1953). — KIRWIN, T. J., and G. A. HAWES: Diagnostic value of residual urine estimation. J. Urol. (Baltimore) **41**, 413—430 (1939). — MALAMENT, M., and R. C. BUNTS: Transurethral resection and neurogenic bladder. Virginia med. Monthly **78**, 243—246 (1949). — MARSHALL, V. F.: Relation of preoperative estimate to pathologic demonstration of extent of vesical neoplasms. J. Urol. (Baltimore) **68**, 714—723 (1952). — MELICOW, M. M.: Tumors of urinary bladder; a clinico-pathological analysis of over 2500 specimens and biopsies. J. Urol. (Baltimore) **74**, 498—521 (1955). — MIDDLETON, R. P.: HENRY JACOB BIGELOW and his operation; a major contribution to modern urology comes finally into its own. J. Urol. (Baltimore) **49**, 883—888 (1943). — MOORE, T. D., A. L. HERRING and D. A. McCANNELL: Some urological aspects of endometriosis. J. Urol. (Baltimore) **49**, 171—177 (1943). — O'BRIEN, H. A., J. D. MITCHELL and E. C. MARTIN: Transurethral resection of large prostates. J. Urol. (Baltimore) **62**, 225—230 (1949). — PEYTON, A. B.: Bladder neck obstruction in young male adult. J. Urol. (Baltimore) **69**, 109—117 (1953). — SHIVERS, C. H. DE T.: Medical findings in benign prostatic hyperplasia; a new method of grouping cases for operation. J. Urol. (Baltimore) **49**, 847—856 (1943). — THOMPSON, G. J., and J. H. KAPLAN: Advantages of transurethral removal of certain bladder tumors. J. Urol. (Baltimore) **73**, 270—279 (1955). — WESSELL, M. S., A. B. KURITZ, R. A. BURGER and G. W. REAGAN: Mucinous carcinoma of urachus invading bladder. J. Urol. (Baltimore) **67**, 523—525 (1952).

Chapter XVIII

BERGMAN, R. T., R. TURNER, R. W. BARNES and H. L. HADLEY: Comparative analysis of 1000 consecutive cases of transurethral prostatic resection. J. Urol. (Baltimore) **74**, 533 to 545 (1955). — BUMPUS jr., H. C., and B. D. MASSEY: Transurethral resection: Does it require as exacting a preoperative preparation as prostatectomy? Calif. west. Med. **46**, 89—92 (1937). — CREEVY, C. D., and M. J. FEENEY: Routine use of antibiotics in transurethral prostatic resection: A clinical investigation. J. Urol. (Baltimore) **71**, 615—623 (1954). — FERRIER, P. A.: Personal communications. — GAUDIN, H. J., H. A. ZIDE and G. J. THOMPSON: Use of sulfanilamide after transurethral prostatectomy. J. Amer. med. Ass. **110**, 1887 to 1890 (1938). — GRAVES, C. L., F. M. SELLERS and M. KARP: Study of anesthesia for 1176 transurethral prostatectomies. J. Amer. med. Ass. **156**, 1045, 1048 (1954). — KREUTZMANN, H. A. R.: Improved suprapubic trocar and cannula. J. Urol. (Baltimore) **40**, 341—342 (1938). — LAWSON, J. D., A. L. SCHNEEBERG and W. B. TOMLINSON: Observations of dynamics of acute urinary retention in man. J. Urol. (Baltimore) **67**, 951—956 (1952). — LUNDY, J. S.: Choice of anesthetic methods for urologic operations in various age groups. J. Urol. (Baltimore) **67**, 745—749 (1952). — O'HEERON, M. K., A. W. MILES and M. G. RAPE: Intraprostatic local anesthesia for transurethral prostatic resections. J. Urol. (Baltimore) **62**, 231—244 (1949). — RAINES, S. L., and R. POPPER: Anesthesia in prostatic surgery. Sth. Med. Surg. **47**, 1092—1093 (1954). — WISHARD, W. N., H. G. HAMER and H. O. MERTZ: Local infiltrative anesthesia of prostate preliminary to resection. J. Amer. med. Ass. **102**, 32—35 (1934).

Chapter XIX

BARNES, R. W.: Method and rhythm in transurethral prostatic resection. J. Urol. (Baltimore) **65**, 603—607 (1951). — BARNES, R. W., and R. T. BERGMAN: Transurethral diverticulotomy; preliminary report. Urol. cutan. Rev. **42**, 15—17 (1938). — BARNES, R. W., R. T. BERGMAN and S. FARLEY: Histopathological study of prostatic tissue following endoscopic prostatic resection. J. Urol. (Baltimore) **57**, 755—757 (1947). — BARNES, R. W., and I. E. MARTIN: Endoscopic identification of tissue during transurethral prostatic resection. J. Urol. (Baltimore) **62**, 730—735 (1949). — BAURYS, W.: Hemostasis in transurethral resections: the use of adrenalinhyaluronidase mixture. J. Urol. (Baltimore) **66**, 265—269 (1951). — BLEAKNEY, P. A., and S. J. PACKARD: Factors that expedite transurethral prostatic resection. J. Urol. (Baltimore) **76**, 115—122 (1956). — CREEVY, C. D.: Intraprostatic injection of pitressin and adrenalin in control of bleeding during transurethral resection — a preliminary report. J. Urol. (Baltimore) **50**, 593—596 (1943). — McDONALD, H. P.: Care of patient following prostatic resection. Sth. med. J. (Bgham, Ala.) **38**, 260—264 (1945). — NESBIT, R. M.: Advantages of perineal urethrotomy in prostatic resection. Sth. Surg. **7**, 501—504 (1938). — Transurethral prostatic resection: a discussion of some principles and problems. J. Urol. (Baltimore) **66**, 362—372 (1951). — REYNOLDS, L. R., T. L. SCHULTE and H. J. HAMMER: Bladder tumors — clinical evaluation of radical transurethral management. J. Urol. (Baltimore) **61**, 912—916 (1949). — SENGER, F. L., J. J. BOTTONE and S. H. ROTHFELD: Bladder diverticulum in female. J. Urol. (Baltimore) **68**, 699—702 (1952). — THOMPSON, G. J.: Simultaneous litholapaxy and prostatic resection. Proc. Mayo Clin. **10**,

689—692 (1935). — WEAR, J. B.: Some observations on the technique of transurethral prostatic resection. J. Urol. (Baltimore) **62**, 470—473 (1949). — WEYRAUCH, H. M., and M. L. ROSENBERG: Avoiding poor results in transurethral prostatectomy. Trans. Amer. Ass. gen.-urin. Surg. **45**, 166—179 (1953).

Chapter XX

BARNES, R. W.: Endoscopic prostatic surgery. St. Louis: C. V. Mosby Comp. 1943. — BUMPUS jr., H. C.: Advantages of punch resection of removing obstructing portions of prostate. J. Urol. (Baltimore) **38**, 322—326 (1937). — EMMETT, J.: Transurethral resection with cold punch operative technique. J. Urol. (Baltimore) **49**, 815—839 (1943). — NESBIT, R. M.: Transurethral prostatectomy. Springfield: Ch. C. Thomas 1943.

Chapter XXI

BAIRD, S. S., and H. M. SPENCE: Transurethral resection of female bladder neck; a rebuttal. Urol. cutan. Rev. **52**, 658—661 (1948). — BARNES, R. W.: Endoscopic prostatic surgery. St. Louis: C. V. Mosby Co., 1943, p. 75. — CAULK, J. R.: Contracture of the vesical neck in female. J. Urol. (Baltimore) **6**, 341—343 (1921). ~ Obstruction at bladder neck in men, women and children. Int. Clin. **4**, 136—148 (1937). — CAULK, J. R., and J. F. PATTON: Cautery punch operation for removal of obstructive lesions at vesical orifice in women and children. J. Urol. (Baltimore) **33**, 504—520 (1935). — COUTTS, W. E., and R. VARGAS-ZALAZAR: Contribution to etiology of acquired fibrosis of bladder neck. Brit. J. Urol. **17**, 136—141 (1945). — DAVIS, D. M.: Vesical orifice obstruction in women and its treatment by transurethral resection. J. Urol. (Baltimore) **73**, 112—116 (1955). — EMMETT, J. L., S. P. R. HUTCHINS and J. R. MCDONALD: Treatment of urinary retention in women by transurethral resection. J. Urol. (Baltimore) **63**, 1031—1041 (1950). — ENGLES, C. F.: Bladder neck resection in women and children. Trans. west. Sect. Amer. urol. Ass. **12**, 94—95 (1944). — EVATT, E. J.: Contribution to development of prostate gland in human female and a study of homologies of urethra and vagina of the sexes. J. Anat. Physiol. (Lond.) **45**, 122—130 (1910). — EVERETT, H. S.: Condemnation of resectoscopic procedures upon female vesical neck. Urol. cutan. Rev. **52**, 121—124 (1948). — FITE, E. H.: Vesical neck obstruction in female treated by resection with McCarthy resectoscope. Urol. cutan. Rev. **38**, 163—164 (1934). — FOLEY, F. E. B.: Diagnosis and classification of various forms of bladder neck obstruction. Minn. Med., **12**, 137—145 (1929). — FOLSOM, A. I.: Female urethra; A clinical and pathological study. J. Amer. med. Ass. **97**, 1345—1351 (1931). — FOLSOM, A. I., and J. C. ALEXANDER: Referred pain from female urethra. J. Urol. (Baltimore) **31**, 731—739 (1934). — FOLSOM, A. I., and H. A. O'BRIEN: Female obstructing prostate. J. Amer. med. Ass. **121**, 573—580 (1943). — Female urethra. J. Amer. med. Ass. **128**, 408—414 (1945). — HICKS, J. B.: Bladder neck contracture in women. Surg. Clin. N. Amer. **14**, 1219—1223 (1934). — HOOVER, S. R.: Prostatism in women. Urol. cutan. Rev. **50**, 326—328 (1946). — HOUTUM, G. VAN: Six cases of chronic retention of urine in women caused by dysectasia of bladder neck. Proc. roy. Soc. Med. **28**, 1511—1514 (1935). — HOWARD, T. L., and H. A. BUCHTEL: Resection of vesical neck in children; Indications and results. Amer. J. med. Ass. **146**, 1202—1206 (1951). — HUTCHINS, S. P. R.: Vesical neck obstruction in women. J. Urol. (Baltimore) **69**, 102—108 (1953). — HYAMS, J. A., and S. R. WEINBERG: Hyperplastic change at vesical neck in female. J. Urol. (Baltimore) **51**, 149—161 (1944). — JACOBSON jr., C. E.: Unrecognized vesical neck obstruction in women. New Engl. J. Med. **235**, 645—648 (1946). — JOHNSON, F. P.: The homologue of prostate in female. J. Urol. (Baltimore) **8**, 13—27 (1922). — KEYES, E. L.: The clinical course of bladder neck abstruction attributed to sclerosis or bar (discussion). Amer. J. Surg. **19**, 215—229 (1933). — LOEB, M. J.: Bladder neck obstruction congenital in young males: inflammatory in females. Urol. cutan. Rev. **44**, 455—460 (1940). — MACKENZIE, D. W., and S. BECK: A histopathologic study of female bladder neck and, urethra. Trans Amer. Ass. gen.-urin. Surg. **29**, 227—253 (1936). — MATHÉ, C. P.: Vesical neck obstruction in female. J. int. Coll. Surg. **21**, 146—159 (1954). — MCDONALD, H. P., W. E. UPCHURCH and C. E. STURDEVANT: Vesical neck obstruction in children. J. Urol. (Baltimore) **70**, 94 (1935). — MELVILLE, B.: Resection of bladder neck in female report with of a case. Aust. N. Z. J. Surg. **15**, 299—303 (1946). — MILLS, W. G. Q.: Treatment of chronic urinary retention in women by transurethral resection of bladder neck. Brit. J. Urol. **24**, 236—237 (1952). — MIRABILE, C.: Resection of bladder neck for obstruction in women. New Engl. J. Med. **228**, 751—753 (1943). — NEFF, J. H.: Resection of prolapsed mucosa at vesical neck for retention of urine in female. Trans. Amer. Ass. gen.-urin. Surg. **31**, 263—270 (1938). — NEGLEY, J. C., and E. M. BERKERY: Bladder neck obstruction: Relief by surgical methods. Urol. cutan. Rev. **55**, 327—331 (1950). — NELSON, N. M., R. W. BARNES, H. L. HADLEY and R. T. BERGMAN: Transurethral resection of bladder neck in female. J. Urol. (Baltimore)

77, 198—213 (1957). — NESBIT, R. M.: Vesical neck contracture in female with urinary obstruction. Urol. cutan. Rev. **37**, 291—293 (1933). — O'BRIAN, J. H., and B. D. PINCK: Vesical neck obstruction in female. J. med. Soc. N.J. **46**, 400—403 (1949). — PATTON, J. F.: Bladder neck obstruction in women and children. Rocky Mtn. med. J. **46**, 540—544 (1949). — POWELL, N. B., and E. B. POWELL: Female urethra; a clinico-pathological study. J. Urol. (Baltimore) **61**, 557—570 (1949). — RANDALL, A.: Prostatisme sans prostate. N. Y. med. J. **102**, 1123—1132, 1177—1186 (1915). — RENNER, M. J.: So called female prostate and concretion formation in female urethra. Surg. Gynec. Obstet. **52**, 1087—1092 (1931). — RITTER, J. S., and L. A. SHIFRIN: Vesical neck obstruction in female. Urol. cutan. Rev. **52**, 147—149 (1948). — SHUMAKER, L. B., and F. C. HENDRICKSON: Contracture of vesical neck in females. Urol. cutan. Rev. **52**, 205—206 (1948). — THOMPSON, G. J.: Transurethral operations for relief of dysfunction of vesical neck in female. Proc. Mayo Clin. **10**, 598—600 (1935). — Urinary obstruction of vesical neck and posterior urethra of congenital origin. J. Urol. (Baltimore) **47**, 591—601 (1942). — WINSBURY-WHITE, H. P.: Two cases of retention of urine in women. Lancet **1936**, 1008. — YOUNG, H. H.: Pathology and treatment of obstructions at vesical neck in women. J. Amer. med. Ass. **115**, 2133—2136 (1940).

Chapter XXII

BIORN, C. L., W. H. BROWNING and L. THOMPSON: Transient bacteremia immediately following transurethral prostatic resection. J. Urol. (Baltimore) **63**, 155—161 (1950). — GOLDSTEIN, A. E., M. R. GOLDEN and H. E. SILBERSTEIN: Postoperative blood loss in prostatectomy. J. Urol. (Baltimore) **71**, 63—66 (1954). — HARRISON III., R. H., J. S. BOREN and J. R. ROBISON: Dilutional hyponatremic shock: Another concept of transurethral prostatic resection reaction. J. Urol. (Baltimore) **75**, 95—110 (1956). — KENYON, H. R.: Perforations in transurethral operations: Technique for immediate diagnosis and management of extravasations. J. Amer. med. Ass. **142**, 798—802 (1950). — LOMBARDO jr., L. J.: Fibrinolysis following prostatic surgery. J. Urol. (Baltimore) **77**, 289 (1957). — MULHOLLAND, S. W., and H. M. MADONNA: J. Urol. (Baltimore) **68**, 489—495 (1952). — MURPHY, J. J., V. IOB and J. LAPIDES: Effect of transurethral prostatectomy on fluid balance in the elderly patient. J. Urol. (Baltimore) **73**, 860—865 (1955). — NESBIT, R. M.: Advantages of perineal urethrotomy in prostatic resection. Sth. Surg. **7**, 501—504 (1938). — NESBIT, R. M., and K. B. CONGER: Studies of blood loss during transurethral prostatic resection. J. Urol. (Baltimore) **46**, 713—717 (1941). — PILCHER, F., and C. SHEARD: Measurements on loss of blood during transurethral prostatic resection and other surgical procedures, determined by spectrophotometric and photelometric methods. Proc. Mayo Clin. **12**, 209—213 (1937).

Chapter XXIII

BOYD, H. L.: The use of thrombin (tropical) in control bleeding associated with prostatic surgery. J. Urol. (Baltimore) **54**, 385—390 (1945). — BUDDINGTON, W. T., and R. C. GRAVES: Management of catheter drainage. J. Urol. (Baltimore) **62**, 387—393 (1949). — GREENBERGER, A. S. and M. E.: Capsella bursa pastoris as a hemostatic after prostatectomy. J. Lancet **60**, 422—423 (1940). — LEE, L. W.: Clinical use of urecholine in dysfunctions of bladder. J. Urol. (Baltimore) **62**, 300—307 (1949). — McLELLAN, A. M.: Sodium citrate solution for preventing formation of blood clots in bladder. J. Urol. (Baltimore) **30**, 251—252 (1933). — MILLIKEN, L. F.: The use of a new blood coagulant in transurethral prostatic resection J. Urol. (Baltimore) **42**, 75 (1939). — OLWIN, J. H., F. B. PAPIERNIAK and J. W. MERRICKS: Use of anticoagulants following transurethral prostatic resection. J. Urol. (Baltimore) **63**, 303—308 (1950). — SCHNEIDER, D. H.: Aseptic closed urinary tract drainage. J. Urol. (Baltimore) **74**, 158—161 (1955). — STEGEMAN, W.: Total avoidance of postoperative bladder irrigation; results and conclusions from 200 consecutive cases. J. Urol. (Baltimore) **63**, 882—886 (1950).

Chapter XXIV

BARNES, R. W., H. L. HADLEY, R. T. BERGMAN and R. TURNER: Conservative versus radical treatment of bladder tumors. M. J. Aust. **1**, 197—205 (1955). — BARNES, R. W., W. M. WILSON, R. T. BERGMAN, S. E. FARLEY and H. L. HADLEY: Plastic surgery on urinary bladder. J. Urol. (Baltimore) **69**, 641—651 (1953). — BERGMAN, R. T., R. TURNER, R. W. BARNES and H. L. HADLEY: Comparative analysis of one thousand consecutive cases of transurethral prostatic resection. J. Urol. (Baltimore) **74**, 533—548 (1955). — DAVIS, E., and R. M. NESBIT: Comparison of late functional results in perineal prostatectomy and transurethral prostatic resection. Trans. Amer. Ass. gen.-urin. Surg. **33**, 251—255 (1940). — DAVIS, J. P.: Ureteral injury by transurethral electroresection and coagulation. J. Urol.

(Baltimore) **68**, 168—177 (1952). — DEAN, A. L.: Surgical treatment of tumors of genito-urinary organs. J. Urol. (Baltimore) **70**, 246—256 (1953). — DORMAN, H. N.: Transurethral prostatic resection: A statistical study based on 300 consecutive cases. J. Urol. (Baltimore) **45**, 411—427 (1941). — EMMETT, J. L., D. D. ALBERS and R. E. ANDERSON: Statistical and analytic review of final results of transurethral resection for cord bladder. J. Urol. (Baltimore) **65**, 36—59 (1951). — FLOCKS, R. H.: Treatment of patients with carcinoma of bladder. J. Amer. med. Ass. **145**, 295—301 (1951). — HIGBEE, D. R.: Conservative surgery in treatment of large single papillary tumors of the bladder. J. Urol. (Baltimore) **70**, 237—241 (1953). — HUGGINS, C., W. W. SCOTT and C. V. HODGES: Studies on prostatic cancer, III; the effects of fever, of desoxycorticosterone and of estrogen on clinical patients with metastatic carcinoma of the prostate. J. Urol. (Baltimore) **46**, 997—1006 (1941). — JACOBS, F. M., J. L. BERRY and J. C. ORMAN: Transurethral resection for paralyzed bladder: Preliminary results. J. Urol. (Baltimore) **70**, 615—619 (1953). — LANDES, R. R.: Painless dilatation of ureteral meatal strictures following transurethral prostatic resection. J. Urol. (Baltimore) **70**, 626 (1953). — LEWIS, B., and G. CARROL: Prostatic resection without moonlight and roses. Urol. cutan. Rev. **37**, 1—7 (1933). — LIVERMORE, G. R.: Etiology of unsatisfactory results following prostatic resection. Trans. Amer. Ass. gen.-urin. Surg. **33**, 243—250 (1940). — MATHE, C. P.: Cystitis: Classification and treatment: discussion of type occurring after transurethral resection. J. Urol. (Baltimore) **62**, 308—316 (1949). — McCARTHY, J. F.: Critique of present methods in the surgery of the prostate gland. J. Urol. (Baltimore) **45**, 428—431 (1941). — MERCIER, O.: End results of 900 cases of transurethral prostatic resections of the prostate. J. Urol. (Baltimore) **49**, 665—674 (1943). — MORTENSEN, H.: Morbidity after prostatectomy by transurethral resection. Med. J. Austr. **2**, 919—921 (1955). — O'CONOR, V. J.: Bilateral intramural strictures of the ureters after transurethral prostatic resection of the prostate. J. Amer. med. Ass. **145**, 1249—1251 (1951). — ORR, L. M., P. R. KUNDERT and F. J. PYLE: Transurethral prostatic resection — ultimate results. J. Urol. (Baltimore) **49**, 840—846 (1943). — ROLNICK, H. C., and L. A. RISKIND: Mortality in prostatic surgery. J. Urol. (Baltimore) **37**, 12—17 (1937). — STRATTE, J. J., and J. STRATTE: Strictures following transurethral resection: Comments on indication for operation. Amer. J. Surg. **73**, 503—509 (1947). — THOMPSON, G. J.: Transurethral surgery in 1939. Proc. Mayo Clin. **15**, 780—784 (1940). — VERGES-FLAQUE, A.: Urinary incontinence following prostatectomy: its cure by nonoperative treatment. J. Urol. (Baltimore) **61**, 96—101 (1949). — VEST, S. A.: Mortality in surgery of prostate. J. Urol. (Baltimore) **45**, 439—450 (1941).

Grundlegende Änderungen
der Pflege urologischer Instrumente

Reinigen und Sterilisieren im Urologischen Krankenhaus München
Chefarzt Prof. Dr. Ferd. May

Die Pflege des urologischen Instrumentariums bedarf einer besonderen Sorgfalt, da das Ergebnis der Untersuchung von dem guten Funktionieren der Instrumente abhängig ist. Beschädigungen durch unsachgemäße Behandlung führen zu einem vorzeitigen Verschleiß der wertvollen Instrumente und damit zu erhöhten Kosten des Betriebs. Die Beseitigung der Krankheitskeime, die bei

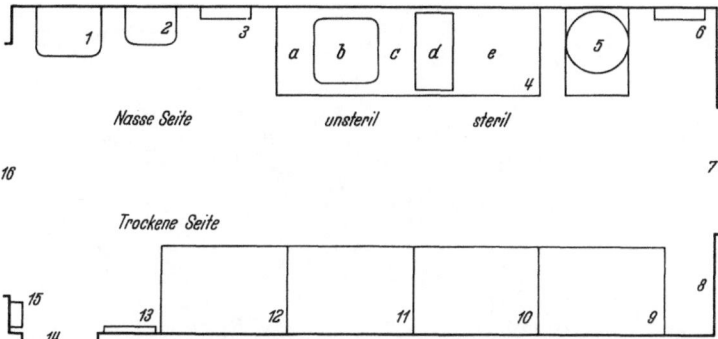

Abb. 1. Instrumentarium einer urologischen Klinik schematisch dargestellt (5,6 m lang, 4,6 m breit). *Nasse Seite, unsteril:* 1 Waschbecken; 2 Schmutzwasserausguß; 3 Katheter-Reiniger; 4 großes Gerät zum Reinigen, Kochen, Trocknen; *a* Ablage für unsaubere Instrumente; *b* Reinigungsbecken (subaquale Hochdruckspülung); *c* Ablage für gereinigte Instrumente; *d* Instrumentenkocher. *Nasse Seite, steril:* e Wärmeplatte zum Trocknen der sterilen Instrumente; 5 Sterilisator und Speicher für Blasen-Spülwasser; 6 Trockner und Härter für Ureter-Katheter; 7 Tür zum Flur (Gang); 8 Schürzen, Mäntel u. dgl. *Trockene Seite:* 9, 10, 11, 12 Arbeitstische zuzüglich Schränke zum Aufbewahren steriler Instrumente (über der Tischfläche: Arbeitsleuchten, optische Prüffläche, Lupe zum Prüfen); 13 Halter für Kabel und Schläuche; 14 Tür zu den Untersuchungsräumen; 15 Telefon; 16 Fenster

urologischen Untersuchungen mit den Instrumenten in Verbindung kommen, verlangt eine einwandfreie Sterilisation, der aber eine gründliche Reinigung vorausgehen muß. Die urologischen Instrumente mit ihren engen Kanälen bieten dabei Schwierigkeiten erheblicher Art.

Bei der Einrichtung des Urologischen Krankenhauses München wurde diesen Forderungen besondere Beachtung geschenkt. Es wurde von der die Instrumente liefernden Firma verlangt, daß die Cystoskope kochfest sein müssen, eine Forderung, die im Laufe der Zeit auch auf die Optiken ausgedehnt wurde. Zur Pflege, Reinigung, Sterilisierung und Aufbewahrung der Instrumente wurde ein eigener Raum, Instrumentarium genannt, eingerichtet, in dem die einzelnen Arbeitsvorgänge in einer zwangsläufigen Reihenfolge vor sich gehen, so daß Fehler nach Möglichkeit ausgeschaltet werden. Die Arbeitsplätze sind eingeteilt in „naß — „trocken — „unsteril" — „steril" (Abb 1).

Auf der nassen Seite werden die Instrumente nach der Benutzung abgegeben und erst einer gründlichen Reinigung unterzogen. Während die Außenseite vor allem bei der Benutzung von in Wasser löslichen Gleitmitteln ohne wesentliche Schwierigkeiten gereinigt werden kann, ist die Entfernung von Verunreinigungen in den engen Kanälen (Blut, Eiter und sonstigen Sekreten) nur durch eine gründliche Spülung mit Wasser unter hohem Druck möglich. Um das Umherspritzen von Wasser dabei zu vermeiden, wird diese Druckspülung in einem mit Wasser gefüllten Becken, also unter Wasser, vorgenommen. Anschließend erfolgt in einem Instrumentenkocher die Sterilisierung der Cystoskope mit ihren Optiken. Um

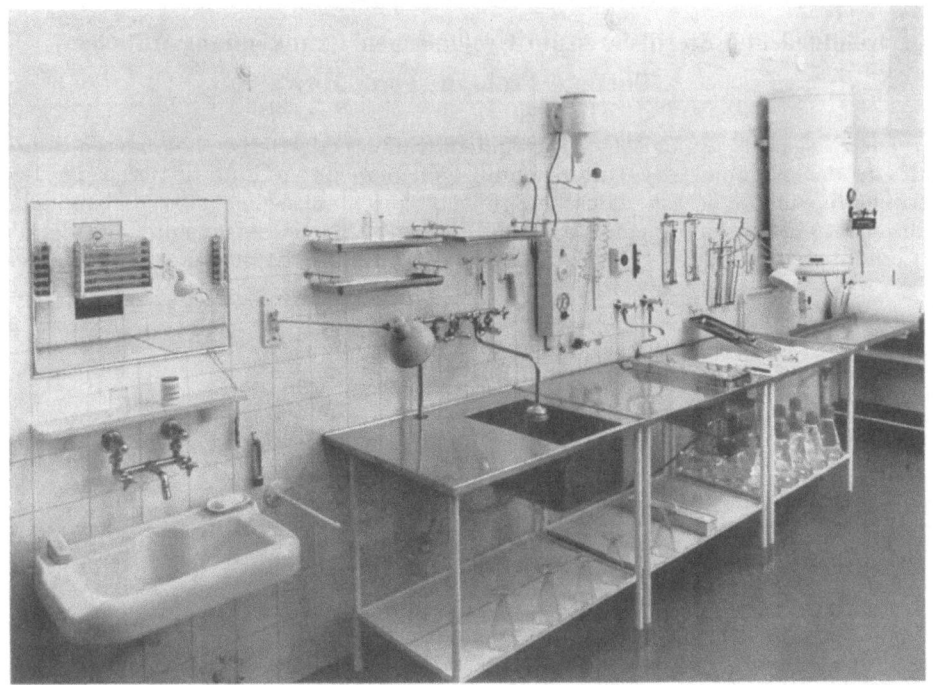

Abb. 2.

Beschädigungen zu vermeiden, befinden sich diese auf einem eigenen Halter, mit dem sie nach dem Kochen, ohne selbst berührt werden zu müssen, zum Trocknen an die Wand gehängt werden. Alle Zusatzteile, Katheter, Bougies usw. werden ebenfalls durch Kochen sterilisiert (Abb. 2).

Einige Spezialeinrichtungen, wie eine Trockenplatte für Zusatzteile und Katheter, ergänzen die Einrichtung der nassen Seite. Die Ureterkatheter werden erst einer Vordesinfektion mit einer Oxycyanatlösung, die lange Zeit durch das Lumen läuft, unterworfen und nach Einlegen in Schutzhüllen (gewebte Schläuche aus plastischem Material) durch Kochen sterilisiert. Um ihnen den genügenden Härtegrad wieder zu geben, werden sie in gestrecktem Zustand unter sterilen Kautelen mit Heißluft getrocknet und so zum nächsten Gebrauch aufbewahrt. Die gefährliche Verwendung von Drähten zur Einführung von Kathetern ist damit ausgeschaltet.

Auf der „trockenen Seite" werden die Instrumente nach vollkommener Trocknung steril in luftdichten Kästen unter der Einwirkung von Formaldehydgas aufbewahrt und bei Bedarf an die Ärzte, die in den direkt an das „Instrumentarium" angrenzenden Untersuchungsräumen arbeiten, ausgegeben (Abb. 3).

Abb. 3.

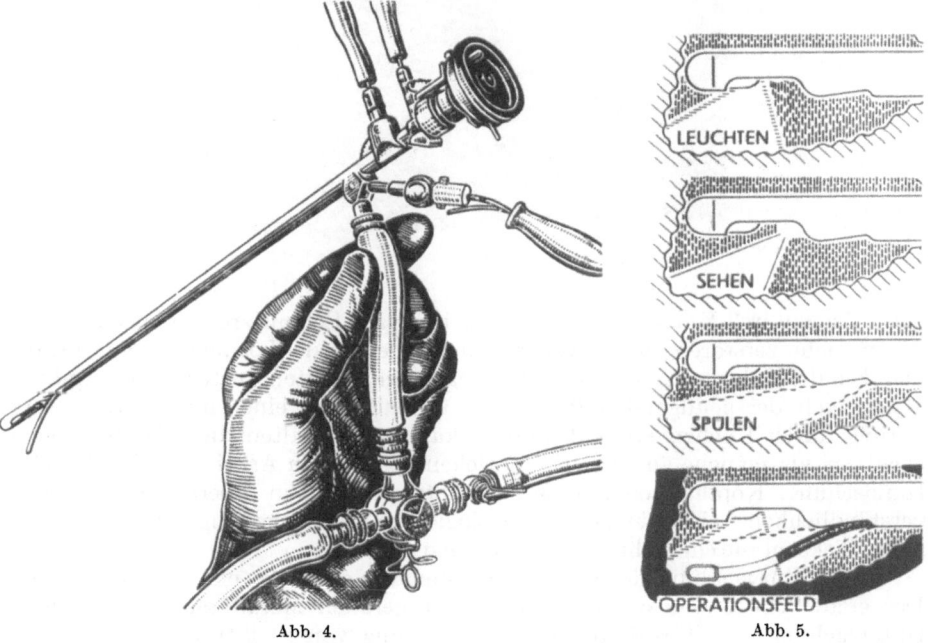

Abb. 4.   Abb. 5.

Mit diesen Maßnahmen, die seit 1938 immer wieder verfeinert und verbessert wurden, ist es uns gelungen, die Lebensdauer unserer Instrumente wesentlich zu verlängern und die durch Fehler bei der Instrumentenpflege auftretenden

Schäden auszuschalten. Eine wesentliche Bedingung muß aber dabei eingehalten werden: die Pflege der Instrumente muß einer gut eingearbeiteten Operations-Schwester anvertraut werden, die gründliche Kenntnis der Instrumente besitzt und die gegebenen Arbeitsvorschriften genau befolgt.

Wenn man die Kanäle lange gebrauchter und dabei schlecht gereinigter Cystoskope betrachtet, so kann man feststellen, daß alle Innenräume mit großen Mengen von Detritus, d.h. eingetrockneten Sekretresten, altem Blut usw. bedeckt

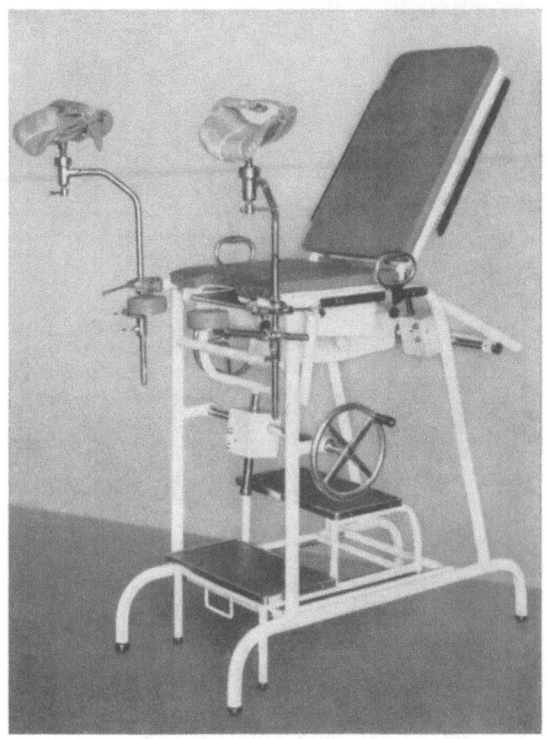

Abb. 6.

sind. Auch durch Kochen lassen sich die in diesen Schmutzresten liegenden Bakterien nicht zerstören, noch weniger durch germicide Lösungen. Nur kräftige Durchspülung der von der Untersuchung noch feuchten Instrumente ergibt die Möglichkeit der genügenden Reinigung, so daß dann eine einwandfreie Sterilisierung durch Kochen erreicht werden kann. Wir halten diese Maßnahme für wichtiger als manche äußeren Kennzeichen des sterilen Arbeitens wie Abdecken, Handschuhe, Kopfhauben und Mundschutz, wie sie im Operationssaal Selbstverständlichkeit sein müssen, im urologischen Untersuchungsraum aber doch nicht wirksam durchgeführt werden können.

In der Entwicklung der Cystoskope sind wir ebenfalls neue Wege gegangen. Der erste Schritt war die Einführung der geraden Metallbougies, die wir zur Untersuchung der Harnröhre und zur Dehnung von Strikturen benutzen. Die Möglichkeit, die das gerade Metallinstrument bei der Austastung der hinteren Harnröhre gibt, überzeugte uns davon, daß der gebogene Schnabel des Cystoskops überflüssig ist. Deshalb erfolgte die Entwicklung des geraden Cystoskops, das auch durch Verkleinerung der „inneren schädlichen Länge", durch Vergrößerung der Helligkeit des Bildes und durch besondere Spülmöglichkeiten verbessert wurde. Die Blase kann mit diesem Instrument während des Beobachtens sehr wirksam gespült werden, so daß die Cystoskopie bei verschiedenen Füllungszuständen möglich ist. Die Spülflüssigkeit wird durch eine entsprechende Anordnung der Kanäle und genügende Weite aller durchflossenen Stellen in einer bisher noch nicht möglichen Menge in die Blase geleitet, außerdem wird sie so konzentriert gelenkt, daß ein kräftiger Strahl die in der Blase beobachtete Stelle

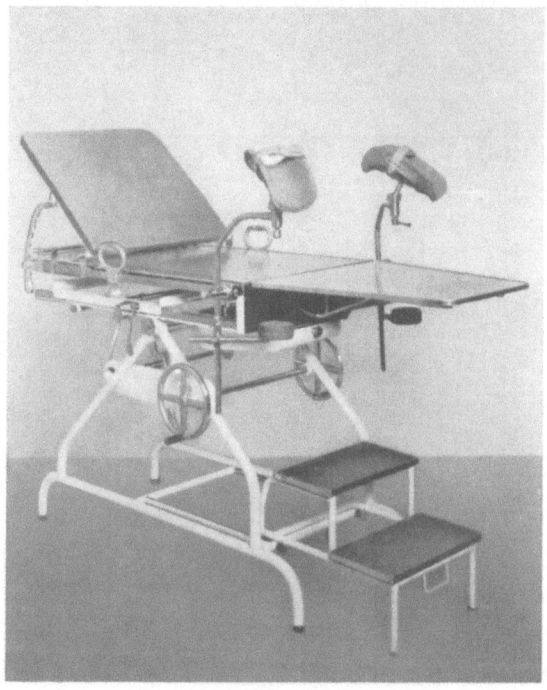

Abb. 7.

bespült. Eine wesentliche Verbesserung der diagnostischen und operativen Möglichkeiten bei der Cystoskopie wurde dadurch erreicht. Durch wechselweise Anwendung der Operations-Optik (135°) und der Weitwinkel-Optik (Panor) sind die Beobachtungsmöglichkeiten weiter verbessert (Abb. 4 u. 5).

Die von uns entwickelten Tische (Abb. 6 zum Untersuchen und transurethralen Operieren, Abb. 7 für Röntgen-Aufnahmen) sind einfach, genügend beweglich und lassen sich leicht und sicher bedienen.

Literatur

ARNHOLDT, F.: Das neue Instrument. — Das Strahl-Cystoskop, ein Fortschritt in der Technik des Cystoskopierens. Med. heute **1953**, 1, 4. — ARNHOLDT, F., u. P. WESTERBERG: Über die Bougierung der Harnröhren-Striktur. Z. Urol. **44**, 744 (1951). — DÜTTMANN, G.: Z. Urol **32**, 695 (1938). — FUNFACK, M.: Beitrag zur konservativen Beseitigung von Harnleitersteinen und festsitzenden Nierenbeckenkelchsteinen. Z. Urol. **34**, 205 (1940). — MOHENFELLNER, R.: Die Behandlung der männlichen Urethral-Striktur mit Fischer-Urethroskop und Heywalt May Bougies. Urol. int. (Basel) **3**, 243—246 (1956). — MAY, F.: Ein neues Katheter-Gleitmittel. Z. Urol. **35**, 170 (1941). — Zwei Jahre Urol. Krankenhaus München. Z. Urol. **35**, 297 (1941). — Verh. dtsch. Ges. Urol. **15**, 49, 295 (1948). — Erfahrungen mit einem verbesserten Spül-Cystoskop. Z. Urol. **46**, 586 (1953). — Du Nouveau dans la technique du cystoscope. Soc. Suisse d'Urologie, 23. 10. 1953. Méd. et Hyg. (Genève) 1953. — Neues zur Cystoskop-Technik. Helv. chir. Acta 21 (1954). — MORGENSTERN, A.: Über das Zerkleinern und Herausspülen von Blasensteinen durch die Harnröhre unter Sicht. Z. Urol. 28, 235 (1934). — RAVASINI, G.: Die kontrollierbare urethroskopische Elektrotomie für die Behandlung der Harnröhren-Strikturen. Urologia 24, H. 3 (1957). — STAEHLER, W.: Ein neues, verbessertes Blasenhals-Elektro-Schneidgerät. Z. Urol. 33, 245 (1939). — Operative Cystoskopie. Leipzig: Georg Thieme 1941. — WICHER, W.: Über das Reinigen, Entkeimen, Trocknen, Härten und Aufbewahren von Ureter-Kathetern. Z. Urol. 34, 173 (1940). — WILLI, H.: Urologische Untersuchungen im Säuglingsalter. Fortschr. Röntgenstr. 57, 2 (1938). — ZEISS, L.: Über eine neue Methode der konservativen Harnleiterstein-Behandlung. Z. Urol. 33, 121 (1939).

Author Index

Subject Index